D0081937

Human Heredity

THE COLLEGE LIBRARY OF BIOLOGICAL SCIENCES
Edited by Ralph Buchsbaum

Human Heredity

By

JAMES V. NEEL, Ph.D., M.D.

AND

WILLIAM J. SCHULL, Ph.D.

Heredity Clinic, University of Michigan

 THE UNIVERSITY OF CHICAGO PRESS

THE UNIVERSITY OF CHICAGO COMMITTEE
ON PUBLICATIONS IN BIOLOGY AND MEDICINE

EMMET B. BAY · LOWELL T. COGGESHALL
LESTER R. DRAGSTEDT · PETER P. H. DE BRUYN
THOMAS PARK · WILLIAM H. TALIAFERRO

THE UNIVERSITY OF CHICAGO PRESS, CHICAGO 37
Cambridge University Press, London, N.W. 1, England
The University of Toronto Press, Toronto 5, Canada

Copyright 1954 by The University of Chicago. All
rights reserved. Published 1954. Second impression 1957.
Composed and printed by THE UNIVERSITY OF CHICAGO
PRESS, Chicago, Illinois, U.S.A.

To

PRISCILLA

and

VICKI

Table of Contents

CHAPTER 1

Man as an Object of Genetic Study

1.1. *Special problems in the study of human heredity.*—Man as an object of genetic study presents both peculiar advantages and disadvantages. We will consider the disadvantages first. Although the beginnings of genetics go back much further, it is customary to place the emergence of genetics as an exact science in the year 1900. This was the year in which three European biologists independently rediscovered the basic principles of genetics, which had been enunciated by Gregor Mendel in 1866 but which had gone almost entirely unnoticed by the scientific world for the next thirty-four years. Geneticists in the first several decades following the rediscovery of Mendel's laws in 1900 were largely occupied with what we may term the "mechanisms" of heredity, i.e., the definition of how the determiners of inherited characteristics, the *genes*, were transmitted from one generation to the next, and the criteria for deciding whether a given trait was due to one or several genes and whether these genes were located on the same or different chromosomes. For a variety of reasons, man is not a favorable subject for studies of this nature: (1) Controlled matings between two individuals of known genetic background, followed by the production of large numbers of offspring from a single mating—upon which the plant and animal geneticists rely so heavily—are, of course, not characteristic of human societies. (2) A further disadvantage of human material is the practical impossibility, in the study of the genetics of a particular trait, of standardizing or altering the environment at will; this is a particularly serious drawback where the manifestations of the gene are influenced by nutrition, training, etc. (3) The length of the interval between birth and reproduction in man, combined with the poor records which society generally maintains with respect to its deceased members, are scarcely favorable to genetic inquiries. (4) Lastly, the rather considerable number of human chromosomes (twenty-four different pairs, as contrasted, e.g., to *Drosophila melanogaster* with four) renders it difficult to establish the chromosomal relations of the genes responsible for the various traits known to be inherited in man.

Because of these shortcomings of human material, students of human heredity in the early decades of this century were for the most part led to

1

confine their observations to the collection of extensive pedigrees in which a given obvious trait appeared repeatedly in successive generations. There was a tendency to judge the value of the study in terms of the number of individuals in the pedigree. Under these circumstances the bulk of the attention was diverted to traits exhibiting simple dominant heredity. It was this early search for—and preoccupation with—families containing many persons affected with a given trait that has given rise to the belief, still encountered in some circles, that the study of human heredity consists in the collection of unusual, striking, and/or quaint pedigrees. As we shall see, this is far from the truth.

1.2. *Special advantages of human material.*—We turn now to the advantages of man for genetic studies. A generation of geneticists trained on animals as easily manipulated as the fruit fly and the mouse has been quick to recognize the difficulties involved in working out the exact genetic basis of the more complex human traits. This has led to frequent statements regarding man's unsuitability for genetic studies. This unsuitability is not so serious as is frequently pictured. It is obvious that human beings are not to be crossed like cattle or flies. But while controlled matings are impossible in human societies, we now realize that man in his time and numbers has contrived to enter into many of the matings desired by the geneticist; it remains only to locate these matings for study. Furthermore, the combined efforts of many investigators have resulted in the development of a powerful set of mathematical techniques specifically designed to extract as much information as possible from human genetic data. These techniques at least partly offset the problems created by the small size of human families and the difficulty in obtaining reliable data extending over many generations.

There are two subdivisions of genetics which are currently under intensive investigation in which as much can be learned from the study of man as from any other animal. These are physiological genetics and population genetics. Physiological genetics is concerned with the problem of how the genes work, i.e., with the definition of the developmental sequence between the presence in an individual of a particular gene and the appearance in that same individual of a particular morphological or biochemical trait. We are more familiar with the detailed anatomy, physiology, and biochemistry of man than of any other animal. Small departures from the norm which might go undetected in a fruit fly or a mouse are relatively much more apparent in man. There is therefore the possibility, once an inherited trait has been identified in man, of bringing a great many correlated observations to bear on the ultimate nature of the inherited defect. On the debit side, on the other hand, is the impossibility of deriving pure strains with which to work and the relative inaccessibility of embryological material. There is growing evidence, as we shall see later, that genes work through controlling the many biochemical reactions which occur during the development and the adult life of every organism. In chapter 12 we shall consider some of the many biochemical reactions of man which have been shown to be under genetic control.

Population genetics deals with the nature of the genetic differences between groups of individuals. There are large numbers of readily accessible members of the human species, most of whom will be found to be sufficiently co-operative to yield data of value to the geneticist. On the basis of various morphological characteristics, the anthropologist has separated the human species into a number of distinct stocks. Recent advances in serology have put into the hands of the geneticist a number of easily identified, inherited serological differences between individuals. The study of the frequency with which these and other inherited traits are represented in different groups of people is a powerful supplement to the more traditional anthropological methods of studying the relationships between peoples. Chapter 15 will be devoted to this development.

Man is a curious creature, and he is particularly curious about man. Aberrant human individuals—as well as differences between groups of individuals—have excited comment from earliest times. Some of this comment has been preserved in the incredibly voluminous medical and anthropological literature. Although the descriptions are sometimes biased and incomplete and often omit data considered important by the geneticist, there is no denying the existence of a large literature pertinent to the problems of human genetics. This is in striking contrast to the situation which obtains with respect to many of the important common domestic animals. Commercial breeders of horses, cows, pigs, and sheep have usually done their best to conceal the occurrence of off-types in their herds, since information as to the occurrence of such off-types lowered the financial value of the herd. Consequently, aside from man, we have extensive data concerning aberrant individuals among mammalian species only for certain laboratory animals—mice, rats, and guinea pigs—where the occurrence of such abnormalities may be studied in an environment free of special bias.

Certain organisms lend themselves particularly well to the study of special aspects of genetics. Thus the fruit fly, Drosophila, approximates the ideal laboratory animal for the study of the "mechanisms" of inheritance. The mold Neurospora has proved excellent material for the study of genetically controlled biochemical reactions. The inheritance of quantitative traits is well studied in various agriculturally important plants—corn, beans, wheat. But while a great deal is known about various specific phases of genetics in particular plant and animal species, it is probable that, in the aggregate, more is known about the heredity of man than of any other form, with the possible exception of Drosophila and corn. And yet, as we shall see, only the barest sort of start has been made on the subject of human heredity.

1.3. *Purpose of this book.*—In a small book like this, it would be impossible to attempt to review all that we know about human inheritance. Although the study of human genetics is still in its infancy, it would require several large volumes for even a skeleton summary of the techniques and knowledge in this field. We shall therefore attempt to introduce the reader to some of the landmarks of past work in human heredity and some of the signposts for future development. It is hoped that, with these reference points clearly in mind, the reader will be oriented with respect to a variety of problems which may arise in the future. We shall, then, emphasize the methodology of human genetics far more than the established facts of human inheritance. Even thus limiting ourselves, we shall be forced to slight some topics. Some of our readers will be disappointed at certain omissions and emphases in this book. To this we can only say that we have attempted to include those matters which in our experience most need treatment.

The reader who has not had college courses in the calculus and biometry may experience difficulty in places. We offer no apologies for this. The complexities of the study of human heredity are such that knowledge of certain branches of mathematics is no less essential to the serious student of human heredity than to the astronomer or the physical chemist. The text which attempts to disguise this fact is, in the long run, doing the student a disservice. But in recognition of the fact that some "serious students" may acquire their mathematical background later than others, the book is designed so that the two chapters which draw heaviest on a mathematical background, chapters 13 and 14, may be omitted without great loss of continuity.

1.4. *References.*—The student will undoubtedly profit by referring from time to time to other textbook presentations of general genetics as well as of human heredity. The bibliography which follows gives suggested references.

Bibliography

MULLER, H.; LITTLE, C. C.; and SNYDER, L. H. 1947. Genetics, medicine, and man. Ithaca: Cornell University Press.
ROBERTS, J. A. F. 1940. An introduction to medical genetics. London: Oxford University Press.
SINNOTT, E. W.; DUNN, L. C.; and DOBZHANSKY, TH. 1950. Principles of genetics. 4th ed. New York: McGraw-Hill Book Co.
SNYDER, L. H. 1941. Medical genetics. Durham, N.C.: Duke University Press.
————. 1951. The principles of heredity. 4th ed. Boston: D. C. Heath & Co.

Sorsby, A. (ed.). 1953. Clinical genetics. London: Butterworth & Co., Ltd.
Srb, A., and Owen, R. 1952. General genetics. San Francisco: W. H. Freeman & Co.
Stern, C. 1949. Principles of human genetics. San Francisco: W. H. Freeman & Co.

There are a number of journals especially devoted to the publication of material on human inheritance. The two English-language journals which the student will find interesting are the *Annals of Eugenics* (Cambridge: At the University Press) and the *American Journal of Human Genetics* (Baltimore: Waverly Press).

1.5. *Acknowledgments.*—It is a pleasure to acknowledge the constructive criticism of many of our friends and colleagues. Drs. J. N. Spuhler, Earl Green, Madge Macklin, Duncan McDonald, Newton Morton, J. H. Renwick, and T. E. Reed generously went over the entire manuscript, while Drs. C. C. Li, Brian McMahon, Allan Fox, Sidney Cobb, and Leonard Kurland read selected portions. We are deeply indebted to Mrs. Jane Schneidewind for outstanding secretarial assistance and to Miss Grace Yesley for the preparation of most of the illustrative material.

It is also a pleasure to acknowledge our indebtedness to the numerous publishers and our colleagues for their permission to reproduce copyright items. We are indebted to Professor Sir Ronald A. Fisher, Cambridge, and to Messrs. Oliver and Boyd, Limited, Edinburgh, for permission to reprint Table 13.9 from *Statistical Methods for Research Workers.*

CHAPTER 2

The Physical Basis of Heredity in Man

IN MAN, as in plants and other animals, the key to an understanding of the laws of heredity lies in an understanding of the behavior of the chromosomes at the time of cell division. This is because the units of heredity, the genes, are, with very few exceptions, an integral part of the chromosomes.

2.1. *Chromosomes.*—The chromosomes are small bodies of various sizes and shapes located in the cell nucleus. They take up certain dyes with avidity and appear quite dark in the usual stained preparation of a cell. With few exceptions, every cell of the body contains a set of chromosomes. The average number of chromosomes per cell varies widely from one species to the next. The cells of some animals contain as few as 2 chromosomes, while the cells of other animals contain as many as 200.

In man, the usual number of chromosomes is 48. These consist of twenty-four pairs, one member of each pair derived from the father and the other member from the mother. We refer to the members of a pair of chromosomes as "homologous" chromosomes.

2.2. *Genes.*—The term "gene" has been used to designate the submicroscopic, intracellular determiners of the inherited characteristics of an organism. Suitable experiments have revealed that the genes are arranged in a linear sequence along the length of the chromosomes. The sequence in which the genes are arranged tends to be the same for all the individuals of a particular species. At one time in the development of genetics it was customary to envision the genes, even adjacent ones, as sharply separated from one another both morphologically and functionally. It is now realized that, in the chromosomal continuum, adjacent genes may not be independent of one another in their functioning. In other words, in contrast to the earlier point of view, genes are currently envisioned as "much more loosely defined parts of an aggregate, the chromosome, which in itself is a unit and reacts readily to certain changes in the environment" (Demerec, 1951).

2.3. *Mitosis.*—In preparation for the ordinary (*mitotic*) cell divisions of the body, each of the chromosomes contained within the nucleus of the cell reduplicates itself, in a fashion not now understood. The reduplication remains incomplete at one point, termed the "centromere." The membrane which bounds the nucleus then disintegrates, and each reduplicated chromosome, with the two strands still attached at the centromere, migrates toward the center of the cell. The intracellular forces which are responsible for this movement of the chromosomes are poorly understood. One of the chief visible manifestations of these forces is the so-called "spindle" which appears at cell

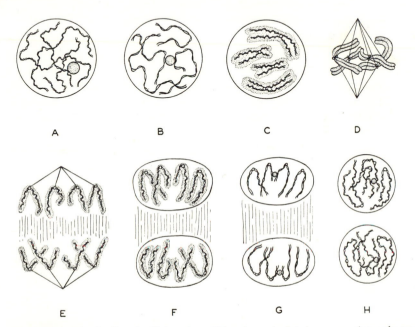

FIG. 2-1.—Mitosis. For simplicity, the cell has been depicted as possessing only two pairs of chromosomes. Each of the two daughter-cells has the same chromosomal complement as the mother-cell. For convenience in discussing the events of mitosis, it is customary to recognize the prophase (A–C), metaphase (D), anaphase (E), and telophase (F–H). (Modified, by permission, from Fundamentals of cytology, by Dr. L. W. Sharp, copyright 1943, McGraw-Hill Book Co.)

division. When the chromosomes reach the equatorial position of the spindle, the centromere divides, and the products of the reduplication move away from each other to opposite poles of the spindle. Usually each daughter-cell receives an exact replica of each chromosome present in the original cell, with the result that there are 48 chromosomes in each daughter-cell. Figure 2-1 is a diagram of this sequence of events. From the beginning of the process to the arrangement of the chromosomes on the spindle, there is a progressive shortening or contraction of the chromosomes.

2.4. *Meiosis.*—The two cell divisions (*meiotic*) which precede the formation of the germ cells follow a somewhat different pattern. At an early stage homologous chromosomes pair lengthwise in a manner designed to bring corresponding parts of the two chromosomes together. Just before or just after this pairing, each chromosome reduplicates as before, the process again remaining incomplete at the centromere. At this stage the products of the reduplication, termed "chromatids," remain in close contact with one another. While the chromosomes are thus paired, there may be an exchange of segments between chromatids of homologous chromosomes. Again the chromosomes undergo a progressive shortening. There follow in rapid succession two cell divisions. The first of these divisions separates the two centromeres, each with two attached chromatids. The centromere now divides, and the next cell division separates the two chromatids. Each of the four resulting daughter-cells receives one representative of each pair of chromosomes, so that the number of chromosomes is now 24 rather than 48. However, as a result of the above-mentioned exchange of segments, termed "crossing-over," this single representative of the original pair of chromosomes may be composed of elements derived from both members of the original pair. The sequence of events is shown in Figure 2-2.

The behavior of a given pair of chromosomes at meiosis is independent of the behavior of any other pair. Thus, if with respect to chromosome pair 1, a germ cell receives the member of the pair derived from the father, with respect to chromosome pair 2 the germ cell may receive either a paternally or a maternally derived chromosome, and so on for chromosomes 3, 4, etc. The genes located on the chromosomes will, of course, exhibit the same behavior.

2.5. *Cytological maps of chromosomes.*—Figure 2-3 is a photomicrograph of the chromosomes of man as they appear during a mitotic cell division just before the split in each chromosome becomes readily apparent. During the earlier stages of cell division ("prophase"), the chromosomes are much less compact than this, appearing as slender strands of darkly staining material, along the length of which there occur unequally sized aggregations of even more darkly stained material, termed "chromomeres." In various plant and animal species, certain chromosomes are regularly associated with collections of peculiarly staining material termed "nucleoli" (singular: "nucleolus"). In the nuclei of human cells there are two nucleolus-bearing chromosomes. By careful attention to the chromomere pattern and the presence or absence of a nucleolus and its characteristics, a start has been made on a cytological map of the human chromosomes (Schultz and St. Lawrence, 1949; Kodani, unpublished). Figure 2-4 is a drawing of the prophase appearance in testicular material of what may arbitrarily be designated chromosomes 1 and 2 of man, the chromosomes associated with nucleoli. It is apparent that the chromomeres of each of these chromosomes differ from one another in size and in the relative distance from one to the next. These are constant differences, appearing in cell after cell. With patience and practice, a chromomere "pattern" can be recognized, which, together with the presence of the nucleolus, makes the identification of each of these two chromosomes possible in properly prepared material. The possibility exists that, with

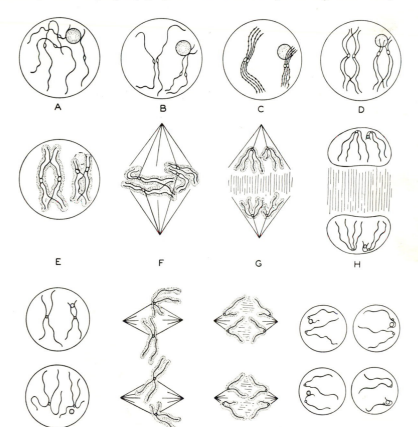

FIG. 2-2.—Meiosis. Again for the sake of simplicity, only two pairs of chromosomes have been depicted. As described in the text, the cells which result from the meiotic divisions receive only one representative of each original pair of chromosomes. The globular body within the nucleus depicted in this and the preceding figure is a nucleolus, of use in preparing cytological maps of chromosomes. (Modified, by permission, from Fundamentals of cytology, by Dr. L. W. Sharp, copyright 1943, McGraw-Hill Book Co.)

time, a similar map could be constructed for each of the 22 other chromosomes. However, even in the best preparations the chromosomes overlie one another in such a fashion as to make it difficult to get a clear view of any particular chromosome. Furthermore, the chromosomes often become fragmented in the course of preparing slides. This renders the task of preparing chromomere maps difficult.

This constancy in the sequence of chromomeres along the length of a chromosome is the cytological basis for the constancy in the sequence of the genes which was referred to earlier. Each chromomere in reality may be associated with one or several genes. The intimate, detailed structure of a

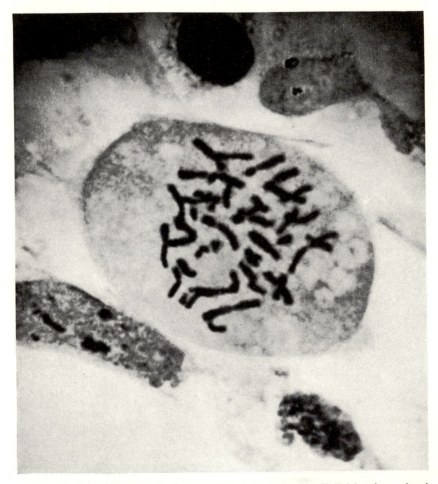

FIG. 2-3.—The chromosomes of man as they appear at mid-cell division (metaphase). (Photograph courtesy of Dr. T. C. Hsu.)

chromosome is still poorly understood. It is composed of special proteins, termed "nucleoproteins"; but how these are arranged is not known.

2.6. *Sex determination.*—The members of each of the 24 pairs of chromosomes are generally identical in appearance. There is one exception to this finding. Males possess one unequal pair of chromosomes, termed the "XY-pair." Females, on the other hand, possess two X-chromosomes. The X-

FIG. 2-4.—A semidiagrammatic representation of the appearance of chromosomes 1 and 2 of man (the two nucleolar-bearing chromosomes) as they appear in the early stages of meiosis. (Chromosome 1 by permission of Dr. Jack Schultz and the Journal of Heredity; chromosome 2 by courtesy of Dr. Masuo Kodani.)

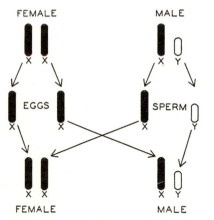

FIG. 2-5.—The chromosomal basis for the determination of sex in man. Further explanation in text.

chromosome is considerably larger than the Y. The sperm cells of a male are of two types, half containing an X-chromosome, and half a Y. All the egg cells, on the other hand, are X-bearing. Random union of the two types of sperm with eggs gives rise to two kinds of zygotes in equal numbers—those with an X- and a Y-chromosome, and those with two X's. The former develop into males, the latter into females. This is a self-perpetuating mechanism. The situation is as depicted in Figure 2-5.

2.7. The number of genes.—Considerable effort has been expended in estimating the probable number of genes present in Drosophila. The most reliable estimates thus far available suggest that the (diploid) nucleus of a female Drosophila contains a minimum of 10,000–20,000 genes. The nucleus of a male Drosophila—because of the presence of a Y-chromosome, which is, for the most part, genetically inert, rather than a second X-chromosome— would, of course, contain somewhat fewer genes. The single attempt thus far made to estimate the number of genes present in man has yielded a (diploid) figure of 40,000–80,000, with the difference between males and females relatively less than in Drosophila because of the larger number of autosomes[1] in man (Spuhler, 1948). These estimates are very approximate, the evidence that man has a greater number of genes than Drosophila being especially circumstantial. However, even if man possesses no more genes than Drosophila, the enormity of the task involved in the development of a detailed familiarity with the kinds and functions of human genes is obvious.

Bibliography

SPECIFIC REFERENCES

DEMEREC, M. 1951. Annual report of the director of the department of genetics, Carnegie Institution of Washington Year Book, No. 50, pp. 167–74.

SCHULTZ, J., and ST. LAWRENCE, P. 1949. A cytological basis for a map of the nucleolar chromosome in man, J. Hered., **40**:31–38.

SPUHLER, J. N. 1948. On the number of genes in man, Science, **108**:279–80.

GENERAL REFERENCES

Cold Spring Harbor symposia on quantitative biology, Vol. IX: Genes and chromosomes. Lancaster: Science Press, 1941.

Cold Spring Harbor symposia on quantitative biology, Vol. XVI: Genes and mutations. Lancaster: Science Press, 1951.

DARLINGTON, C. D. 1937. Recent advances in cytology. 2d ed. Philadelphia: P. Blakiston's Son & Co.

MIRSKY, A. E. 1953. The chemistry of heredity, Scient. Monthly, February, 1953, p. 47.

RILEY, H. P. 1948. Introduction to genetics and cytogenetics. New York: John Wiley & Sons.

SHARP, L. W. 1943. Fundamentals of cytology. New York: McGraw-Hill Book Co.

WHITE, M. J. D. 1942. The chromosomes. 2d ed rev. London: Methuen & Co., Ltd.

---

1. All the chromosomes which are not sex chromosomes are referred to as "autosomes."

CHAPTER *3*

Man's Genetic Diversity

IT IS difficult for those unfamiliar with genetic concepts to appreciate the enormous possibilities for genetic diversity which exist within the human species. Lack of appreciation of these possibilities lies behind some of the misunderstandings, to be discussed later, concerning human inheritance. A consideration of what is now known concerning the genetics of the human red blood cell will illustrate the extent of the possible genetic differences between men.

3.1. *The number of possible different genotypes with respect to the inherited anemias.*—Roughly speaking, the inherited characteristics of the red blood cells are of two types. We recognize, on the one hand, a series of inherited defects in the red cell which often result in an anemia and, on the other hand, a series of inherited serological reactions. The hereditary anemias with a simple genetic basis include hereditary spherocytosis, thalassemia, sickle-cell anemia, ovalocytosis, and Fanconi's syndrome. Each of these anemias is due to an abnormal gene situated at some particular point on a chromosome ("genetic locus"). With respect to any particular one of the five different genetic loci involved in these anemias, an individual may have received the abnormal gene from both parents or from only one parent, or, as is much more likely, he may have received the corresponding normal gene from both parents. If as regards a particular genetic locus an individual has received from both parents an identical form of the gene occupying that locus, we refer to him as "homozygous" at that genetic locus. Where, on the other hand, the members of a gene pair differ with respect to their characteristics, we speak of the individual as "heterozygous" at that locus. The alternative forms of a gene which may occur at any genetic locus are termed "allelomorphs" or, more simply, "alleles." With reference to each of the inherited anemias, an individual may be homozygous with respect to the normal gene, heterozygous for the abnormal gene, or homozygous for the abnormal gene. There are thus three genetic alternatives. Assuming that the findings at one locus are not related to those at another, the number of theoretically possible different genetic combinations as regards these five anemias is $3^5 = 243$.

13

3.2. *The number of possible different genotypes with respect to serological traits.*
—In addition to the inherited anemias, there are inherited serological differences between red cells. These differences determine the manner in which
the red cells react with certain test sera. They will be discussed at some length
in subsequent chapters. These serological differences appear for the most part
to have a rather simple genetic basis, the presence of a particular reaction
corresponding to the presence of a particular gene and the absence of the reaction to the absence of that gene. For some of these serological reactions
there are multiple forms which the reaction may take, corresponding to multiple alternative forms of the responsible gene, i.e., "multiple alleles." The
occurrence of multiple alleles greatly increases the possibilities for genetic
differences between individuals. Thus, where there are three alternative
forms of the gene (only two of which can be represented in an individual at
one time), the number of possible genetic constitutions ("genotypes") is six.
Where there are four alleles, the number of genotypes is ten. The general
formula for the number of possible genotypes where a series of multiple
alleles exists is $n + (n - 1) + (n - 2) + (n - 3) + \ldots + 1 = n(n + 1)/2$,
where $n = $ the number of alleles. This corresponds to $n(n - 1)/2$ heterozygotes and n homozygotes.

There are now recognized at least seventeen apparently independently inherited types of serological differences between individuals. Eight of these
differences are quite rare, and the positive type is very seldom encountered.
One of the seventeen, the so-called "Rh reaction," is represented by at least
eight different alleles (or closely linked gene complexes; see Sec. 8.6, p. 86).
There are thirty-six combinations in which these eight alleles may occur in any
one individual. Another difference, upon which the common A_1, A_2, B, and O
blood groups are based, may be traced to four alleles, which may occur in
ten different combinations. The remaining fifteen differences depend, so far
as is now known, on a single pair of alleles; but it is quite possible that further
studies will reveal multiple alleles at some of these loci. Even if only a single
pair of genes were postulated at each genetic locus associated with a serological reaction, the total number of genetic combinations possible with reference
to the genes responsible for the already recognized serological reactions would
be $3^{17} = 129,140,163$.

3.3. *The total number of possible genotypes with respect to the red blood cell.*—
If we now consider the number of possible genotypes with respect both to
serology and to anemia, we must recognize well in excess of $129,140,163 \times
243 = 31,381,059,609$. *It is obvious that very many of these possible combinations have never been realized.* Some of the anemias have restricted racial distributions, being more common in one part of the world than in another.
Thus individuals homozygous for the thalassemia gene occur about once in

each 2,400 births to Italians and Greeks but are very rare among Negroes. Individuals homozygous for the sickle-cell gene are found about twice in each 1,000 births to American Negroes but very rarely among non-Negroid groups. Although the genes responsible for the other anemias appear to have a more generalized distribution among the various races of man, the probability of homozygosity for all five of the genes which we have mentioned in this chapter as being associated with an anemia is certainly very much less than

$$\frac{1}{2.4 \times 10^3} \times \frac{1}{5 \times 10^2} \times \frac{1}{1 \times 10^5} \times \frac{1}{1 \times 10^5} \times \frac{1}{1 \times 10^5} = \frac{1}{12 \times 10^{20}}$$

Certain of the serological characteristics referred to above are also quite uncommon. Accordingly, while we may write that there are in excess of one billion possible genotypes with respect to the known inherited characteristics of the red cell, we must recognize that the statement represents a genetical tour de force, a majority of these being so uncommon that they may never have occurred in the history of the world. Furthermore, some of the necessary testing sera are so rare that no one laboratory, even the finest, has ever assembled all of them. Race and Sanger, working in one of the best-equipped serological laboratories in the world, have estimated that, using eight groups of sera readily available, they can distinguish 29,952 serological combinations. Since the test sera will not always indicate whether an individual is heterozygous or homozygous for a given gene, some of these combinations include several different genotypes. In a test of 475 Londoners in which these sera were used, 296 different blood-group combinations were encountered. But, despite these qualifications, the fact remains that there exist, and we are already in a position to recognize, a perfectly enormous array of different human genotypes classified on this basis alone.

The study of the genetically controlled characteristics of the red blood cell must, like the study of human heredity in general, be in a very early stage. Only a fraction of the genetic differences between the red cells of individuals has been recognized. In view of the fact that the discovery of each new genetic locus, with a minimum of three possible genotypes, at least triples the number of possible combinations and in view of the further fact that multiple allelic series are not at all uncommon, it seems not unlikely that within ten or twenty years the number of possible recognizable genetic combinations with respect to the red blood cell will be greatly increased, with several million of these combinations actually being represented among the present world's inhabitants.

3.4. *The biochemical basis of the inherited morphological and serological differences between erythrocytes.*—The ultimate biochemical basis for all the various inherited morphological and serological differences between erythrocytes is unknown. However, it has been found that the complex protein known as "hemoglobin," which imparts to erythrocytes their red color and is responsible for oxygen transport, is abnormal in some of the inherited anemias characterized by unusual-appearing erythrocytes (see sec. 12.7, p. 169). The A, B, O blood-group reactions are apparently due to substances which consist in part of carbohydrate-lipid complexes, these substances in blood cells normally being organically combined with proteins. It seems reasonable to postulate that, with the passage of time, all these inherited variations will be found associated with complex differences in the composition of the proteins and associated groups of the red blood cell. There emerges, then, a picture of the genetic control of an immense range of biochemical variation in the erythrocyte.

3.5. *The genetic diversity of man.*—The red blood cell is but one of the many specialized cells of the body. From the standpoint of visible cellular differentiation, it is one of the simpler of the specialized cells. There is far more microscopic complexity to a nerve cell, a white blood cell, a muscle cell, or a cell from a secreting gland. It seems certain that only a start has been made in defining the genetically controlled differences between the erythrocytes of various individuals. There is no reason to suppose that the erythrocyte is peculiar in its genetic complexity. Rather, its accessibility and serological reactions enable us to make relatively rapid progress in defining that complexity. If this is so, then, by extrapolation from what is now known concerning the genetics of the red blood cell, we are led to the concept of vast potentialities of gene-controlled differences between individuals. The estimate that the chromosomes of each adult contain between 20,000 and 40,000 different gene pairs provides the theoretical basis for these differences; the facts already known concerning the red blood cell provide a tangible bridge between theory and fact.

Bibliography

NEEL, J. V. 1953. Inherited abnormalities of the cellular constituents of the blood. *In:* SORSBY, A. (ed.), Clinical genetics. London: Butterworth & Co.
RACE, R. R., and SANGER, R. 1950. Blood groups in man. Springfield: Charles C Thomas.
WIENER, A. S. 1943. Blood groups and transfusion. 3d ed. Springfield: Charles C Thomas.

CHAPTER 4

Nature and Nurture

THE PROBLEM of the extent to which environmental factors may influence the expression of the genotype is one of the oldest in biology. It is unfortunate that, as seen in retrospect, this problem has often been approached with more bias than perspicacity. So vast is the field of knowledge encompassed in the study of biology that it is difficult for any one individual to have a grasp of the significance of all the developments occurring in all the various fields of research. The result has inevitably been a tendency to emphasize (and overemphasize) the significance of the material with which the observer is most familiar. Nowhere has this tendency been more obvious than in the debates which have flourished over the problem of heredity versus environment.

4.1. *The interaction of gene with environment.*—Much of our present thinking about this problem as it concerns man is strongly colored by the results of animal experimentation. When dealing with animals, the experimenter can create situations which lead to clearer answers than is usually the case for man. A single example will suffice. The fruit fly, Drosophila, has, scattered over its body, thin hairs or bristles. The position of each bristle is generally well defined. In one particular region of the body 4 bristles are ordinarily present. However, when flies homozygous for the *polychaetoid* gene are grown under the usual laboratory conditions, the number of bristles in this area averages 6.5. Although, on the average, there are 2.5 more bristles than normal, there is considerable variability from fly to fly in the total number and size of such bristles, some having as few as 5 and others as many as 9. This variability occurs in spite of the fact that the flies have been closely inbred for many generations, so that any one fly of the strain is genetically very similar to any other. Presumably, then, in this particular inbred line the differences between flies in bristle number are due to circumstances effective during the development of the fly. Studies have been made of some of the environmental factors which may influence bristle number. For instance, flies which develop at 14° C. average 9 bristles, whereas genetically identical flies developing at 29° C. average only 6 bristles. Flies raised at room temperature under marginal food conditions, so that they are constantly on the verge of starvation, average 5 bristles, whereas those raised under very favorable food conditions average 7 bristles. The conclusion seems justified that part of the variation from fly to fly in bristle number which is seen when this strain of flies develops under "natural" conditions is due to differences in the temperature and the food to which individual flies are exposed. But there are other factors involved as well, the nature of which is unknown.

We may, if we wish, make a rough comparison between variations in bristle number in flies and variations in tooth number and structure in man. There are significant differences between people in the number of teeth present. Some individuals lack one or more incisor teeth, others lack third molars, and some few develop no teeth at all. These differences tend to be inherited. There are also structural differences between such teeth as are present. These are in part inherited but are in part due to the occurrence of certain types of illness or dietary deficiencies at critical periods in the development of the tooth. The state of one's dentition at, say, age twenty-one is thus due to both genetic and environmental factors.

A second comparison with the bristle situation in Drosophila might involve the degree of muscular development in man. People differ greatly in their muscularity. These differences in muscular development are at least in part inherited. There are slender persons who can exercise day after day to the point of exhaustion and still their muscular development remains poor, in contrast to thickset individuals who, on a minimum of exercise, retain their muscularity. However, these differences can express themselves only at a certain nutritional level. When the available food supply is reduced to a bare minimum, as happened in certain concentration and prisoner-of-war camps during World War II, these differences in muscle mass tend to disappear. Likewise, in Drosophila, the differences in bristle number between wild-type flies and the particular strain discussed above tend to be minimized at low nutritional levels.

4.2. *The approach to the nature-nurture problem in man.*—While it is true that various subdivisions of the human species may be recognized, genetically homogeneous strains of man, in the sense that the term "genetically homogeneous" is used by the plant and animal breeder, do not exist. Even if such strains did exist, the organization of society is scarcely such as to permit carefully controlled observations concerning the effect of environmental modifications on human growth and development. It is therefore practically impossible when one is dealing with human populations to create situations which throw a sharply critical light on the relative importance of heredity and environment. Thus, despite the fact that dental differences between men have been recognized since the dawn of history, in some ways we know less about the genetic control of dentition and the role of environmental factors than we do about the genetics of bristle number in the fruit fly and the modification of the expression of the genotype by environmental variables. However, in spite of the difficulties in evaluating the role of environmental effects in man, there are at least three different ways in which one can begin to arrive at a more precise opinion as to how environmental factors influence human traits than is possible in the examples discussed above.

1. While there are no inbred (i.e., genetically homogeneous) lines in man, the birth of identical twins always provides a "strain" of two genetically identical individuals. A careful study of similarities and dissimilarities between such twins provides very important information concerning the role of heredity in a wide variety of traits. Occasionally, identical twins become separated at an early age and are reared in quite dissimilar environments. The differences which develop between such twins under these circumstances are one measure of the extent to which environment may modify the expression of the genotype. Since identical twins occur only once in each 240 births and since, for sentimental reasons, society exerts more effort to keep identical twins united than is the case with ordinary siblings, only a relatively few instances of separated twins have been carefully studied in this country. The principal findings from these studies, as well as the study of twins in general, will be considered in chapter 16.

2. A second approach to the evaluation of the heredity-environment problem stems from studies of the characteristics of groups of people whose environment is drastically changed. Immigration has brought the peoples of many lands to the United States. By and large, these immigrants have been young adults. Once here, immigrants have tended to marry other individuals from the same country. The children of such marriages are genetically similar to their parents but are reared under different environmental conditions. Actually, because of the tendency of an immigrant group to preserve its customs, the difference between the environment of the "old country" and the country of adoption is far less than the range of environmental differences to be encountered in the United States proper. A comparison of the physical characteristics of immigrants and their children has brought to light some interesting facts. One of the earlier and best-known studies in this field was that of Boas, who compared immigrants from a variety of localities in Europe with their children born in this country and demonstrated differences in the form of the head, a characteristic then usually regarded by the anthropologist as being quite constant for any particular racial group and relatively little affected by environment. Somewhat later, Shapiro (1939) demonstrated that the children of Japanese immigrants to Hawaii were heavier and taller than their parents. The difference in height amounted to

$1\frac{5}{8}$ inches in males and $\frac{5}{8}$ inch in females. In addition, there were changes in head form, as well as a tendency toward narrower hips, chests, and faces. The children of Chinese immigrants exhibit somewhat similar changes (Lasker, 1946). Finally, Goldstein (1943) has compared the children of Mexican immigrants to Texas with their parents and with the children of nonimmigrant Mexicans from the same locality as the immigrants. Again there were significant size differences, but, in addition and in confirmation of Boas' findings on the descendants of Europeans and Shapiro's findings on Japanese immigrants, the heads and faces of the adult children tended to be somewhat smaller than those of their parents of like sex.

While it is true that in the past century there appears to have been a gradual increase in the size of man in many different countries, for reasons not entirely clear, the increases demonstrated by these studies are clearly over and beyond the increase expected on the basis of the general trend. The obvious inference to be drawn from this is that the environment of the United States is more favorable for growth than that of the countries from which these individuals were derived. The fact that Sicilians tend to be shorter than Swedes may reflect a difference in diet between the two groups, as well as a different genetic endowment. That environment may alter final size in man is not surprising, in view of the many animal experiments on the effects of various feeds on growth. It is, however, interesting that this should be accompanied by changes in proportions.

As noted above, the American-born children of immigrants still tend to resemble the parental group closely in nutritional habits. In large cities, where most of these studies have been carried out, the cultural continuity with the "old country" was often striking. If even under these circumstances environmental effects are apparent, then to what extent may the later cultural assimilation of these groups and the resultant environmental changes be accompanied by even more marked changes in stature and bodily proportions? Only the corner of the curtain concealing a very interesting line of observation has been lifted. Studies on the second-generation descendants of immigrants are greatly to be desired.

3. A third line of evidence bearing on the question of how environmental factors may alter the expression of the genotype comes from medicine. The treatment of many of the various kinds of inherited diseases consists essentially of an environmental change. In some cases the change is relatively minor; in other cases, quite severe. By observing a patient prior to treatment and then following this same patient closely while he is under treatment, a physician reaches certain conclusions about environmental factors. This "before and after" approach is illustrated by the following examples:

a) As certain individuals grow older, they lose the ability to use blood sugar as a source of energy because of a deficiency of insulin, a hormone secreted by the pancreas. The resulting disease is known as "diabetes mellitus." Many studies have clearly indicated the importance of heredity in this disease. When, now, a physician is called upon to treat diabetes mellitus, the first step almost always consists of a restriction on the amount of sugar and starch in the patient's diet. With this change in the "external environment," the blood sugar in milder cases of the disease falls to normal or near-normal levels, and no further treatment is necessary. For the more severely affected, these simple measures are not enough. Such individuals must receive injections of insulin in order to control the level of blood sugar. Thus by a combination of methods involving alterations in both the "external" and the "internal" environments in which the genes responsible for diabetes function, it is possible to modify greatly the effects of these genes.

b) A second example of the environmental modification of genetic endowment concerns xeroderma pigmentosa. In this disease, which is determined by a recessive gene, the skin is unusually sensitive to light. Exposure to the sun in amounts usually readily tolerated by normal persons results, in this disease, in the development of an extreme degree of "freckling." Not uncommonly, skin cancers develop in these people. However, as long as people with the disease avoid exposure to light, they get along relatively well.

c) As a final illustration of the interplay between heredity and environment, we may refer to the development of allergies. Some people are abnormally sensitive to small amounts of protein substances which have no effect on most people. Thus when some individuals are exposed to the pollens of certain plants, they sneeze and develop runny eyes and a moist nose, a condition commonly referred to as "hay fever." Other people develop hives when they eat strawberries. Still others develop a skin eruption when they come in contact with various substances which do not harm most of us. There is good evidence that the ease with which one becomes sensitized has a hereditary basis. However, this genetic predisposition can scarcely become apparent unless the individual is exposed to the appropriate environmental agent. For instance, the pollen of the common ragweed is one of the worst offenders as a cause of hay fever. Ragweed is found all over the United States, but more commonly in the eastern and central portions. The plant does not grow well in the more northerly portions of this continent. In the regions where there is little or no ragweed pollen in the air, ragweed hay fever cannot develop. Conversely, the disease is most prevalent where the weed is commonest. A person reared in Alaska, no matter how susceptible he may be to developing ragweed hay fever, will not express his genotype so long as he remains in the north.

4.3. *Mental abilities.*—The problem of the relative contribution of innate endowment versus experience and training is particularly complex with reference to mental abilities and traits. No biologist will deny the great plasticity of the human mind. Nevertheless, there are clear indications that, in mental as well as in physical traits, heredity supplies the raw materials on which environment works and that the nature of these raw materials differs from one individual to the next. A limiting factor in the conduct of studies on this problem has been the relative crudeness of our measures of innate, as contrasted to acquired ability. We will return to this problem later.

4.4. *The treatment of inherited disease.*—It would be well at an early stage in this book to dispel a common misconception. This is the concept that what is inherited is fixed and immutable, and nothing can be done about it. At the very least, this may become a convenient excuse for all one's shortcomings. Carried to an extreme, it leads to a philosophy of "biological determinism" which may stifle initiative and justify all manner of personal failures. That this concept is spurious is quite apparent from a brief consideration of everyday medical practices. We have already quoted several illustrations of the environmental modification of the expression of the genotype. Many more instances could be quoted from the field of medicine, involving the successful treatment of inherited disability. Thus the cloudiness of the lens of the eye known as "cataract" is often inherited as if due to a single gene. While this cloudiness cannot be dispelled by any known medical treatment, it is possible to remove the affected lens surgically and substitute an appropriate type of eyeglasses. While such glasses will not restore vision completely to normal, they can rehabilitate affected persons from near-blindness to the point of very useful vision. As another example, let us consider the type of rheumatism known as "gout." This is due to a gene-determined defect in the way the body handles the breakdown of certain foodstuffs. While it is not yet possible to correct this deficiency, it is possible by the use of certain drugs and hormones (colchicine and cortisone) to control the symptoms of the disease quite well. As a final illustration, consider pernicious anemia. Because of a genetically determined defect in the lining of the stomach, certain persons are unable to utilize their food properly and, as one consequence, develop a severe anemia known as "pernicious anemia." Without treatment this anemia is fatal. If, however, these persons are treated with vitamin B_{12}, their blood becomes entirely normal.

4.5. *The interaction of gene with gene.*—Thus far in this chapter we have been concerned exclusively with the effect of environmental factors on the expression of specific genes or combinations of genes. It is appropriate at this point to consider the effect of gene upon gene. The manner in which one gene, by its presence in the organism, may modify the functioning of another gene is no less important than the manner in which the environment may influence the action of a gene. For a simple illustration of this interrelationship of the genes in their functioning—a phenomenon which is usually termed "genetic interaction"—we may turn again to the genetic control of bristle number in Drosophila (Neel, 1941). We have already discussed the effect of the *polychaetoid* gene on bristle number. Two other genes which also affect bristle number in Drosophila are known as *hairy* and *Hairy wing*. The *Hairy wing* (*Hw*) gene adds bristles to the same region as does the *polychaetoid* (*pyd*) gene. The *hairy* (*h*) gene by itself adds no extra bristles to this region but is responsible for extra bristles elsewhere. By standard genetic techniques it is possible to syntheize strains of flies in which one, two, or all three of these genes are homozygous but which are otherwise genetically very similar. In one experiment in which this was done, the number of extra bristles present in the various genotypes was as shown in the accompanying tabulation.

Genotype	*pyd*	*h*	*Hw*	*pyd h*	*pyd Hw*	*h Hw*	*pyd h Hw*
Average no. of extra bristles..........	1.76	0.00	3.42	1.68	5.81	4.37	8.68
Expected no. of extra bristles if gene effects additive.....	....	....	....	1.76	5.18	3.42	5.18

If the effects of these genes in combination were simply additive, i.e., if the number of extra bristles in genotype *ab* was simply equal to the number in *a* plus the number in *b*, then the results would be as shown in the second row of the table. The results are quite otherwise. The *h* gene, which by itself appears to do nothing, when combined with *Hw* results in *more* bristles than in *Hw* alone but, combined with *pyd*, appears to result, if anything, in *fewer*. The combination of *pyd* and *Hw* has more bristles than a simple additive effect, while flies with all three genes have many more bristles than a simple additive effect.

Similar complex interaction phenomena have been noted in many other investigations on plant and animal material. It is difficult in man to study the interactions of specific genes, because, as noted above, gene action is best studied when attention can be focused on the effects of only a few of the many thousands of genes present, with the rest of the genetic background held relatively constant. This situation, while easily realized in various inbred strains of experimental animals, does not usually exist in man; and so it is rarely possible in man to obtain clear examples of the modification of the effect of one gene or genes by another. These difficulties notwithstanding, several examples of the interdependence of genes in their expression are already known in man. Perhaps the most familiar of these concerns red hair color. This trait appears usually to be determined by homozygosity for a single gene. Black hair color is also genetically determined. There exist many gradations in color between jet-black hair and the "towhead," and it seems probable that a number of different genes co-operate to produce the extreme degrees of black hair, just as there are several genes co-operating to produce a deeply pigmented skin. The gene or genes responsible for black hair are not allelomorphs of the gene responsible for red hair. In the presence of a large amount of black pigment, the occurrence of red pigmentation in the hair may, without very special tests, appear to be completely hidden. In other words, a person who is homozygous for the gene for red hair color will, as a rule, not show the effect of this gene if he also possesses the genes responsible for black hair. There is a masking effect, sometimes referred to as "epistasis."

A second illustration of factor interaction in man comes from the field of hematology. In chapter 3 the inherited anemia known as "thalassemia" was mentioned. Individuals heterozygous for the thalassemia gene exhibit a relatively minor defect in red-cell formation, whereas homozygous individuals suffer from a very severe illness. Figure 4-1, a, is a photomicrograph of normal blood. Figure 4-1, b, is a photograph of blood from a person heterozygous for the thalassemia gene. It will be noted that some of the cells are oval in shape. In Figure 4-1, c, are pictured blood cells from a person heterozygous for the gene responsible for the sickling phenomenon. This is an extremely interesting inherited anomaly to be treated in greater detail later (sec. 12.7, p. 169). It is sufficient to state here that when the red cells of persons heterozygous for this gene are deprived of oxygen, they tend to send out projections and assume very bizarre shapes. But when not deprived of oxygen, as was the case for the erythrocytes pictured in Figure 4-1, c, they appear normal. Finally, Figure 4-1, d, is a photomicrograph of the blood of an individual who is heterozygous for both these genes, which, so far as is now known, are not allelomorphs. The abnormal appearance of the blood is readily apparent. The cells differ markedly from one another in size and shape, and now almost every cell has the target appearance, this being due to an abnormal thinness of the cell. The effect on the health of the person who receives both genes is just as striking as the effect on the appearance of the cells. Whereas persons heterozygous for only the sickling gene or only the thalassemia gene do not appear to be significantly inconvenienced in any way, persons who receive both these genes suffer from a severe chronic anemia. Two genes, each of which has by itself a minor effect, produce severe disease in the double heterozygote.

4.6. *Combined genetic and environmental modifications of gene expression.—* The genetical literature contains many examples of families in which a disease whose development is apparently conditioned by a single major gene varies significantly in its manifestations from one affected member of the family to the next. An interesting pedigree in this connection comes from the field of ophthalmology (Lutman and Neel, 1945). Figure 4-2a is a pedigree of hereditary cataracts in which the condition appears to be due to a single

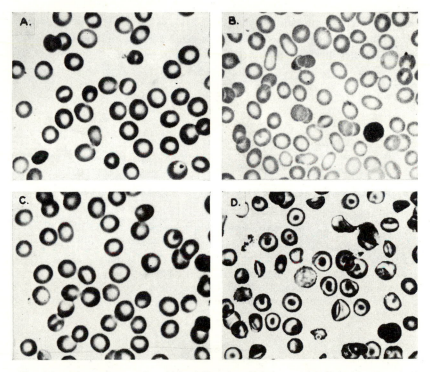

Fig. 4-1.—Photomicrographs of blood films from (a) a normal individual, (b) an individual heterozygous for the thalassemia gene, (c) an individual heterozygous for the gene responsible for the sickling phenomenon, and (d) an individual heterozygous for both the thalassemia and the sickling genes. Further explanation in text.

gene. Within the pedigree there is considerable variation in the exact expression of the trait. Three more or less distinct types of cataract can be recognized, as illustrated in Figures 4-2b, 2c, and 2d. These figures represent the appearance of the lens of the eye when magnified and examined by the appropriate ophthalmological instruments. In the first and most common type the central portion of the lens, the so-called "fetal nucleus," is uniformly cloudy, and scattered throughout are semiopaque flakes, each centered by a dense white spot. In the second type the cataract is due to flake-like, iridescent opacities which are especially numerous in the anterior portion of the lens. The third type presents the picture of coalescing feathery masses. The second type of cataract was restricted to one branch of the family, while the third was seen in only one individual; the mother and two daughters of this latter individual showed the more common first type of cataract. It is not clear to what extent the variability in the type of cataract present is due to segregating genetic modifiers and to what extent to poorly understood environmental variables effective at an early stage of development. The rather clear-cut differences suggest genetic modifiers. If so, then here is another illustration of factor interaction in man.

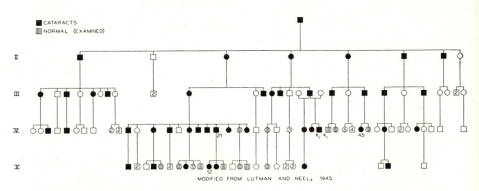

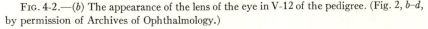

Fig. 4-2.—(*a*) A pedigree of inherited cataract (modified after Lutman and Neel, 1945). In this and all subsequent pedigrees, squares (□) represent males, circles (○), females. Affected individuals are indicated by the solid black symbols. The spouses of members of the kindred have been omitted because in this case they were all normal and so contribute no pertinent information.

Fig. 4-2*b*　　　　　　Fig. 4-2*c*　　　　　　Fig. 4-2*d*

Fig. 4-2.—(*b*) The appearance of the lens of the eye in V-12 of the pedigree. (Fig. 2, *b–d*, by permission of Archives of Ophthalmology.)

Fig. 4-2.—(*c*) The appearance of the lens of the eye in IV-45 of the pedigree

Fig. 4-2.—(*d*) The appearance of the lens of the eye in IV-21 of the pedigree

4.7. *The synthesis.*—It would be possible to quote many further examples, drawn from the fields both of experimental genetics and of human heredity, of the interaction of genes with environment and gene with gene. However, the examples already quoted should be sufficient to make the point at issue. When we speak of the effect of any particular gene, this is really an abbreviated way of referring to the result of the introduction of a particular gene into a particular genetic background and a particular environment. Genes do not operate in a vacuum, but, rather, each has its role to play in the complex machinery of development. In some instances the effect of the gene is such that it cannot be masked or hidden by any known combination of other genes and environmental agencies compatible with life. In other cases, however, the results of receiving a particular deleterious gene may be entirely disguised by the right combination of genetic and environmental modifiers. Conversely, the effects of a gene contributing to the development of a favorable trait may be hidden by an unfortunate environmental situation. In the following chapters we shall sometimes refer to the effect of a particular gene. It must be understood that this is an oversimplification of the true situation, adopted to save space. What is really meant in every instance is the effect of a particular gene interacting with a certain genotype under certain environmental conditions.

In the previous chapter we have considered the genetic complexity of man. That the environment in its way is no less complex can scarcely be doubted. One has only to look briefly at the different climates and cultures under which man flourishes. The study of the interaction of these two sets of variables—heredity and environment, nature and nurture—constitutes an awe-inspiring task. Only the barest start has been made. J. B. S. Haldane (1946) has pointed out that if we have individuals of two different genotypes and if we rank them in two different environments with respect to some quantitative characteristic for which they can be graded (weight, intelligence, muscular strength, etc.), then there are six fundamentally different responses possible, as shown in Table 4-1. In cases 1 and 4, genotype *a* is superior to genotype *b* is both environments; but in case 1, genotype *b* does best in the environment in which genotype *a* does best, whereas, in case 4, genotype *b* does best in the environment where genotype *a* does least well. In cases 2 and 3, environment *Y* is inferior to environment *X* for both genotypes *a* and *b;* but, in case 2, genotype *a* does better than *b* in environment *Y*, while in case 3 the reverse is true. Finally, in cases 5 and 6 the best performance is by genotype *a* in environment *X*, the second best by genotype *b* in environment *Y*, with differences in the responses of the two genotypes to the two remaining environmental situations. Actual examples from the fields of agriculture and animal husbandry can be quoted for most of these cases.

If we consider three genotypes being tested in three different environments, the number of ways in which the genotypes may respond increases sharply, to 10,080. The general formula for calculating the number of different patterns of response is $(mn)!/m!n!$, where m equals the number of genotypes and n equals the number of environments. The number of different possible types of response is multiplied in a truly astonishing fashion as either m or n increases, so that, for example, when $m = n = 10$, there are 7.09×10^{144} types.

In the foregoing, we have scored the different genotypes in the different environments with respect to one single characteristic. Usually we are in-

TABLE 4-1

THE DIFFERENT POSSIBLE RANKINGS OF TWO GENOTYPES
TESTED IN TWO DIFFERENT ENVIRONMENTS

CASE 1	ENVIRONMENT		CASE 2	ENVIRONMENT	
	X	Y		X	Y
Genotype a........	1	2	Genotype a........	1	3
b........	3	4	b........	2	4
CASE 3	ENVIRONMENT		CASE 4	ENVIRONMENT	
	X	Y		X	Y
Genotype a........	1	4	Genotype a........	1	2
b........	2	3	b........	4	3
CASE 5	ENVIRONMENT		CASE 6	ENVIRONMENT	
	X	Y		X	Y
Genotype a........	1	3	Genotype a........	1	4
b........	4	2	b........	3	2

terested not in one characteristic but in a number of them simultaneously. Thus in man we might be interested in intelligence, longevity, and freedom from disease. In the simplest case of two genotypes, two environments, and two criteria, there are 72 different types of interaction. The general formula for the number of different types of interaction is $[(mn)!]^k/m!n!k!$, where m and n have the same meaning as previously and k represents the number of different criteria under consideration. The student can satisfy himself concerning the fashion in which the number of different types of interaction increases as the situation becomes more complex. The complexity of the problem of reaching generalizations concerning the "superiority" of one group or person compared with another is readily apparent.

In closing this chapter, one further complication of the nature-nurture problem should be mentioned. In Table 4-1 we have depicted the possibility that, with respect to a given quantitative trait, genotype a in environment X is better than genotype b in environment Y. We must recognize the possibility, however, that the two may actually not differ significantly. In other words, genetically different groups may, under different environmental conditions, arrive at the same end-point. There is not one but several different ways of achieving a desired end-result.

Bibliography

SPECIFIC REFERENCES

BOAS, F. 1940. Race, language, and culture. New York: Macmillan Co.

GOLDSTEIN, M. S. 1943. Demographic and bodily changes in descendants of Mexican immigrants. ("Publications of the Institute of Latin-American Studies, University of Texas.) Austin, Texas.

HALDANE, J. B. S. 1946. The interaction of nature and nurture, Ann. Eugenics, **13:** 197–205.

LASKER, G. W. 1946. Migration and physical differentiation, Am. J. Phys. Anthropol., **4:**273–300.

LUTMAN, F. C., and NEEL, J. V. 1945. Inherited cataract in the B. genealogy, Arch. Ophth., **33:**341–57.

NEEL, J. V. 1941. Studies on the interaction of mutations affecting the chaetae of *Drosophila melanogaster*. I. The interaction of hairy, polychaetoid, and Hairy wing, Genetics, **26:**52–68.

SHAPIRO, H. L. 1939. Migration and environment. New York: Oxford University Press.

GENERAL REFERENCES

HOGBEN, L. 1939. Nature and nurture. Rev. ed. New York: W. W. Norton & Co., Inc.

CHAPTER 5

Probability

ONE OF the most powerful tools in the armamentarium of the human geneticist is probability theory. The basic tenets of classical genetics, whether of plants or of animals, are, by their very nature, probability statements. We shall repeatedly be called upon to make such statements in describing the children which may be expected from a hypothetical mating. In view of this intimate relationship between the theories of probability and genetics, a review of some of the basic rules of probability theory has become an integral part of any textbook on genetics. We shall, of necessity, confine our attention to those basic rules and concepts which will find application in subsequent chapters.

5.1. *The concept of statistical probability.*—The stimulus which has given rise to the theory of probability has been man's desire to predict events or to provide himself with a means of appraising various courses of action in the face of uncertainty. Historically, probability theory developed around games of chance, such as card playing. Its successful application to such games, as evidenced by the long history of Monte Carlo, has led to a more general theory applicable to a greater number of situations.

The intuitive background on which rests the application of probability theory to genetics may be illustrated as follows: We may visualize certain experiments or observations with a number of different possible results as being repeatable a large number of times under similar circumstances. In such situations we find that if there is more than one possible result, we are unable to predict exactly the outcome of any individual experiment. The outcome of the experiment tends to vary in an irregular fashion which eludes prediction. Such variation we call "random variation" and the experiment a "random experiment." Despite this irregularity in the outcome of a single experiment, if we focus our attention on a long sequence of such experiments, we may find a striking regularity in the average results. For example, consider a random experiment, say A. Assume that there are two possible results to this experiment which we shall call events A_1 and A_2. The random experiment A might be, for instance, the sex of a child at a single delivery. Events A_1 and A_2 could then be the birth of a male child and the birth of a female child. Imagine, now, that this experiment is repeated a large number of times under similar circumstances and that at each experiment the result is noted. At the end of N such repetitions, we could form the ratio of the total number of times the event A_1 occurred to the total number of experiments. This we call the "relative frequency" of the event A_1 in N experiments. Now if we observe this relative frequency as N increases, we shall generally find that it will tend to become more and more stable, that is, it will vary less and less. We feel, then, that if it were possible to repeat this experiment a large enough number of times, the ratio would eventually approximate some fixed number. The fixed number we call the "probability" of the event A_1. This leads us to one definition of probability, namely: "If (a) whenever a series of many trials is made, the ratio of the number of times the event A_1 occurred to the total number of trials is nearly p, and if (b) the ratio is usually nearer to p when longer series of trials are made, then we agree in advance to define the probability of A_1 as p, or more briefly $P(A_1) = p$" (Wilks, 1949). This definition may be termed the "frequency definition of probability."

The use of this definition in genetics may be illustrated by the investigations of Gregor Mendel. He was interested in a number of contrasting characteristics of the garden pea, among which was the color of the seed. He noted two seed colors, yellow and green; furthermore, he noted that peas with green seeds always bred true, whereas some peas with yellow seeds bred true and others did not. Mendel crossed two of these true-breeding plants, one with yellow and one with green seeds, and obtained a generation of hybrids, the first filial generation, or F_1. Plants of this generation were permitted to self-fertilize, and the seeds so obtained were planted. The generation of plants which was produced, the F_2, was found to consist of 428 plants with yellow seeds and 152 plants with green seeds, a ratio of 2.82:1. Mendel concluded that the probability of obtaining a plant with yellow seeds from the crossing of two hybrids was 3/4. Subsequent workers, in verifying Mendel's results, found that, among 195,477 F_2 plants, 146,802 had yellow seeds and 48,675 had green, a ratio of 3.016:1. This experiment follows exactly the conditions laid down in the definition of experimental probability. The experimental approach to probability is widely used in human genetics, particularly in determining the genetic risks in those cases in which the genetic theory behind the occurrence of a particular event is unclear. A number of examples in which such "experimental" probabilities are of value will be cited in subsequent chapters.

The experimental determination of probabilities is applicable to any random experiment; however, there exists an alternative approach to fixing probabilities in many of these situations. This alternative approach has as its basis the mathematical theory of arrangements. In general, this method is as follows: We visualize the conceptual counterpart of an experiment and then enumerate the possible results of this experiment (generally these possible results are dictated by genetic theory). The array of possible results enumerated is said to be exhaustive if at least one of the results necessarily occurs with each single experiment; the results are said to be mutually exclusive if no two results can occur simultaneously. "Now the fundamental assumption underlying the assigning of probabilities is that the results of an experiment may be subdivided into a number of exhaustive, mutually exclusive and *equally likely cases*, say $a_1, \ldots, a_n$, to each of which, since they are assumed to be equally likely, is attached the probability $1/n$" (Uspensky, 1937). In probability problems, however, interest is often centered not upon individual cases but upon *events*, when an event may materialize as a result of any of a number of cases so long as these cases all possess some common attribute of interest. The probability of the event is then defined as "the ratio of the number of cases favorable to the event to the total number of equally probable cases," or, more specifically, if event A_1 may materialize as the result of any m among n equally likely cases, then $P(A_1) = m/n$. As an illustration of this method consider the following problem: Only one member of a family of seven is to be interviewed for a family history. This individual is to be drawn "at random" from a family known to consist of four males and three females. What is the probability that the individual to be interviewed will be male? There are seven ways of drawing one individual from a family of seven; and if the drawing is "at random," it is implied that each of the seven possibilities is equally likely. Of the seven ways, four are favorable to the drawing of a male. The probability, then, that the individual to be interviewed will be a male is, by definition, $4/7$.

The primary objection raised to this definition by the mathematical theorist is leveled at the inclusion of the phrase "equally probable cases," the objection being that the reasoning is circular, that is, that the probable is defined in terms of the probable. Such a theorist might suggest that this difficulty could be overcome by defining probability as "a property of sets of points." Thus, for example, we might develop a probability theory in terms of a fundamental set, finite or infinite, of elements. This fundamental set, per se, may be made up of points, individuals, observations, etc. We may postulate that to each element in the set there corresponds a number, called the "probability" of that element, such that the sum of these numbers for all elements of the set equals 1. We could then imagine some portion of this set, say the subset ω, as possessing some property, X, not possessed by all elements of the fundamental set. The probability of the event X could then be defined as the sum of the probabilities of the elements of the subset ω. This definition does not necessarily imply that all elements of the fundamental set have the same probability; however, in point of fact, this assumption is frequently made. The niceties of the various definitions of mathematical probability need not concern us, but the student should be aware that the definition of probability given here does not have universal acceptance. This does not detract, however, from its applicability to most, if not all, genetic problems involving probability statements.

5.2. *The basic rules of probability.*—Implicitly understood in the definition of probability are the assumptions that probability is measurable on a continuous scale and expressible as a real number. Furthermore, it should be noted that the definition delineates the end-points of the scale of probability; for, if all the n mutually exclusive and equally likely cases are favorable to the event A_1, that is, if $m = n$, then $P(A_1) = 1$. Conversely, if none of the n mutually exclusive and equally probable cases is favorable to the event A, that is, if $m = 0$, then $P(A_1) = 0$. These probabilities may be interpreted, in the first instance, as certainty that the event A_1 *will* occur and, in the second, as certainty that the event A_1 *will not* occur.

The basic rules of probability are operational procedures which permit the assigning of probabilities to a variety of problems. Frequently, for example, it is of interest to know the probabilities of complementary events, such as the events A_1 and not-A_1. In this case we note that if among n mutually exclusive and equally probable results, m are favorable to the event A_1, then $(n - m)$ must be favorable to the event not-A_1. The probability of the event not-A_1 is then merely $(n - m)/n$. It follows directly, then, that

$$P(\text{not-}A_1) = 1 - P(A_1). \qquad (5.2.1)$$

This is sometimes termed the "rule of complementation." The application of this rule may be illustrated by the following simple example: Assume that among 100 deliveries, it is known that 88 terminated in the birth of white infants, 10 in Negro infants, and 2 in Mongolian infants. If a termination is selected at random, what is the probability that the infant so determined will be white? Not-white? Since the termination is to be drawn at random, there are 100 equally likely results, of which 88 are favorable to the birth of a white infant. The probability, then, that the infant determined by a termination drawn at random will be white is 88/100, and, conversely, the probability that the infant will be "not-white" is 12/100.

A second rule arises when we are interested in the occurrence of either of two mutually exclusive events. Now if among n equally likely and mutually exclusive cases, m_1 are favorable to the event A_1 and m_2 are favorable to the event A_2, then $(m_1 + m_2)$ are favorable to the event "either A_1 or A_2." The probability of the occurrence of either A_1 or A_2 is consequently $(m_1 + m_2)/n$, from which it follows that

$$P(A_1 \text{ or } A_2) = P(A_1) + P(A_2). \tag{5.2.2}$$

This is called the "rule of addition of probabilities of mutually exclusive events." The following trivial example will serve to illustrate the use of this rule: Suppose that we are interested in the probability of the occurrence of either a male child or a female child in a single delivery. We assume two equally likely outcomes of a single delivery, one favorable to the occurrence of a male child and one favorable to the occurrence of a female child.[1] Thus the number of results favorable to the occurrence of either a female child or a male child is 2; hence the probability of the occurrence of either a male or a female child at a single delivery is 1—this is precisely the result which experience would lead us to expect.

Still another class of problems of frequent occurrence permits the formulation of a third operational rule. Assume that, among n mutually exclusive and equally probable results, m are favorable to the event A and that, among N equally likely results, M are favorable to the event B. We note that $P(A) = m/n$ and $P(B) = M/N$. When events A and B are uninfluenced by each other, that is, when each of the n cases pertinent to A may be combined with each of the N cases pertinent to B so that each of the nN combined cases are equally likely, events A and B are said to be "statistically independent," or just "independent." If this obtains, we note that each case favorable for A combines with each case favorable for B to give a case favorable to their joint occurrence. It follows, then, that the probability of the joint event A and B is

$$P(AB) = \frac{mM}{nN} = P(A)P(B). \tag{5.2.3}$$

This is called the "rule of the multiplication of probabilities for independent events." As an illustration, let us determine the probability that a single delivery will terminate in a male infant born on either Saturday or Sunday. For simplicity, we shall assume that a delivery will terminate in a male or a female equally frequently and that births are uniformly distributed throughout the week. Now the probability that a single delivery will terminate in a male is, by the definition of probability, 1/2; similarly, the probability that a pregnancy will terminate on either Saturday or Sunday is 2/7. If the day of delivery is independent of the sex of the offspring, in the proba-

1. In this and several following examples we assume that a female birth is as likely as a male birth. Actually, male births are slightly, but definitely, more common than female; the reason is unknown.

bility sense, then there are 14 equally likely outcomes of a single birth corresponding to a male born on Sunday, a female born on Sunday, a male born on Monday, a female born on Monday, etc. Of these 14 equally probable results, only 2 are favorable to the birth of a male child on Saturday or Sunday; hence the probability of the joint occurrence of a male and a termination occurring on either Saturday or Sunday is 2/14, or the product of the events taken singly.

In assigning probabilities, the number which is assigned to a particular event is a function of the possible outcomes of the experiment. This number of possible outcomes of an experiment may be altered by prior knowledge, real or fancied, of what has happened. We may illustrate this by an example. Consider families of three children. Eight different types of such families may be visualized on the basis of the order of birth and the sex of the children. These eight cases are BBB, BBG, BGB, GBB, BGG, GBG, GGB, and GGG, where B and G are a boy and a girl, respectively. If each of these eight types is assumed to be equally probable, then we may determine the probability of a family of three boys, one boy and two girls, etc. The probability, for example, that a family of three children will consist of one boy and two girls is 3/8, since three of the eight equally likely cases are favorable to this event. Suppose that we knew that the family had at least one child which was a girl and were asked to calculate the probability that the family consists of two girls and one boy. Now of the original equally probable results, only those which consist of at least one girl are pertinent to this problem. We observe that seven of the eight original outcomes satisfy this restriction. Of these seven, three are favorable to the event "one boy—two girls"; hence the probability of this event among families of three children, given the occurrence of at least one girl, is 3/7. This is termed the "conditional probability of the event" "one boy—two girls," given the occurrence of at least one girl among the three-child families. In general, the conditional probability of event A under hypothesis H, that is, given H, is denoted by $P(A \mid H)$ and is defined as

$$P(A \mid H) = \frac{P(AH)}{P(H)}. \tag{5.2.4}$$

If this definition is applied to the example just given, we note that $P(AH)$, the probability of the event "one boy—two girls," is 3/8, and that $P(H)$, the probability that a family of three children chosen at random will consist of at least one girl, is 7/8. It follows, then, from the definition of conditional probability, that $P(A \mid H)$ is 3/7. It should be noted that if the events A and H are independent, that is, if $P(AH) = P(A)P(H)$, then

$$P(A \mid H) = P(A).$$

Accordingly, events A and H are statistically independent if the absolute probability of A, $P(A)$ is equal to the conditional probability of A under hypothesis H. In the example given it is clear that A is not independent of H.

5.3. *Repeated trials.*—We have observed that when two events are independent in the probability sense, the probability of their joint occurrence is the product of their probabilities taken singly. The example chosen involved the sex of a child and the day of its birth. We might have elected, however, to consider repetitions of the same experiment by visualizing the successive outcomes of the experiment as independent of one another.

Consider a series of repetitions of an experiment in which there are only two possible outcomes in each experiment and in which their probabilities are constant throughout the trials. The sex of successive children in a family may be assumed to fill these requirements. Let p be the probability of a male child in one trial (birth) and let $q = 1 - p$ be the probability of a female child, the other possible outcome in one trial. Now two independent trials define 2^2 arrays of outcomes with probabilities defined, by multiplication, as pp, pq, qp, and qq. Similarly, three independent trials define 2^3 arrays of outcomes, with probabilities ppp, ppq, pqp, qpp, pqq, qpq, qqp, and qqq. If the sequence of events is ignored, we note that the eight different combinations correspond to four different sets of events, namely, in terms of the example, three boys, two boys and one girl, one boy and two girls, and three girls. Furthermore, we observe that the probabilities of these various sets of events are p^3, $3p^2q$, $3pq^2$, and q^3, respectively.

In N independent trials there exist 2^N different combinations, if order and number of the different events are considered, with the probabilities of these arrays defined by multiplication. If we consider order, then the probability of any specified order of one female child among N is $p^{N-1}q$. If order is ignored, that is, if all combinations of $N - 1$ boys and one girl are assumed to be essentially the same set of events, then the probability that among N children one will be a female is given by $Np^{n-1}q$, since there are N different combinations corresponding to this set of events—that is, to a family of N children only one of whom is a girl. Similarly, it may be shown that the probability that N children will contain k girls is $\binom{N}{k} p^{N-k}q^k$ where $\binom{N}{k}$ is merely $N!/[k!(N - k)!]$. This latter probability will be recognized as the general term in the binomial $(p + q)^N$. It is apparent from this that the probability that the number of females will be k, where $k = 0, 1, \ldots, N$, is given by the successive terms in the binomial expansion.

As an illustration of this method we may ask, "What is the probability that among five children two will be girls and three will be boys?" We can solve this problem either by the binomial or by enumerating the various possibilities. Let us assume that p is the probability of a single birth terminating in a male infant, and $p = q$, where q is the probability of a birth terminating in a female child, the only other possible outcome. The appropriate term in the binomial is that term with the exponent of p equal to 3, or $\binom{5}{3} p^3 q^2$. Now $p^3 q^2$ equals $1/32$, and the coefficient $\binom{5}{3}$ reduces to 10; hence the probability of a family of five consisting of three boys and two girls is $5/16$. If we had elected to enumerate exhaustively the different combinations, the following 32 arrays would have been found:

BBBBB

GBBBB, BGBBB, BBGBB, BBBGB, BBBBG

GGBBB, GBGBB, GBBGB, GBBBG, BGGBB, BGBGB, BGBBG, BBGGB, BBGBG, BBBGG

BBGGG, BGBGG, BGGBG, BGGGB, GBBGG, GBGBG, GBGGB, GGBBG, GGBGB, GGGBB

BGGGG, GBGGG, GGBGG, GGGBG, GGGGB

GGGGG

Of these 32 equally probable combinations of events, we note that 10 are favorable to the set of births involving three boys and two girls. The binomial expansion, therefore, leads us to the same conclusion as an exhaustive enumeration of the various alternatives, but with considerably less effort than must be expended in the latter enumeration.

As a second and slightly different illustration of the use of the binomial, consider the following problem: "What is the probability that five repetitions of the experiment, the sex of a child at a single delivery, will yield sets of events consisting of either three, four, or five boys?" The results of a single birth, or any similar experiment dealing with attributes, may be expressed in numerical form by the simple expedient of assigning, in this case, 0 to the birth of a female child and 1 to the birth of a male child. The sum, say S, of these values over a number of repetitions of the experiment would be equal to the number of experiments which terminated successfully if we assume that the birth of a male represents a successful experiment. Our problem, then, is, essentially, "What is the probability that in five trials $S \geq 3$?" If, again, p is the probability of a male at a single trial and q is the probability of a female at a single trial, then the probability that $S \geq 3$ is merely the sum of the probabilities that $S = 3$ or 4 or 5, that is,

$$P\,(S \geq 3)\ = \binom{5}{3}\,p^3\,q^2 + \binom{5}{4}\,p^4\,q + p^5\,.$$

Since $p = q$, we note that $p^3 q^2 = p^4 q = p^5 = 1/32$ and that the sum of the coefficients of these three terms is 16. It follows, then, that there is one chance in two that a family of five children, selected at random, will contain three or more boys.

5.4. *Mathematical expectation.*—In genetic problems where probability theory is applicable, we are frequently interested in the probabilities for a set of events, where the events may be simply described by numbers. For example, in the preceding problem we were interested in various values of S, the total number of males, and in their probabilities. We were interested in the probabilities of S being 3, 4, or 5; but we might equally well have been interested in the probabilities that S was 0, 1, or 2. These numbers, 0, 1, . . . , 5, can be considered as possible values of a chance variable S; and we realize that for each of these values of S there is a probability that S will be equal to the given value. In general, if $f(s)$ corresponds to the probability that $S = s$, that is, of getting s males, then we can call $f(s)$ the "probability distribution of S." In our example, since S can take on only integral values, for obvious reasons, we term S a "discrete variable."

It is often convenient to describe the probability distribution of a random variable in terms of a few typical values. The values most frequently chosen because of their usefulness are the mean or mathematical expectation and the variance. The term "mathematical expectation" is a legacy from the origin of probability theory in games of chance, where it was of interest to know the gambler's "mathematical expectation of gain." The mean value, or the mathematical expectation of a variable S, is indicated symbolically by $E(S)$ and is defined as the sum of the products of the values of the variable and their probabilities. For example, if the variable S assumes the values $s_1, s_2, \ldots, s_n$, with probabilities $p_1, p_2, \ldots, p_n$, then

$$E(S) = s_1 p_1 + s_2 p_2 + \ldots + s_n p_n . \tag{5.4.1}$$

Similarly, the mean value of the variable, say S^2, which assumes the values $s_1^2, s_2^2, \ldots, s_n^2$, with probabilities $p_1, p_2, \ldots, p_n$ is, by definition,

$$E(S^2) = s_1^2 p_1 + s_2^2 p_2 + \ldots + s_n^2 p_n = \sum_{i=1}^{N} s_i^2 p_i .$$

In the future the limits over which summation is to occur will not be indicated unless unnecessary confusion will result from their omission, i.e.,

$$\sum_{i=1}^{n} \text{ will be written as } \sum_{i} .$$

In the binomial distribution the mean number of children with a particular trait may be shown to be equal to Np as follows: The variable S in any trial may assume the values 0 or 1 with probabilities q and p. In the ith trial, the mean is

$$E(S_i) = p_i(1) + q_i(0) = p ;$$

however, there are N independent trials, and p is constant throughout; hence

$$E(S_N) = \sum_i E(S_i) = Np. \qquad (5.4.2)$$

The variance, previously mentioned, is defined as the mathematical expectation of the squared deviation of the variable from its mean, that is to say, as $E[S - E(S)]^2$. Now, to determine the variance, we note that

$$E[S - E(S)]^2 = E\{S^2 - 2S \cdot E(S) + [E(S)]^2\}.$$

When the braces on the right are removed, this becomes

$$E[S - E(S)]^2 = E(S^2) - E[2S \cdot E(S)] + E[E(S)]^2 ;$$

but, since the expectation of a constant is that constant,

$$E[E(S)]^2 = [E(S)]^2 \quad \text{and} \quad E[2S \cdot E(S)] = 2E(S)E(S).$$

Therefore,

$$E[S - E(S)]^2 = E(S^2) - [E(S)]^2 .$$

In the ith trial the variance is merely

$$E[S^2] - [E(S)]^2 = p - p^2 = pq .$$

Again, however, there are N independent trials, and p is constant throughout; hence the variance of S, the number of successes in N trials, is

$$E(S^2) - [E(S)]^2 = Npq . \qquad (5.4.3)$$

We may illustrate the calculation of the mean and variance by considering the distribution of the sexes among families of six children. Suppose that p is the probability of a boy and q is the probability of a girl. We observe that, since $p = q = 1/2$ and $N = 6$, the mean, Np, of this distribution is 3, and the variance, Npq, is 1.5.

Bibliography

DAVID, F. N. 1951. Probability theory for statistical methods. Cambridge: At the University Press.

FELLER, W. 1950. An introduction to probability theory and its applications. New York: John Wiley & Sons.

NEYMAN, J. 1950. A first course in probability and statistics. New York: Henry Holt & Co.

USPENSKY, J. V. 1937. Introduction to mathematical probability. New York: McGraw-Hill Book Co.

WILKS, S. S. 1949. Elementary statistical analysis. Princeton: Princeton University Press.

Problems

1. What is the probability that a family of seven children will consist of three boys and four girls? Of four boys and three girls? Of at least two girls?
2. If from a given marriage the probability that a child will have a specific inherited disease is 1/4, what is the mean number of affected children to be expected in a family of four? Of seven? What are the variances in these two instances?
3. From a given marriage, the probability that a child will have a specific defect is known to be 1/2. Given that at least two children in a four-child family are affected, what is the probability that all four children will have the defect?
4. The letters *e, e, g,* and *n* are placed in a box. From this box a letter is to be drawn at random. Once drawn, the letter is not to be returned to the box. What is the probability that the four draws necessary to empty the box will result in the word "gene" when the letters are arranged in the order of their draw? If the letters *c, e, h, m, m, o, o, o, r,* and *s* were placed in the box, what is the probability that ten successive draws would result in the word "chromosome" when the letters are arranged in the order in which they were drawn?
5. At the present time, approximately one death in every eight is due to cancer. If we assume that this is the "true" frequency of cancer deaths and that cancer deaths are randomly distributed, what is the probability that in a family of six children, four of whom are dead, three of the four will have died of cancer?

CHAPTER 6

The Dominant Gene in Man

IT IS customary to refer to a gene, A, as being "dominant" to an allele, a, when it is impossible by any known test to distinguish between AA and Aa individuals. This is not to say that differences do not exist but rather that they cannot now be detected. The term "dominant," then, refers not to any inherent property of the gene but to the state of our present knowledge concerning that gene. In other words, the term "dominant" describes what a geneticist observes with respect to the action of a gene, but it does not describe a specific, fixed, and immutable property of the gene. A gene which appears to be a dominant in a particular genetic background and environment may not be so in another environment and genetic background.

Most of the published reports concerned with heredity in man involve simple dominant inheritance. The techniques necessary to identify a particular trait as being due to a dominant gene differ with the frequency of the trait. Thus, if a trait is rare, the procedures necessary to test the hypothesis that it is due to a dominant gene are not in all respects the same as those applied when the trait is relatively common. We shall consider first the case of the rare trait and then the case of the common trait.

6.1. *Criteria of inheritance due to a single, completely dominant, rare, autosomal gene.*—If a trait is due to a single, completely dominant, rare autosomal gene, then individuals homozygous for the gene will be so infrequently encountered that in the analysis of pedigrees the possibility of their occurrence can be neglected, and all matings of affected and normal assumed to be of the $Aa \times aa$ type. Under these circumstances, one will observe that (1) the trait is transmitted directly from affected person to affected person, without skips or breaks in continuity; (2) the two sexes are affected in equal numbers; and (3) approximately half the children of an affected individual show the trait.

Figure 6-1 is a hypothetical pedigree of simple dominant heredity, from which the reasons for the foregoing statements will be apparent. An Aa-type individual produces two kinds of eggs or sperm, as the case may be, half containing the A gene and half the a gene. His or her normal mate can produce

41

only *a* gametes. Random union of eggs and sperm gives rise to two types of zygotes in equal numbers, *Aa* and *aa*. Figure 6-2 is an actual pedigree of dominant heredity, illustrating the three above-enumerated characteristic of this type of inheritance.

The literature contains many pedigrees tracing the inheritance of a dominant gene through six or even seven generations. Once a dominant gene makes its appearance in a family, it may persist for an indefinite period of time. Perhaps the best-documented example of this persistence of a trait due to a dominant gene involves a condition characterized by a fusion of the the first and second phalangeal bones of the fingers (Drinkwater, 1917). In

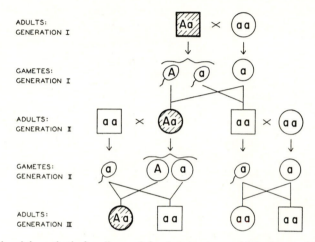

Fig. 6-1.—A hypothetical pedigree of dominant autosomal inheritance. Affected individuals are indicated by the heavily outlined symbols. Further explanation in text.

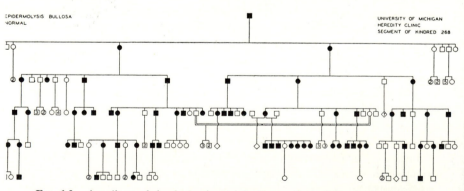

Fig. 6-2.—A pedigree of simple dominant inheritance, the trait in question being epidermolysis bullosa, a tendency of the skin to form large blisters following minor trauma.

1914, in the course of repairs on the Shrewsbury Cathedral in England, it became necessary to disinter the body of John Talbot, the first Earl of Shrewsbury, who had been killed in battle in 1453. A fourteenth-generation descendant of the Earl was in charge of the repairs. This latter individual had the abnormality of the fingers referred to above. The same abnormality was observed in the Earl's skeleton. The defect had persisted almost five centuries.

6.2. *The cytology of the human X- and Y-chromosomes.*—As noted earlier, in man, as in other mammals, sex determination is associated with an unequal-appearing pair of chromosomes, the male possessing an XY-pair and the female an XX-pair. The cytological studies of Koller and Darlington (1934) and Koller (1937) raised the possibility that in various mammals, including

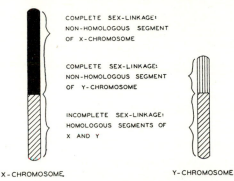

FIG. 6-3.—The assumed situation regarding homologous and nonhomologous segments of the X- and Y-chromosomes.

man, the X- and Y-chromosomes consist in part of morphologically and genetically similar segments which pair and may exchange sections at meiotic prophase. If this is so, then we must speak of "homologous" and "nonhomologous" portions of the sex chromosomes. The situation which on this basis may be postulated to exist in the male is shown in Figure 6-3.

The concept of homologous segments of the X- and Y-chromosomes has become widely disseminated throughout the literature on human genetics. However, Mathey (1951) and Sachs (1954) have questioned the accuracy of the original observations on which the concept is based. So much has been written about the "partial sex-linkage" exhibited by genes located on the homologous segments of the X and Y that no introductory text is complete without a description of the possibility. However, it is suggested that the student not only regard as conjectural the spatial relationships of the homologous and nonhomologous segments shown in Figure 6-3 but, indeed, con-

Solution if all daughters of men & all children of women are affected. Not good model.

44   HUMAN HEREDITY

sider the very existence of a homologous portion of the X- and Y-chromosomes as not definitely proved.

Assuming for the moment that the chromosomal relationships are as depicted in Figure 6-3, then there are three different possible patterns of inheritance which may be exhibited by dominant genes located in the sex chromosomes. These will be considered in sections 6.3–6.5.

6.3. *Criteria of inheritance due to a single, completely dominant, rare, sex-linked gene.*—Consider, first, a rare dominant gene located in the nonhomologous segment of the X-chromosome. The locus at which such a gene occurs in represented only once in the male, but twice in the female. Males with a particular sex-linked gene at a particular locus are frequently referred to as "hemizygous" for that gene. Since we are assuming that the trait is rare, the probability that an affected female is homozygous may be disregarded. A trait due to a completely dominant gene located in this segment will exhibit a distinctive pattern of inheritance, as follows: (1) the trait will be transmitted directly from affected person to affected person, without skips or breaks in continuity; (2) affected males married to normal females will transmit the trait only to their daughters, not to their sons; (3) affected females married to normal males will transmit the trait to sons and daughters in equal numbers, half the offspring of each sex being affected; and (4) in the population as a whole, there will be approximately twice as many females who show the trait as males (why?).

Reference to Figure 6-4 will establish the reasons for this pattern of inheritance. Very few pedigrees illustrating the complete dominance of a sex-linked gene located in the nonhomologous segment of the X-chromosome are known in man. That is to say, genes so located are, by and large, either recessive or have differing effects in the male and the female (cf. chap. 8).

6.4. *Y-linked (holandric) inheritance.*—We will consider next the behavior of a rare dominant gene located in the nonhomologous segment of the Y-chromosome. A few examples of what appears to be inheritance due to such genes are known. A trait due to a gene in the nonhomologous portion of the Y-chromosome will be transmitted from an affected male to all his male offspring; females will not be affected. Thus a pedigree of webbed toes has been described in which fourteen males in four generations showed the trait, all the male offspring of each affected male showing the webbing (Schofield, 1922).

6.5. *Criteria of inheritance due to a single, rare, completely dominant, partially sex-linked gene.*—This brings us to the third possible type of dominant, sex-linked inheritance. Shortly after recognition that the X- and Y-chromo-

$$\lim \frac{a_n}{b_n} = \gamma$$

where γ is solution of

$$\frac{1-\gamma}{2-\gamma} = \gamma \qquad \gamma = \frac{3-\sqrt{5}}{2}$$

somes of mammals might consist in part of cytologically similar portions, Haldane (1936) investigated the question of whether by the appropriate genetic tests it would be possible to recognize in man genes which exhibited the behavior one might expect of genes located in this so-called "homologous" segment. Such genes could be spoken of as exhibiting partial or incomplete sex-linkage, since, in any particular family, they would tend to follow the nonhomologous (differential) portion of either the X- or the Y-chromosome, showing a change from association with one to the other only if crossing-over had occurred. In his search for such genes, Haldane was un-

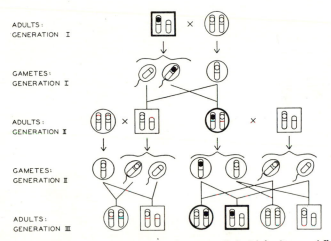

Fig. 6-4.—A hypothetical pedigree of dominant, sex-linked inheritance. Affected individuals are again indicated by the heavily outlined symbols. Further explanation in text.

doubtedly stimulated by the fact that in 1921 Aida had demonstrated the occurrence of just such a type of inheritance in certain fish.

Several means of identifying heredity due to a rare, dominant, partially sex-linked gene are available (cf. Macklin, 1944). We shall cite but two.

1. If one has at his disposal a collection of *extensive* pedigrees segregating for a particular trait, it will be apparent that in some pedigrees the trait tends to follow the Y-chromosome, resulting in a marked excess of affected males, whereas in other pedigrees the gene will tend to behave like an ordinary sex-linked dominant. It is necessary, however, to have *extensive* pedigrees before one can distinguish this phenomenon from chance fluctuations in the sex ratio among affected persons.

2. Another method of identifying dominant partial sex-linkage is the one used originally by Haldane (1936). If there is such sex-linkage, then the *affected children of affected males* will, in the absence of crossing-over, be of

the same sex as the affected paternal grandparent, as shown in Figure 6-5. Even if a certain amount of crossing-over does occur, involving a transfer of the gene in question from the X- to the Y-chromosome, there will still be a tendency for the foregoing statement to be correct, unless, as can be assumed to be the case only rarely, crossing-over is very common. On the other hand, if there is no partial sex-linkage, the children of an affected male will be expected to be of the same sex as the affected grandparent in half the cases, and not in the other half. Haldane tabulated all the pedigrees of dominantly inherited retinitis pigmentosa in the literature in which the findings in both

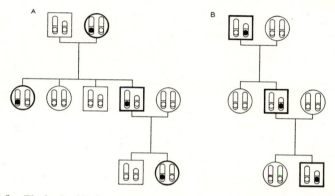

FIG. 6-5.—The basis of Haldane's test for dominant partial sex-linkage. Note how, in the absence of crossing-over, the affected individual is of the same sex as his or her affected paternal grandparent. Even if crossing-over does occur, the tendency to resemble the affected paternal grandparent will persist as long as their is less than 50 per cent recombination.

the parents and children of an affected father were known; his observations are shown in the accompanying tabulation. The departure from equality in

Children	Affected	Unaffected
Of like sex with affected paternal grandparent.....	81	60
Of unlike sex with affected paternal grandparent...	64	83

the number of affected and unaffected, for both kinds of children, can be shown to be significant, thus suggesting that, in at least some pedigrees, retinitis pigmentosa is inherited as if due to a dominant, partially sex-linked gene. Although many other dominantly inherited diseases were investigated in this respect, none showed similar sex-ratio deviations consistent with dominant partial sex-linkage. It should be pointed out that if there is considerable crossing-over between a partially sex-linked gene and the nonhomologous segments, it is very difficult to distinguish this type of inheritance from true autosomal heredity.

6.6. *Criteria of inheritance due to a single, completely dominant, common, autosomal gene.*—Thus far we have restricted ourselves to expectation when the trait in question is rare, and individuals homozygous for the responsible gene are not encountered. When the trait is more common, the analysis of pedigrees is somewhat more difficult. Affected individuals now may be either AA or Aa, and matings between affected and normal persons are thus either $AA \times aa$ or $Aa \times aa$. In the first type of mating, all the children will be affected; in the second type, only half. Considering the results of both types of marriage together, something over half the children will be affected. Furthermore, we will now encounter marriages between two affected persons, and these marriages can be of three types, namely, $AA \times AA$, $AA \times Aa$, and $Aa \times Aa$. In the first two types, all the children will be expected to be affected, while in the third, only three-quarters (why?). Considering the results of all the types of marriage involving two affected persons, something over three-quarters of all the children will be affected.

In testing a body of genetic data for agreement with the hypothesis that the trait in question is due to a common, dominant gene, one first inspects the pedigree or collection of pedigrees to see whether the facts are consistent with dominant inheritance. If they are, then the next step in the analysis is possible only if one has accurate figures on the frequency of the trait in the population. If one has such figures, then it is possible to calculate the expected frequency of AA, Aa, and aa individuals. From this, one can estimate the expected relative frequency of the various types of matings in the population under study, and the expectation with reference to the children of these matings. As the final step, one compares the observed with the expected results on the basis of hypothesis. Significant discrepancies mean that the hypothesis must be re-examined. This problem will be treated in some detail in section 13.6 (p. 197).

6.7. *Criteria of inheritance due to a single, completely dominant, common, sex-linked gene.*—Where a trait due to a completely sex-linked gene is relatively common, affected females may be of two types, either AA or Aa. Affected males, as before, will all be AY. The first two conditions mentioned on page 44 will again be satisfied. Since, however, affected females may now be either AA or Aa, condition 3 on page 44 will no longer hold. Matings of affected females with normal males will be of two types, $AA \times aY$ and $Aa \times aY$. All the sons and daughters of the first type of mating will be affected, but only half the sons and daughters of the latter type. Taking all the marriages of affected females $\times$ normal males together, something over half the sons and daughters of such marriages will show the trait. If we know the exact frequency of the trait in the population, then the exact proportion of affected sons and daughters expected can be calculated. We shall also return to this problem in chapter 13.

When a trait due to a sex-linked dominant gene is rare, then, as noted in condition 4 on page 44, affected females are approximately twice as common as affected males. But where the trait is relatively common, so that the occurrence of AA females cannot be disregarded, although there will still be an excess of affected females, these will now be less than approximately twice the number of affected males.

6.8. *Certain items of interest in connection with dominant inheritance.*—There is the general impression, difficult to document with real accuracy, that more of the inherited traits known in man seem to be due to dominant genes than is the case for other animals, such as Drosophila and the mouse (Levit, 1936). This difference between man and other animals may be more apparent than real. The characteristics of man are both more familiar to us and more readily detected than are those of a mouse or a fly. Any genetically determined departure from normality is thus more readily detected in man than in the fly. For instance, the trait brachydactyly, characterized by a shortening of the fingers, was one of the very first in man shown to be inherited as if due to a dominant gene. A similar morphological abnormality would be very easily missed in the mouse. In other words, it may be that the same proportion of genes have effects when heterozygous in man, the mouse, and the fruit fly; but we are better able to detect these effects in man, in consequence of which there are more *apparently* dominant genes in man. A second factor in the apparently greater number of dominant genes in man may be the relative ease with which dominant heredity may be identified, as contrasted with recessive.

There are two commonly encountered misapprehensions concerning dominant (and recessive) inheritance which can be disposed of at this point. The first is that a trait which is truly inherited should be present at birth. The medical literature contains many pedigrees of dominant inheritance in which the trait in question fails to appear until age twenty or thirty or later. Huntington's chorea will serve as an example. This disease is characterized by the development, at about age thirty or forty, of mental deterioration accompanied by involuntary movements of the face, body, and extremities. These symptoms are due to an actual degeneration of certain brain centers. Figure 6-6 is a pedigree of this disease, meeting all the requirements of dominant heredity. Hereditary disease may develop at any time during life: It is not the disease as such but the predisposition to the disease which is inherited. The reason why individuals of certain genetic backgrounds initially appear to function normally and then, sometimes gradually, sometimes suddenly, develop a given disease is not clear. There would appear to be a genetically determined "weakest link," which sooner or later must snap. The possibility exists that in time, with further knowledge, we shall be able to locate that link and, by the appropriate medical approach, reinforce it before the damage is done.

A second very elementary misconception has to do with the relation between the outcome of successive pregnancies. When one tosses a coin, the fact that the preceding flip terminated in "heads" has no relation to the outcome of the present flip. Just so in dominant inheritance: the fact that the

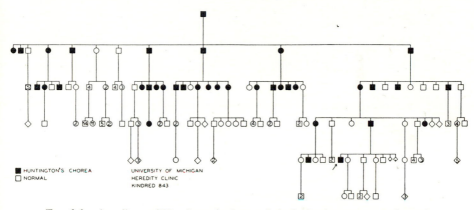

FIG. 6-6.—A pedigree of Huntington's chorea. A single dominant autosomal gene is responsible for the onset at about age thirty to forty of the degeneration of various areas in the brain. In many pedigrees there appears to be a deficiency of affected persons because of the death of carriers of the gene before the age of onset.

preceding child exhibits a dominantly inherited trait has no bearing on whether the next child will or will not show the trait. The results of each pregnancy are stochastically independent of any other. If one observes a sufficiently large collection of families, all possible proportions and sequences of affected children will be observed, in a manner predictable from the binomial expression $(p + q)^n$, where p = probability of normality = $1/2$, q = probability of showing trait = $1/2$, and n = number of children in the family.

6.9. *The concept of penetrance.*—There are traits which appear to be inherited as if due to a dominant gene but for the fact that occasionally, in a family in which segregation for the gene is occurring, a normal-appearing offspring of a person with the trait will produce affected offspring. Figure 6-7 illustrates just such a pedigree, the trait in question being polydactyly.

The most reasonable explanation of such a pedigree is that the normal trans-
mitters have really received the gene in question but for some reason it has
failed to find expression. We refer to such genes as "incompletely penetrant."
Presumably, the fact that such a gene is penetrant in one individual and not
in another is related to environmental variables, including factors operative
in early uterine development, and genetic modifiers of the type discussed in
chapter 4. When the penetrance of a dominant gene falls to 20 or 30 per cent,
the pattern of affected individuals within a family is such as to make conclusions concerning the type of heredity difficult. The reader interested in the
problems involved in estimating penetrance will find a good discussion in
Allen (1952).

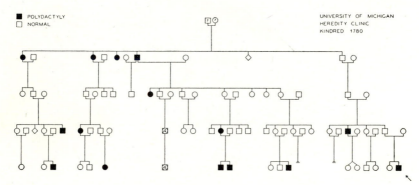

Fig. 6-7.—A pedigree of polydactyly apparently due to an irregularly dominant
gene.

6.10. *Hereditary syndromes.*—Thus far we have concerned ourselves with
genes with a single known effect. Not infrequently in human inheritance we
encounter constellations of characteristics which tend to be inherited together. For instance, in certain rare families one regularly encounters a
combination of distichiasis (a double row of eyelashes) and lymphedema of
the lower extremities (marked edematous swelling of the legs). Figure 6-8
summarizes one such family to come to the authors' attention. On the basis
of what we know concerning human embryology, it is difficult to visualize
any developmental connection between the two traits. Two explanations
have been advanced for such constellations; one is that two different, closely
linked genes are involved, and the other is that a single gene with multiple
effects is responsible. If it is a matter of close genetic linkage, than both the
traits in question should be encountered as single genetic entities far more
frequently than the combination of the two. As a matter of fact, in this particular instance, both traits have been encountered separately, both distichiasis and hereditary lymphedema appearing in some families as dominant

traits; from the medical literature it would appear that the two traits are encountered separately considerably more often than they are found in association with each other. At the present time there is no way in this particular syndrome to reach a firm decision between the two possibilities, i.e., linkage versus multiple effects of a single gene.

Figure 6-9 is concerned with a more involved problem. Marfan's disease in its complete form is characterized by a triad of findings consisting of an extremely slender body build, a dislocation of the lens of the eye, and signs of congenital heart disease. Such a patient is shown in Figure 6-10. There is a wide variation in the extent to which each aspect of the triad is expressed. Body build, although usually slender, may be almost normal. There

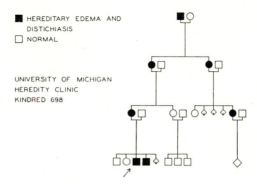

FIG. 6-8.—A pedigree of distichiasis associated with Milroy's disease (hereditary edema). The combination behaves in this family as if due to a single dominant gene.

may be minimal or no dislocation of the lens. The heart murmur which indicates congenital heart disease may be absent. In any particular affected individual, then, there is a wide variation in the manifestations of the syndrome—mildly affected individuals are sometimes spoken of as *formes frustes*. Some authors have attempted to explain this variation in the disease as the result of the dissociation, through crossing-over, of a complex of three linked genes, each of the three genes being associated with one of the cardinal findings of the syndrome. Such an explanation fails to account for such parent-child combinations as *II-2* and *III-3* or *II-3* and *III-9* of Figure 6-9; for on this explanation, once an aspect of the triad has been lost (through crossing-over), it should not reappear in subsequent generations. Rather, it seems more likely that a single gene whose effects are subject to environmental and genetic modification is responsible. In this connection it should be pointed out that there are known in both the fruit fly and the mouse a number of "syndromes" which, by the most rigorous genetic tests, appear to be due to a single gene (summary in Caspari, 1952).

Some of the hereditary syndromes now recognized in man are represented by even more striking constellations of abnormalities than is the case for Marfan's disease. For instance, the so-called "Laurence-Moon-Biedl syndrome" in its full expression is characterized by pigmentation of the retina of the eye, polydactyly, mental retardation, obesity, and sexual underdevelopment. Hurler's syndrome (gargoylism) is a combination of mental retardation, opacities of the cornea, partial claw hand, abnormal body proportions, changes in the appearance of the white blood cells, and, less constantly, congenital heart disease. In osteogenesis imperfecta the characteristic findings are abnormally fragile bones, blue sclera, hyperelasticity of the skin and joints, and early deafness. These and other syndromes are described in a chapter in Gates (1946).

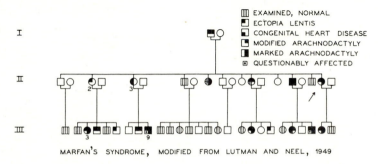

MARFAN'S SYNDROME, MODIFIED FROM LUTMAN AND NEEL, 1949

FIG. 6-9.—A pedigree of Marfan's syndrome, illustrating the variability in the expression of various aspects of the syndrome. (Modified from Lutman and Neel, Arch. Ophth., **41**:276–305, 1949.)

It is difficult to visualize how one gene can be responsible for so diverse a collection of abnormalities. However, recent findings with respect to still another syndrome, known as "Wilson's disease," illustrate in striking fashion how apparently unrelated abnormalities may have a common basis. In Wilson's disease there is the onset, usually in the second decade, of the degeneration of certain brain centers, the replacement of the normal liver cells by fibrous tissue, and the appearance of a peculiar pigmented ring in the cornea of the eye. More recently, the excretion in the urine both of (1) abnormally large amounts of peptides with terminal dicarboxylic amino acids and (2) amino acid residues (aminoaciduria) has been found to be characteristic of the disease. The syndrome is due to a recessive gene (Bearn, 1953).

As long ago as 1922 Siemerling and Oloff suggested that the disease might be due to an abnormal accumulation of copper in the affected tissues. Subsequent investigations did, in fact, demonstrate a marked increase in the copper content of the brain, eye, and liver. It was then shown that patients with Wilson's disease had abnormally large amounts of copper in their urine. The next important development in this field came from an unexpected quarter in a manner which illustrates at best what is sometimes termed the "unity of science" and, at the least, the unexpected ramifications of scientific discovery. BAL is the commonly used medical abbreviation for a drug whose

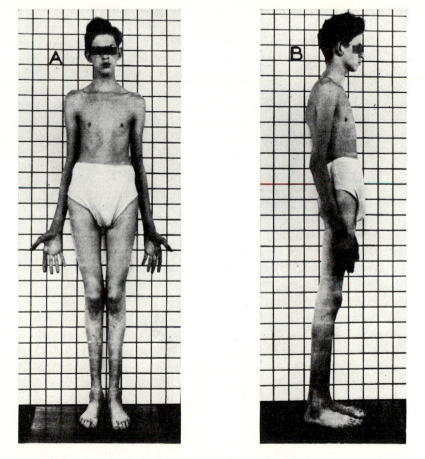

FIG. 6-10.—A young man with Marfan's syndrome. (From Lutman and Neel, 1949, by permission of Archives of Ophthalmology.)

full name is "British anti-lewisite," a compound developed by British investigators during World War II as an antidote for poisoning with arsenic-containing gases. Subsequently it was found that this compound was effective in the treatment of poisoning with such other metals as mercury, gold, and silver. In 1948 Cummings (see also Denny-Brown and Porter, 1951) reported that when patients with Wilson's disease are treated with BAL, there is a striking increase in the amount of copper excreted in the urine, followed in a period of several weeks by improvement in the clinical condition of the patient.

In recent years there has been a great deal of interest in determining the precise nature of the defect in copper metabolism. It has been shown that the amount of copper present in the serum is abnormally low in this disease (Bearn and Kunkel, 1952). Moreover, Scheinberg and Gitlin (1952) have found affected persons to be deficient in ceruloplasmin, the serum protein to which most, if not all, of the copper normally present in serum is bound. Finally, Zimdahl, Hyman, and Cook (1953) have shown that patients with the disease absorb from the digestive tract much more of the copper ingested as part of the diet than do normal persons. It has been suggested that this increased absorption of copper may be in response to the abnormally low ceruloplasmin level. While it appears that an abnormality in copper metabolism is basic to many of the various superficially unrelated manifestations of Wilson's disease, the final details are still obscure.

That the altered copper metabolism may not be the ultimate defect is suggested by the fact that the aminoaciduria may precede the other manifestations of the disease by some years. It has been suggested that this aminoaciduria, with the resultant chronic loss of amino acids, rather than the abnormal copper metabolism may be the cause of the liver disease. It has further been suggested that there is a deficiency in certain enzymes concerned with the peptide breakdown, with the aminoaciduria actually secondary to the kidney's inability to handle peptides (Uzman and Hood, 1952). How this is related to the abnormal copper metabolism is not clear; but one could surmise, for instance, that the chronic loss of certain amino acids and/or peptides interferes with the system responsible for copper metabolism. But although some of the details are still obscure, the studies thus far have clearly shown how superficially unrelated defects may actually have a common metabolic basis.

The hereditary syndromes are a striking demonstration of a fact which has been slowly emerging from work on lower plants and animals, namely, that it is the rule rather than the exception for one gene to have a variety of effects. Why this should be the case will become more apparent in chapter 12. Suffice it to say here that so interrelated are the developmental sequences of the body that any deviation from the normal may have extensive ramifications, the more so, the earlier in development the deviation occurs.

Bibliography

SPECIFIC REFERENCES

AIDA, T. 1921. On the inheritance of color in a fresh-water fish *Aplocheilus latipes* Temmick and Schegel, with special reference to sex-linked inheritance, Genetics, 6:554–73.

ALLEN, G. 1952. The meaning of concordance and discordance in estimation of penetrance and gene frequency, Am. J. Human Genetics, 4:155–72.

BEARN, A. G. 1953. Genetic and biochemical aspects of Wilson's disease, Am. J. Med., 15:442–49.

BEARN, A. G., and KUNKEL, H. G. 1952. Biochemical abnormalities in Wilson's disease, J. Clin. Investigation, 31:616.

CASPARI, E. 1952. Pleiotropic gene action, Evolution, 6:1–18.

CUMMINGS, S. N. 1948. The copper and iron content of brain and liver in the normal and in hepatolenticular degeneration, Brain, 71:410.

DENNY-BROWN, D., and PORTER, H. 1951. The effect of BAL (2,3-dimercaptopropanol) on hepatolenticular degeneration (Wilson's disease), New England J. Med., 245:917–25.

DRINKWATER, H. 1917. Phalangeal synostosis anarthrosis (synostosis, ankylosis) transmitted through fourteen generations, Proc. Roy. Soc. Med. (sec. path.), 10:60–68.

HALDANE, J. B. S. 1936. Search for incomplete sex-linkage in man, Ann. Eugenics, 7:28–57.

———. 1941. Data on incomplete sex-linkage of spastic paraplegia, J. Genetics, 41:141.

KOLLER, P. C. 1937. The genetical and mechanical properties of sex chromosomes. III. Man, Proc. Roy. Soc. Edinburgh, 57:194–214.

KOLLER, P. C., and DARLINGTON, C. D. 1934. The genetical and mechanical properties of the sex chromosomes. I. *Rattus norwegicus*, J. Genetics, 29:159–73.

LEVIT, S. G. 1936. The problem of dominance in man, J. Genetics, 33:411–34.

MACKLIN, M. T. 1944. Xeroderma pigmentosum, Arch. Dermat. & Syph., 49:157–71.

MATHEY, R. 1951. The chromosomes of the vertebrates, Advances in Genetics, 4:159–79.

SACHS, L. 1954. Sex-linkage and the sex chromosomes in man, Ann. Eugenics, 18:255–61.

SCHEINBERG, I. H., and GITLIN, D. 1952. Deficiency of ceruloplasmin in patients with hepatolenticular degeneration (Wilson's disease), Science, 116:484–85.

SCHOFIELD, R. 1922. Inheritance of webbed toes, J. Hered., 12:400–401.

UZMAN, L. L., and HOOD, B. 1952. Familial nature of amino-aciduria of Wilson's disease (hepatolenticular degeneration), Am. J. M. Sc., 223:392–400.

ZIMDAHL, W. T.; HYMAN, I.; and COOK, E. D. 1953. Metabolism of copper in hepatolenticular degeneration, Neurology, 3:569–76.

GENERAL REFERENCES

GATES, R. R. 1946. Human genetics, chap. 18, Hereditary syndromes. New York: Macmillan Co.

Problems

1. Assume that the following pedigree involves a trait due to a rare, dominant, partially sex-linked gene. Which affected individual in the pedigree must be attributed to crossing-over, and is the gene in question located on the X or Y in the affected individual in generation 1?

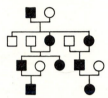

2. How would you interpret the following pedigree of diabetes mellitus?

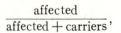

3. In the pedigree of epidermolysis bullosa in Figure 6-2, what is the chance that a child from a marriage of two affected persons will be normal?

4. What normal persons in Figure 6-8 may be assumed to be carriers of the irregularly dominant gene responsible for polydactyly in this family? Using the formula

$$\frac{\text{affected}}{\text{affected} + \text{carriers}},$$

what is the penetrance of the gene in this family? Why is this a maximum estimate of the penetrance?

5. Comment on the value of the concept of penetrance in human heredity.

CHAPTER 7

The Recessive Gene in Man

A GENE a is said to be "recessive" to an allelomorph, A, when in the combination Aa there is no detectable effect of a. The term "recessive," like the term "dominant," refers to what the geneticist observes with respect to the action of a given gene under a given set of circumstances rather than to a fixed and unchanging property of the gene. To each of the three simple modes of transmission of a dominant gene discussed in the preceding chapter, there corresponds a similar form of recessive inheritance; hence, a recessive gene may be autosomal, sex-linked, or partially sex-linked. Just as was true for dominant inheritance, so when one analyzes pedigrees for agreement with a recessive type of inheritance, expectation differs somewhat according to the frequency of the trait in question. We shall consider first the case for the rare, recessively inherited trait and then the case for the more common trait. It must be emphasized that the genetic processes are the same in the two cases and that the distinction is largely one of convenience.

7.1. *Criteria of inheritance due to a single, completely recessive, rare, autosomal gene.*—If a trait is due to a single, completely recessive, rare, autosomal gene, then (1) the father, mother, and more remote ancestors of an affected individual are usually normal; (2) on the average, one-quarter of the siblings of an affected individual will also be affected; (3) both sexes are affected and equally often; and (4) an increase may be observed in the incidence of consanguinity among the parents of affected individuals.

Figure 7-1 illustrates the basis for these findings. In the usual case, both parents of an individual exhibiting a recessively inherited trait are heterozygous for the gene in question, the gene having been transmitted undetected from heterozygote to heterozygote for many generations. Under these circumstances there is a one-in-two chance that either the egg or the sperm will receive the recessive gene, and so a $1/2 \times 1/2$, or one-in-four, probability that the fertilized egg will be homozygous for the gene. The chance that the father will contribute the abnormal gene and the mother its normal allele is $1/2 \times 1/2$, or $1/4$, while the chance that the reverse will be the case, i.e., mother contributing the abnormal gene and father the normal, is also $1/4$. The total chance of a fertilized egg heterozygous for the abnormal gene is thus

57

1/4 + 1/4, or 1/2. Finally, there is a one-in-four probability that the zygote will receive the normal gene from both parents. On the average, then, three out of four children of such a marriage will appear normal; but, of the normal-appearing children, two out of three will be heterozygous carriers of the gene in question. Figure 7-2 is an actual pedigree of recessive inheritance, the trait in question being retinitis pigmentosa, a degenerative disease of the retina characterized by atrophy and pigmentary infiltration.

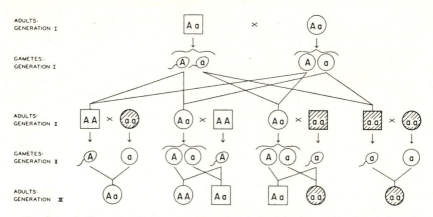

Fig. 7-1.—A diagrammatic representation of recessive inheritance. Further explanation in text.

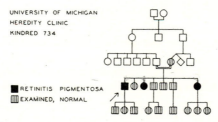

Fig. 7-2.—A pedigree of the recessively inherited form of retinitis pigmentosa

7.2. Criteria of inheritance due to a single, completely recessive, rare, sex-linked gene.—A sex-linked recessive gene, as in the case of a sex-linked dominant, is one which is borne on the nonhomologous section of the X-chromosome. One of the oldest of the recognized familial diseases is hemophilia, a sex-linked recessive trait characterized by a tendency to severe bleeding. The ancients recognized its familial transmission, and the Talmud, which dates from not later than the fifth or sixth century, granted dispensation from circumcision to the male members of a family with a history of profuse bleeding following circumcision; and this dispensation extended to the sons of all the sisters of the mother but not to the sons of the siblings of the father. A study of pedigrees of hemophilia would yield the following distinctive pattern characteristic of all rare sex-linked recessive traits: (1) the trait will show distinct discontinuities in its descent from generation to generation; (2) affected males will not produce affected children (except in the very uncommon event that they happened to be married to carrier females); (3) the affected males will themselves be the offspring of normal females, who will tend to produce as many normal males as they produce affected males; and (4) in the population as a whole, there will be many more affected males than females. When affected females occur, then (5) they will be the off-

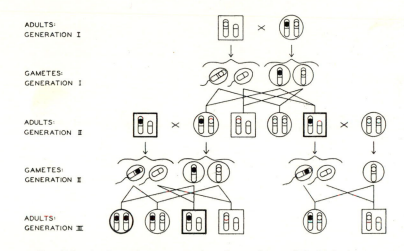

FIG. 7-3.—A schematic representation of recessive sex-linked inheritance

spring of an affected father whose spouse may or may not be affected, and (6) they will tend to occur more frequently in matings involving an affected male and a female relative than from matings involving an affected male and a nonrelated female.

Figure 7-3 is a hypothetical pedigree of sex-linked inheritance, illustrating the cytological basis for the pattern of sex-linked heredity. Figure 7-4 is an actual pedigree of hemophilia.

Some forty-odd sex-linked pathologic traits now are known in man; eighteen different sex-linked genes involving the eye alone have been recognized (Falls, 1952). Of these sex-linked traits, the majority are recessive in their mode of inheritance; a few such traits, in addition to hemophilia, are color blindness, ichthyosis simplex (fish-skin), anhidrotic ectodermal dysplasia (the absence of sweat glands), and megalocornea (abnormally large cornea). By comparison with other mammals, a disproportionately large fraction of the inherited traits of man appear to be sex-linked. The frequency with which sex-linked traits have been recognized in man may be in part due to the fact that the unusual behavior of sex-linked genes leads to their relatively easy recognition. In addition to this, however, it seems that sex-linked traits may actually be relatively more common in man than in many other mammals. Thus, despite the large amount of work done on the genetics of the house mouse, as of this writing only three certain and two possible sex-linked traits have been recognized in this animal (Falconer, 1952; Garber, 1952). Only one sex-linked trait is known in cats (yellow) and only one in dogs (hemophilia).

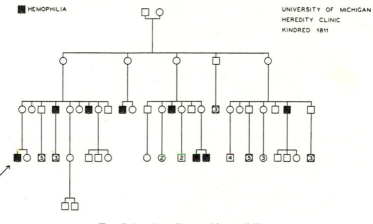

FIG. 7-4.—A pedigree of hemophilia

7.3. Criteria of inheritance due to a single, completely recessive, rare, incompletely sex-linked gene.—In chapter 6 the genetic pattern to be expected from a dominant gene located on the homologous section of the X- and Y-chromosomes was discussed. Recognition of a trait as being due to a partially sex-linked, recessive gene is somewhat more difficult. The criteria are similar to those obtaining in the case of a sex-linked recessive, but they differ in several important respects. Thus, in this type of heredity, the four conditions listed for autosomal recessive inheritance are all met, but, in addition, (1) traits due to such genes tend to appear in one sex only in any one family; (2) the sex of the affected offspring is, in the absence of crossing-over, like that of the paternal grandparent from whom the gene was inherited; and (3) where the trait appears in the offspring of a consanguineous marriage, affected females will tend to be more frequent than males, since the latter can arise only as the result of crossing-over.

Examination of Figures 7-5 and 7-6 will show why these statements are correct. (The reader who is not familiar with the concepts of linkage and crossing-over is advised to delay reading further in this section or studying Figs. 7-5 and 7-6 until he has read chap. 10.) With respect to the second statement, it is obvious that we will not usually know through which paternal grandparent the gene is derived. However, there are two circumstances under which we can be reasonably sure as to the contributory paternal grandparent: (1) If a maternal relative of the father had the same defect that was present in the father's children, it may be assumed with a fair degree of certainty that it was through his mother that he received the gene, whereas if the relative was a paternal one, then it would be through his father that the gene was derived. (2) If the marriage giving rise to an affected child is consanguineous, it is reasonable to assume that the parents have received the gene for the defect through a common ancestor. This assumption is certainly more probable than that they received it through unrelated ancestors. Therefore, if the pedigree gives the information as to the manner in which the husband is related to his wife, one can determine with a fair degree of certainty the parent through whom the husband received his gene for the defect. A compilation of such consanguineous marriages will reveal whether the affected offspring resemble in sex the transmitter paternal grandparent.

Among the traits which have been analyzed by various investigators and suspected of showing partial, recessive sex-linkage are xeroderma pigmentosum, total color blindness, recessive retinitis pigmentosa (some pedigrees), recessive epidermolysis bullosa, hereditary hemorrhagic diathesis, spastic paraplegia, and Oguchi's disease. In view of the fashion in which apparent instances of partial sex-linkage are "selected" out of collections of pedigrees of various diseases, it is difficult to be dogmatic at present concerning the occurrence and relative frequency of partial sex-linkage. Moreover, Macklin (1944, 1952) has pointed out, in connection with an analysis of xeroderma pigmentosum, certain peculiarities in the sex ratio and in the proportions of affected males and females which are unexplained.

7.4. Criteria of inheritance due to a single, completely recessive, common, autosomal gene.—When a trait is due to a common, autosomal recessive gene, heterozygous carriers of the gene will frequently be encountered in the population. Marriages of individuals who exhibit the trait to apparently normal persons will thus often be of the type $aa \times Aa$. Half the children of such marriages will be aa and so affected. This, of course, is the same ratio that would be encountered if the trait were due to a completely dominant gene, a circumstance which in a small collection of pedigrees might give rise to some confusion.

There are three findings whereby, on the basis of data involving two generations, we can distinguish between the hypothesis that a trait is due to a common recessive gene and the hypothesis that it is due to a dominant gene. (1) In contradistinction to the situation with respect to the dominant, we will, in the case of recessive inheritance, also find affected persons arising from the marriage of two normal persons, in the ratio of one affected to three normal. However, as noted in chapter 6, a dominant gene may sometimes be irregularly expressed. In such cases, affected persons may also arise from the marriage of two apparently normal persons. (2) Where a trait is due to a recessive gene, the marriage of two affected persons must

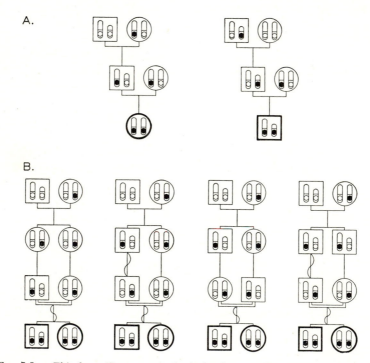

Fig. 7-5a.—This figure illustrates the basis for the assertion made in the text that, in the absence of crossing-over, the sex of an offspring affected with a partially sex-linked recessive trait will be the same as that of the paternal grandparent from whom one of the genes was derived.

Fig. 7-5b.—This figure and Fig. 7-6 illustrate the basis for the assertion made in the text that where a partially sex-linked recessive trait appears among the offspring of a consanguineous marriage, there will be an excess of affected females (modified from Macklin, 1952, by permission from the American Journal of Human Genetics). In this figure we consider the case wherein the gene is introduced through the *female* ancestor common to the cousins. Affected offspring can arise from first-cousin matings as a result either of the com-

always result in similarly affected children if children are produced, whereas this is not so for dominant inheritance. (3) Finally, where a trait is assumed to be due to a recessive gene, we can, if we know the frequency of the trait, calculate what fraction of marriages of affected × normal are $aa \times Aa$, and what fraction $aa \times AA$, only the former yielding affected children. In a collection of such marriages, we can test whether the results agree with hypothesis. This problem will be treated in chapter 13.

7.5. *Criteria of inheritance due to a single, completely recessive, common, sex-linked gene.*—When a trait is due to a common, recessive, sex-linked gene, we may occasionally encounter affected females, and, of course, a high pro-

---

bination of derivative genes, that is, of genes derived from a common gene, or of the fortuitous combination of like but nonderivative genes. If the latter situation obtains, then the sex most frequently affected depends solely upon the paternal grandparent from whom one of the genes was derived (Fig. 7-5a). If, on the other hand, affected individuals are the result of the combination of derivative genes, then an excess of affected females is to be expected among all first-cousin matings. We note, for example, that when first cousins are the offspring of sisters, both cousins may be heterozygous without requiring crossing-over of the gene from the X- to the Y-chromosome (see *left-most mating*). Such a mating can, in turn, give rise to affected females without crossing-over; but for affected males to occur, crossing-over of the gene from the X- to the Y-chromosome in the male parent is a necessity. In the diagram the points of crossing-over are indicated by a wavy line; the base of this wavy line points toward the crossover phenotype. Similarly, if first cousins are the progeny of brothers (*right-most mating*), then, although both brothers may be heterozygous without crossing-over, if they are to pass the gene along to their sons, crossing-over must occur, since the father transmits only a Y-chromosome to his sons. Hence, in this case, if both first cousins are to be heterozygous, at least one crossover must take place; otherwise stated, one of the cousins must be a noncrossover phenotype, and the other (always the male in this instance) must be a crossover phenotype. Such a mating could give rise to affected males without further crossing-over; but, for affected females to occur, another crossing-over in the male parent is necessary. We observe, now, that only four types of first-cousin matings are possible when the gene is introduced through the female ancestor common to the cousins and that these four types depend upon whether the cousins are the offspring of sisters, brothers, or a brother and a sister. If we analyze each of these matings separately in the manner illustrated above, we note that in no instance can affected males occur without a crossing-over having been interposed between the affected male and the common ancestor; whereas affected females can occur without crossing-over in two of the four matings, namely, if the cousins are the offspring of sisters or if the situation is that illustrated in mating 3 (*third from left* in diagram). In a large population with random mating, we should expect each of these four types of matings to be equally frequent in their occurrence. Furthermore, within a family we should expect noncrossover phenotypes to be more frequent than crossover phenotypes. Consequently, since affected males are always either crossover phenotypes themselves or the offspring of crossover phenotypes, we should expect affected males to be less common than affected females when the gene, derivatives of which are combined in the affected person, is introduced through the female ancestor common to the parents of the affected person.

portion of apparently normal women will be heterozygotes. Many matings of affected males × apparently normal females will be of the $aY × Aa$ type, in which case affected males will have affected sons. This, it will be recalled, is not the case when the gene is rare (cf. sec. 7.2, p. 58). There are still two findings which serve to identify heredity due to a common sex-linked recessive, namely, affected females *always* have an affected father, and all the sons of affected females are affected, regardless of the genotype of the father.

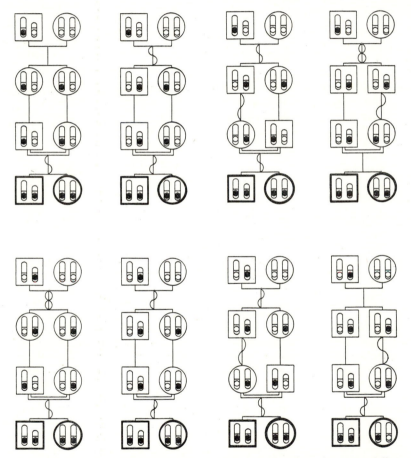

FIG. 7-6.—This illustrates further the basis of the assertion that where a partially sex-linked recessive trait appears among the offspring of a consanguineous marriage, there will be an excess of affected females. In the cases considered in this figure, the gene is introduced through the *male* ancestor common to the cousins. This situation differs from that previously illustrated in Fig.7-5b, in that here we recognize eight different types of matings, depending upon whether the first cousins are the offspring of brothers, sisters, or a brother and a sister and upon whether the gene was introduced via the X- or the Y-chromosome of

7.6. *Multiple modes of inheritance*.—Not infrequently the human geneticist encounters the transmission of the same, or apparently the same, inherited disease in a number of different genetic patterns. One such example is choroidoretinal degeneration. "Choroidoretinal degeneration" is a generic term applied to retinitis pigmentosa and related diseases involving, generally, a degeneration of the choroid and retina of the eye by atrophy and pigmentary infiltration. Choroidoretinal degeneration has a variety of different clinical manifestations and, as might be expected, a number of different modes of inheritance. In the majority of cases, choroidoretinal degeneration follows the pattern of a simple autosomal recessive (cf. Fig. 7-2); but it has been reported as an autosomal dominant, a sex-linked recessive, a partially sex-linked dominant, as well as a partially sex-linked recessive gene. The most probable explanation for this perplexing behavior is that we are dealing with a number of diverse, but as yet undifferentiable, clinical entities, each produced by a different gene.

Even where a disease appears to be inherited as a recessive trait, there is still the possibility that it may have diverse genetic origins. For instance, in the mouse there are known eleven distinct recessive genes which result in head-shaking and erratic, circular movements referred to as "waltzing," generally, but not always, associated with deafness. The effects of some of these genes can be distinguished from others by the occurrence of additional

---

the common male ancestor. Again, points of crossing-over are indicated by a wavy line; the base of this wavy line points toward the crossover phenotype. If these eight matings are analyzed in the same manner as that employed in Fig. 7-5b, we note that in only one case can affected offspring arise without a crossing-over having been interposed between the affected person and the common ancestor. This is the case when the gene is introduced through the X-chromosome of the male ancestor and when the cousins are the offspring of sisters. We observe that in this single instance the affected individuals who arise without crossing-over having occurred are females. We further observe with respect to all matings that whenever the number of crossovers interposed between the affected person and the common ancestor is even, that is, none, 2, etc., the affected person will always be a female; whereas when the number is odd, i.e., 1, 3, etc., the affected person will always be of the male sex. If we now consider conjointly the case when the gene is introduced through the female ancestor or the male ancestor, we see that if mating is at random, there are sixteen equally probable first-cousin matings (since each of the four matings for the case illustrated in Fig. 7-5b can occur in two ways, corresponding to the two different X-chromosomes of the female). Of these sixteen equally probable matings, five can give rise to affected females without crossing-over occurring, whereas none of them can give rise to affected males without at least one crossover having occurred in the direct line of descent of the gene. It is apparent, then, that when consanguineous marriages which give rise to affected offspring are considered collectively, there will be an excess of affected females among the affected offspring. It is also apparent that the majority of this excess stems from those matings wherein the gene is introduced through a common female ancestor.

defects, such as syndactylism; but most of these genes, when homozygous, result in phenotypically indistinguishable mice (Grüneberg, 1947). In man it would only rarely be possible to distinguish between such genes. One possible approach to this problem will be discussed below.

7.7. *The relation of type of inheritance to severity of disease.*—Where a disease which does not make its appearance until sometime after birth may be inherited in any one of several different ways, it is a general rule that when the disease is inherited as if due to a dominant gene, it tends to have a later age of onset than when inherited as if due to a recessive factor. In the event that there is a sex-linked recessive type of the disease, it tends to have a later age of onset than the autosomal recessive type (Macklin, 1932). Table 7-1 con-

TABLE 7-1

RELATION OF AGE OF ONSET OF HEREDITARY DISEASE
TO MODE OF INHERITANCE

DISEASE	AGE OF ONSET (MEAN ± STANDARD ERROR) WHEN APPARENTLY INHERITED AS:		
	Dominant	Autosomal Recessive	Sex-linked Recessive
Peroneal atrophy (Bell, 1935)	18.95 ± 0.62	10.93 ± 0.71	15.90 ± 0.75
Friedreich's ataxia (Bell, 1939)	20.42 ± 1.51	11.75 ± 0.35	. . . .
Spastic ataxia (Bell, 1939)	36.25 ± 1.41	14.72 ± 1.16	. . . .

tains data to illustrate this point. Two overlapping "explanations" are usually given for this finding. The first is that, since the dominant type of disease depends for its persistence on the reproduction of the affected person, whereas the recessive type is propagated by apparently normal heterozygotes, only those dominant genes tend to persist which are associated with a sufficiently late age of onset to permit the affected individual to reproduce. A second "explanation" lies in the hypothesis that where an abnormal gene results in an interference with some physiological process, it seems reasonable to expect that two genes would result in an earlier breakdown than one.

It should also be pointed out that where an inherited disease has a variable age of onset, there is a tendency for the disease to affect the members of a family at the same age. In other words, there is an intra-family correlation as regards age of onset of the disease. This fact is illustrated in Table 7-2 from data which Bell (1939) has assembled regarding the same diseases included in Table 7-1. The similarities between siblings as regards age of onset appear to be about the same for both the dominant and the recessive types of the disease. Such correlations may be explained by the existence of genetic and environmental modifiers which influence the age of onset of the disease. Otherwise stated, in some families there occur, by chance, modifiers which tend to delay the onset of the disease and in other families modifiers which tend to hasten the onset. If this explanation is correct, then we would

TABLE 7-2*

CORRELATION OF AGE OF ONSET OF CERTAIN INHERITED
DISEASES IN PAIRS OF SIBLINGS

Disease	Type of Inheritance	No. of Cases	Correlation Coefficient
Friedreich's ataxia.......	{Dominant	144	0.925
	{Recessive	500	.694
Spastic ataxia..........	{Dominant	198	.812
	{Recessive	164	.845
Peroneal atrophy........	{Dominant	164	.803
	{Recessive	108	0.840

* By permission of Dr. Julia Bell and The Treasury of Human Inheritance.

TABLE 7-3*

CORRELATION COEFFICIENTS FOR AGE OF ONSET OF DISEASE IN TWO CATE-
GORIES OF RELATIVES FOR THREE DOMINANTLY INHERITED DISEASES

RELATIONSHIP	DYSTROPHIA MYOTONICA		HUNTINGTON'S CHOREA		MUSCULAR DYSTROPHY (DOMINANT TYPE)	
	No. of Pairs	Correlation	No. of Pairs	Correlation	No. of Pairs	Correlation
Pairs of siblings.......	270	0.659	442	0.465	240	0.807
Pairs of first cousins...	210	0.365	464	0.223	134	0.696

* By permission of Dr. Julia Bell and The Treasury of Human Inheritance.

expect to find that pairs of affected siblings tended to be more highly corre-lated as regards age of onset than did pairs of less closely related individuals, such as first cousins, since the closer the genetic relationship, the greater the probability that two individuals would possess the same genetic modifiers. Table 7-3 contains data on this point as regards three dominantly inherited diseases where there is adequate data on affected first cousins. In every instance siblings are more similar than first cousins.

7.8. *The ratio of heterozygotes to homozygotes.*—An extremely important problem associated with recessive inheritance is the determination of the ratio of individuals heterozygous for a given gene to the incidence of individuals homozygous for the gene. In the mating of two heterozygous individuals, one-half of the progeny produced from such a mating can be expected to be normal individuals but carriers of the recessive gene, while one-fourth of the progeny will be homozygous for the recessive gene. Does this ratio of two heterozygotes to one homozygote describe the comparative incidence of these two genotypes in the population? Otherwise stated, what are the frequencies of these genotypes in a mixed population? The solution to this problem, granted certain basic assumptions, was given almost simultaneously in 1908 by G. H. Hardy, an English mathematician, and W. Weinberg, a German physician. The argument is as follows:

Suppose that A and a are a pair of alleles and that in any given generation the number of AA, Aa, and aa individuals are as $u : 2v : 1$. Furthermore, suppose that there is no tendency of like genotype to mate with like and that the numbers of the various types of individuals are fairly large, so that mating may be regarded as random. Suppose also that the sexes are evenly distributed among the three genotypes and that all possible matings are equally fertile. Now, since mating is at random, all types of matings will occur and with frequencies proportional to the number of individuals comprising each of the mating types. Consequently, in the next generation the relationship between the numbers of individuals of the various genotypes may be simply determined from the accompanying multiplication table.

	AA (u)	Aa $(2v)$	aa (1)
AA (u)	u^2	$2uv$	u
Aa $(2v)$	$2uv$	$4v^2$	$2v$
aa (1)	u	$2v$	1

From this table, we note that the frequency of $AA \times AA$ matings is u^2, and, since this mating can produce only AA individuals, the contribution of this mating to the next generation is $u^2 AA$ individuals; similarly, we note that the frequency of the mating $AA \times Aa$ is $2uv$, and, since this mating contributes equal numbers of AA and Aa individuals, its contribution to the next generation is $uv AA$ and $uv Aa$. When this argument is used to determine the contributions to the next generation of all the mating types, we find that in the generation which arises from these matings the numbers of the various genotypes are as $AA:Aa:aa$, where

$$AA: \quad u^2 + 2uv + v^2 = (u+v)^2,$$

$$Aa: \quad 2uv + 2u + 2v + 2v^2 = 2(u+v)(v+1),$$

$$aa: \quad v^2 + 2v + 1 = (v+1)^2.$$

Now it is of interest to know under what circumstances this distribution will be the same as the preceding or succeeding generation. The condition under which this is fulfilled is met when the number of AA and aa individuals arising from matings of $Aa \times Aa$ is equal to the number of Aa individuals arising from matings of $AA \times aa$, since all other matings produce offspring genotypically like one or both parents and in numbers equal to those parents. When the original assumptions are met by a generation which arises from one derived by random mating, we see that the numbers of individuals of types AA and aa arising from matings $Aa \times Aa$ are to the number of individuals of type Aa arising from $AA \times aa$ matings as

$$\tfrac{1}{2}[4(u+v)^2(v+1)^2] : 2(u+v)^2(v+1)^2,$$

and this generation will be exactly like the preceding one. Consequently, if the initial premises are fulfilled, a stable distribution in the numbers of AA, Aa, and aa individuals will be reached in one generation of random mating, regardless of the distribution of these individuals in the generation preceding random mating. This is called the "Hardy-Weinberg law."

In terms of the frequencies of the genotypes, the Hardy-Weinberg law states that if p is the frequency of a gene A, and $q = 1 - p$ is the frequency of its allele in a population, then the frequencies of the genotypes are as

$$p^2(AA) : 2pq(Aa) : q^2(aa), \qquad (7.8.1)$$

and, furthermore, that

$$p^2 + 2pq + q^2 = 1. \qquad (7.8.2)$$

The applications of the Hardy-Weinberg law to problems which arise in human genetics are many and varied. Here we shall indicate only its use in the determination of the ratio of individuals heterozygous for a recessive gene to those homozygous for the same gene. When the frequency of a recessive gene is known to be q, then, by the Hardy-Weinberg law, the frequencies of heterozygotes and recessive homozygotes are $2pq$ and q^2, respectively, or the ratio of Aa to aa is as $2pq:q^2$. Suppose, for example, that a recessive gene has a frequency $q = 0.10$, then $p = 1 - q = 0.90$, $q^2 = 0.01$, and $2pq =$

0.18. The ratio of heterozygotes to homozygotes is, in this case, 18:1. Table 7-4 indicates the relative frequencies of homozygotes and heteroxygotes for differing frequencies of a recessive gene; but, since these latter frequencies are based· on the assumption that the conditions for the Hardy-Weinberg equilibrium are satisfied, it would be well to restate these conditions. They are (1) no tendency for like genotypes to mate, that is, mating at random, and all matings equally fertile; (2) the absence of selection for or against any of the genotypes involved; (3) stability of the gene (i.e., no mutation, cf. chap. 11); (4) nonoverlapping generations; and (5) an infinitely large population. Obviously, no one, much less all five, of these criteria is ever completely satisfied in a given human population; however, in many instances

TABLE 7-4*

RELATIVE FREQUENCIES OF INDIVIDUALS HOMOZYGOUS AND HETEROZYGOUS FOR RECESSIVE GENE AT DIFFERENT GENE FREQUENCIES

Gene frequency (per cent)	50	10	3.16	1.41	1.0	0.707	0.316	0.10
Frequency of homozygote (per cent)	25	1	0.1	0.02	0.01	0.005	0.001	0.0001
Frequency of heterozygote (per cent)	50	18	6.12	2.79	1.98	1.40	0.63	0.20
No. of heterozygotes to one homozygote	2	18	61	140	198	281	630	2,000

* By permission of Dr. Nils von Hofsten and Hereditas.

the Hardy-Weinberg law represents a sufficiently close approximation that its use has been justified numerous times in estimating the frequencies of the genes responsible for various inherited diseases.

7.9. *The consanguinity effect in recessive inheritance.*—It has long been observed that individuals affected by the rare, recessively inherited pathologic traits are frequently the products of consanguineous marriages (see Table 7-5). Let us consider why this should be so. We shall begin by defining three terms, namely, relatedness, the coefficient of relationship, and the coefficient of inbreeding.

We shall say that two individuals in a population are related if they have one or more common ancestors. The marriage of two such individuals is termed a "consanguineous marriage." The coefficient of relationship between two individuals, say A and B, is the probability that at a given locus, A and B each possess an identical gene, identical in the sense that at some time in the past these genes had their origin in the same gene. The coefficient of relationship between A and B will be indicated symbolically by r_{AB}. Numerically, r_{AB} can be defined as $\frac{1}{2}$ to the power of the number of links in a chain of descent connecting the two relatives and the common ancestor(s), and summing such terms for all possible different paths (why?). Several simple examples of this method of evaluating r_{AB} are illustrated in Figure 7-7a. It is apparent that $r_{AB} = 0$ if A and B are unrelated and that $r_{AB} \neq 0$ if A and B have at least one common ancestor.

The coefficient of inbreeding for an individual, which may be termed F, we shall define as the probability that an individual, say C, will have at a given locus two genes identical in their origin. Alternatively, we may think

TABLE 7-5*

CONSANGUINITY AMONG PARENTS OF INDIVIDUALS WITH
CERTAIN RECESSIVELY INHERITED DISEASES

Trait	Incidence of First-Cousin Marriages among Parents (This Disease)	Approximate Incidence of First-Cousin among Parents (General Population)
1. *Albinism:*		
Caucasian	0.18–0.24	0.01
Japanese	.37– .59	.06
2. *Infantile amaurotic idiocy:*		
Caucasian	.27– .53	.01
Japanese	.55– .85	.06
3. *Ichthyosis congenita:*		
Caucasian	.30– .40	.01
Japanese	.67– .93	.06
4. *Congenital total color-blindness:*		
Caucasian	.11– .21	.01
Japanese	.39– .51	.06
5. *Xeroderma pigmentosum:*		
Caucasian	.20– .26	.01
Japanese	0.37–0.43	0.06

* Modified from Neel, Kodani, Brewer, and Anderson, 1949, by permission of American Journal of Human Genetics.

of F as the probability that two genes at a given locus are *not independent* in origin. Since the latter obtains only if one of these genes is present in each of the two parents of C, it follows that $F_C = 0$ if the parents of C are unrelated, and $F_C \neq 0$ if the parents of C have at least one common ancestor. Numerically, F can be calculated from

$$F_C = \Sigma \left(\tfrac{1}{2}\right)^{n+n'+1} (1 + F_Z), \qquad (7.9.1)$$

where (1) n and n' are the number of generations in the lines of descent from the common ancestor to the parents of C; (2) summation is over all different combinations of paths; and (3) F_Z is the coefficient of inbreeding of the common ancestor, if inbred. If the common ancestor is not inbred, then $F_Z = 0$ and F_C is merely $\Sigma(\frac{1}{2})^{n+n'+1}$. Several simple illustations of this rule are given in Figure 7-7b. It follows from the definition of F that $(1 - F)$ is the probability that C does not possess identically derived genes at the locus in question. Otherwise stated, $(1 - F)$ is the probability that at a given locus the two genes are independent of each other in a probability sense. They may be similar or dissimilar, but a knowledge of one sheds no light on the other. The

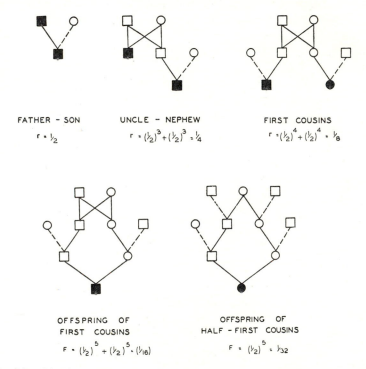

FATHER - SON

$r = \frac{1}{2}$

UNCLE - NEPHEW

$r = (\frac{1}{2})^3 + (\frac{1}{2})^3 = \frac{1}{4}$

FIRST COUSINS

$r = (\frac{1}{2})^4 + (\frac{1}{2})^4 = \frac{1}{8}$

OFFSPRING OF
FIRST COUSINS

$F = (\frac{1}{2})^5 + (\frac{1}{2})^5 = (\frac{1}{16})$

OFFSPRING OF
HALF - FIRST COUSINS

$F = (\frac{1}{2})^5 = \frac{1}{32}$

Fig. 7-7.—(a) The coefficients of relationship for three simple relationships. (b) The coefficients of inbreeding for the cases of offspring of first and of half-first cousins. The individuals or individual whose coefficient of relationship or inbreeding is to be determined are indicated by solid squares or circles. Paths of descent pertinent to the calculation of these coefficients are indicated by unbroken lines.

reader may verify for himself that the coefficient of inbreeding of an individual is merely half the coefficient of relationship between the parents of the individual.

Let us now turn to the effect of inbreeding on the expression of recessive traits. If a given recessive gene occurs in a randomly mating population with a frequency q, then the probability of an affected individual is q^2. These affected individuals will be the products of consanguineous as well as of non-related marriages. From consanguineous marriages, we recognize that an affected individual can arise from (a) the fortuitous combination of like but not identical genes or (b) the combination of identical genes. The probabilities of these two events are readily calculated. In the first instance we have seen that the probability that two genes at a specific locus in an individual derived from a related marriage are independent is $(1 - F)$. Hence the probability that these two genes are independent in origin but alike in function is $(1 - F)q^2$. In the second instance the probability of an affected person is the product of the probabilities that (1) identical genes will be combined in the offspring of a related marriage and (2) these genes will be of the type in question, or Fq. It follows that the frequency with which affected individuals arise from related marriages is the sum of the frequencies with which the two previously mentioned events occur, i.e.,

$$F q + (1 - F)\, q^2 \,. \tag{7.9.2}$$

In the specific case of first-cousin marriage, we note that $F = 1/16$. Accordingly, the frequency with which affected individuals arise from first-cousin marriages is

$$(1/16)\, q + (15/16)\, q^2 \,.$$

One further consequence of consanguineous marriages should be noted. If $Fq + (1 - F)q^2$ is the frequency with which recessive homozygotes arise from related marriages, then $Fp + (1 - F)p^2$ is the frequency with which such marriages give rise to dominant homozygotes. Moreover, since the frequency of heterozygotes is the difference between 1 and the sum of the frequencies of the homozygous classes, the frequency of heterozygotes arising from related marriages must be $2(1 - F)pq$. It is readily apparent that the net effect of related marriages is to augment the probability of an offspring's being homozygous, while diminishing the probability that he will be heterozygous.

Bibliography

SPECIFIC REFERENCES

BELL, J. 1935. On the peroneal type of progressive muscular atrophy, Treas. Human Inher., 4:69–140. London: Cambridge University Press.

———. 1939. On hereditary ataxia and spastic paraplegia, ibid., pp. 141–281.

———. 1947. Dystrophia myotonia and allied diseases, ibid., pp. 343–410.

FALCONER, D. S. 1952. A totally sex-linked gene in the house mouse, Nature, 169: 664.

FALLS, H. F. 1952. The role of the sex chromosome in hereditary ocular pathology, Tr. Am. Ophth. Soc., 50:421–67.

GARBER, E. D. 1952. A dominant, sex-linked mutation in the house mouse, Science, 116:89.

HALDANE, J. B. S. 1938. A hitherto unexpected complication in the genetics of human recessives, Ann. Eugenics, 8:263–65.

HARDY, G. H. 1908. Mendelian proportions in a mixed population, Science, 28: 49–50.

MACKLIN, M. T. 1932. The relation of the mode of inheritance to the severity of an inherited disease, Human Biol., 4:69–79.

————. 1944. Xeroderma pigmentosum, Arch. Dermat. & Syph., 49:157–71.

————. 1952. Sex ratios in partial sex linkage. I. Excess of affected females from consanguineous matings, Am. J. Human Genet., 4:14–30.

VON HOFSTEN, N. 1951. The genetic effect of negative selection in man, Hereditas, 37:157–265.

WEINBERG, W. 1908. Über den Nachweis der Vererbung beim Menschen, Jahresh. Verein f. vaterl. Naturk. in Württemberg, 64:368–82.

GENERAL REFERENCES

GRÜNEBERG, H. 1947. Animal genetics and medicine. New York: Paul B. Hoeber.

Problems

1. In the light of section 7.4, develop a second genetic interpretation of the pedigree of Problem 2, chapter 5.

2. What is the probability that four children produced by a mating of parents heterozygous for the gene responsible for the recessive form of retinitis pigmentosa will consist of a normal male, an affected male, a normal female, and an affected female? That all four children will be affected males?

3. What effect would small average size of family and a small population have on the stability of gene frequencies?

4. Show that in a brother-sister mating, the value of F (cf. sec. 7.9) is $1/4$; that in an uncle-niece union, F equals $1/8$; and that in a second-cousin union, F equals $1/64$.

5. Assume that at a given locus three genes, say a dominant gene A, and two recessive genes, a and a', may occur. Furthermore, assume that aa and aa' individuals, while rare, exhibit a characteristic abnormality, whereas $a'a'$ is lethal at so early a stage in development that the defective embryo cannot be identified. What effect will the lethality of the $a'a'$ genotype have on the frequency with which consanguinity is observed among the parents of affected individuals? (Hint: Haldane, 1938.)

CHAPTER 8

Genes neither Dominant nor Recessive

IN THE preceding two chapters we have considered typical examples of dominant and recessive heredity. In a strict sense, a dominant gene is one whose effects are the same whether the gene is homozygous or heterozygous. By the same token, a recessive gene is one which finds no expression unless homozygous. As our knowledge of heredity in man increases, it is becoming apparent that many genes are neither strictly dominant nor strictly recessive, but somewhere in between. Otherwise stated, we are recognizing more and more instances in which a gene, at first thought to be a recessive, is found to possess some detectable effects when heterozygous, and a gene originally classed as a dominant is found to have even more marked effects when homozygous than when heterozygous. Depending on whether the gene was originally classified as a dominant or a recessive, it is, in the light of this further knowledge, usually referred to as "incompletely dominant" or "incompletely recessive." Genetically speaking, the two terms amount to the same thing.

8.1. *Incompletely recessive genes.*—One example of incompletely recessive genes will suffice. In the preceding chapter the diverse genetic etiologies of the disease known as choroidoretinal degeneration (retinitis pigmentosa) have been emphasized. In some families this disease appears to be inherited as if due to a dominant gene, and in others as if due to a recessive gene; the gene may or may not be sex-linked. Recent intensive studies on certain families which at first appeared to exhibit simple recessive sex-linked inheritance of this disease have revealed an interesting situation. The results of one of the more extensive of these investigations (Falls and Cotterman, 1948) are summarized in Figure 8-1. Had the investigators of this large family restricted their attention to those with visual complaints, this would have appeared to be a family exhibiting simple recessive sex-linked inheritance. Careful studies of the parents of the affected males revealed that the mother in each instance showed very minor changes in the appearance of the retina. Moreover, the daughters of affected males revealed these same changes. When examined with suitable instruments, the retina of the normal eye ordinarily

75

has a uniform or somewhat mottled, salmon-pink appearance. In these women, however, the retina had an unusual "golden" color and reflected light in the dark in much the same way as does a cat's or dog's eye. Thus far, the marriage from which one might expect a homozygous female child has not occurred in this family. It seems reasonable to anticipate that such a female would show a picture at least as extreme as the affected males, and perhaps more so. Here, then, is a gene which is incompletely recessive. As pointed out in chapter 6, completely dominant sex-linked genes in man are rare. On the other hand, as will be apparent from Table 8-1, there are at least seven diseases exhibiting the pattern of ostensibly "recessive," sex-linked inheritance, in which females show minor manifestations of the disease.

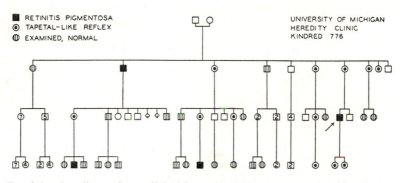

Fig. 8-1.—A pedigree of a sex-linked form of retinitis pigmentosa in which the heterozygous females exhibit a characteristic clinical finding. (Data of Falls and Cotterman, 1948.)

There is no way at the present time of estimating what proportion of the "recessive" genes of man really have some small effect when heterozygous. Even in an organism as well catalogued genetically as Drosophila, considerable doubt surrounds this question. This is because so varied are the possible manifestations of many genes that it is difficult to say with assurance that a given gene has *no* effect when heterozygous. One relatively crude way of approaching this question in Drosophila is to determine the extent to which "recessive" genes which result in the death of the fly when homozygous (i.e., recessive lethal genes) also impair the viability of a fly to an appreciable extent when heterozygous. Stern and his collaborators (1952) found that the average viability of female flies heterozygous for sex-linked lethal genes was 96.5 per cent of normal. There was no difference between the behavior of lethals which had arisen spontaneously and those which had appeared during irradiation experiments (cf. sec. 11.7, p. 144). There was a considerable range in the degree to which the lethals affected the viability of flies, one of the lethals, when heterozygous, reducing viability to 60.2 per cent of normal, while several actually seemed to increase the viability. Muller (1950) and B. Wallace (unpublished) have made similar observations concerning lethal genes located in the second chromosome of Drosophila, although the average reduction in viability appeared to be somewhat less than for sex-linked lethals.

8.2. *Incompletely dominant genes.*—Many of the dominant genes with striking effects which contribute so much to the subject matter of medical genetics are so rare that individuals who are homozygous for these genes have not yet been observed. There are several observations, however, that suggest that in many cases the homozygote would exhibit much more serious disease than the heterozygote. Thus hereditary hemorrhagic telangiectasia is a disease in which one finds scattered all over the body small clusters of abnormally thin-walled blood vessels. These vessels tend to rupture with very slight trauma, sometimes leading to serious hemorrhages. The manifestations of the disease are usually mild in childhood, and even adults are seldom seriously inconvenienced. There is on record in the medical literature one family in which two persons with the disease married. To this union there was born a child who at birth showed a large collection of these abnormal vessels on the chest. Within a few days other such areas began to appear and to bleed. Within a short time the child developed as extreme a case of the disease as has ever been seen, with so much bleeding, both internal and external, that, despite all efforts, death occurred at two and one-half months. It seems reasonable to postulate that the child was homozygous for the gene in question (Snyder and Doan, 1944).

Brachydactyly is a disease characterized by abnormally short fingers and toes, due to the shortening or even complete absence of one of the three phalanges of the fingers and toes. The condition is usually inherited as if due to a dominant gene. The literature contains one report of a marriage between two heterozygous individuals. There resulted a severely crippled child, lacking fingers and toes, who died at the age of one year. Although genetic proof that the child was homozygous is, of course, lacking, it seems reasonable to conclude that this was the case (Mohr and Wriedt, 1919). Here, then, are two probable examples of the use of the term "dominant" in a relative sense.

8.3. *The detection of the genetic carriers of inherited disease.*—One of the problems with which medical genetics is concerned is the detection of the "carriers" of inherited disease. In the broad sense, the term "genetic carrier" may be defined as an individual who possesses the potentialities of transmitting inherited disease while he himself shows either no, or only a slight, departure from normality. Where the carrier shows absolutely no abnormality, he cannot be detected by clinical means but only by the progeny test. But where there exists some deviation from what we call "normality," then he may be diagnosed in a variety of ways. The genetic situations involved in carrier detection may be briefly defined as follows:

1. Where a gene is incompletely recessive (or incompletely dominant), the carrier is the heterozygous individual who, marrying another heterozygote, may produce much more seriously affected offspring. As pointed out in chapter 7, the rarer the recessive gene involved, the higher the ratio of carrier:affected (heterozygote:homozygote).

2. Some inherited diseases do not develop until adolescence or even middle or old age. Individuals with the genetic potentialities for developing such disease of late onset—whether the disease be due to a dominant or a recessive gene or to the interaction of several genes—are also "carriers." Thus, as noted earlier, Huntington's chorea is due to the presence of a single dominant gene. Until the age of thirty-five to forty, when the disease usually first shows its hand, the victim-to-be may be referred to as a "carrier."

3. Some dominant genes have much more serious effects in certain individuals than in others. In any particular family such a gene may pass through several generations with only very minor effects, but then suddenly have much more serious consequences. In such a case the carrier is an individual showing only the mild effects. Gout is due to a dominantly inherited abnormality in the metabolism of uric acid. Individuals heterozygous for this gene regularly show an increase in the amount of uric acid present in the blood stream. However, only about 10 per cent of the men with an elevated level of uric acid in the blood—and even fewer females—actually develop the rheumatic complaints which a physician diagnoses as gout. In this case the individuals who show only the elevated uric acid levels may be referred to as the "carriers."

4. Finally, we may recognize a carrier state for a number of diseases in which a genetic factor is clearly at work but in which the precise genetic mechanism is not understood. Diabetes mellitus is a case in point. Despite the prevalence of the disease, there is not yet agreement as to whether, genetically speaking, there is only one or several different kinds of diabetes mellitus, or whether the usual genetic basis of the disease is a dominant or a recessive gene or the interaction of several genes. Nonetheless, there is good evidence that among the relatives of diabetics there is an increased frequency of persons who, by suitable blood tests, can be shown to have a defect in their ability to handle sugar, even though they do not yet have clinical diabetes. Such persons are also carriers.

In some diseases the carriers may be detected with a high degree of accuracy; in others there is at present no more than a faint hint of a recognizable carrier state. Table 8-1 summarizes much of our present knowledge of carrier states. A few minutes spent in the study of this table will suffice to indicate the many different kinds of defects which may be used to detect a carrier state. Sometimes the departure which provides the clue is a morphological variation, sometimes a biochemical abnormality, or even a peculiarity of the brain waves.

Many of the diseases for which it seems a carrier state can be recognized are rare, and often the stigma of the carrier state are, to say the least, not clear cut. For instance, anophthalmos, or absence of the eyeballs, is often inherited as if due to a recessive gene. It has been reported that the heterozygotes may have small eyeballs. However, the estimation of eye size in man is not always easy because of variations in the structure of the lids and surrounding tissues. Furthermore, there appears to be considerable normal variation in the size of the eyeball. Thus, even if one could accurately estimate eyeball size, by no means all those found to have small eyes would be carriers of the gene in question, while, conversely, some heterozygotes for the gene may have normal eyes.

Epilepsy provides another example of a poorly defined carrier state. Spontaneously developing epilepsy is, at least in part, genetically determined. The brain waves (electroencephalograms) of epileptics are abnormal in 84 per cent of affected persons. It has been reported that both the parents of epileptic children exhibit abnormal brain waves in 24 per cent of the cases studied, while one parent only was abnormal in an additional 58 per cent of the cases studied. But in 18 per cent of the cases both parents were normal in this respect. On the other hand, 16 per cent of a control sample were found to have abnormal electroencephalograms. Thus, although, statistically speaking, there is an unusual frequency of electroencephalographic abnormality among the parents of epileptics, there are enough instances in which both parents are normal, and an abnormal electroencephalogram is sufficiently common in the population at large, to render this finding of limited value in carrier detection.

The most obvious "practical" value of information concerning the carrier states is in the prediction of the matings from which children with hereditary disease may be expected. A second use of this information is in the detection of individuals destined to develop certain diseases, at a stage when only very early changes are present. Such early detection carries with it the possibility of forestalling, by appropriate treatment, some or all of the undesirable effects of the gene.

It is important to recognize the limitations of our present knowledge as a

TABLE 8-1

DISEASES IN WHICH IT MAY BE POSSIBLE TO RECOGNIZE A CARRIER STATE

A grading system of 1–4 has been used to indicate the reliability of recognition of the carrier state, grade 1 being most reliable and grade 4 least reliable. The most probable mode of inheritance has been indicated according to the following system: "a.r." = autosomal "recessive" (incompletely recessive, semidominant, etc.); "s.l.r." = sex-linked "recessive"; "d.v." = autosomal dominant or dominants of variable expression; "d.l." = autosomal dominant of late onset; "s.l.r.l." = sex-linked recessive of late onset; and "un." = mode of heredity not clear.

Disease	Mode of Inheritance	Characteristics of Carrier State May Be:	Genetic Relationship of Carrier to Manifest Disease	Reliability	Reference
1. Afibrinogenemia	a.r.	Fibrinogenopenia	As heterozygote to homozygote	2	Risak (1935), Macfarlane (1938), Schönholzer (1939)
2. Allergic state (early, severe)	a.r.	Development of mild allergies sometime after puberty	As heterozygote to homozygote; but only about 1 in 5 heterozygotes ever develop symptoms	4	Wiener, Zieve, and Fries (1936)
3. Anhidrotic ectodermal dysplasia (sex-linked type)	s.l.r.	Very mildly affected females	Carrier females heterozygous for a gene for which affected males are hemizygous	3	Roberts (1929), Levit (1936)
4. Anophthalmia	a.r.	Small eyeballs	As heterozygote to homozygote	4	Sorsby (1934)
5. Ataxia, hereditary	d.l.	Minimal signs of pyramidal tract involvement	Heterozygous for same genes; but phenotypic effects by which gene usually recognized are not yet apparent	3	Schut and Böök, unpublished
6. Choroidoretinal degeneration (retinitis pigmentosa) of sex-linked type	s.l.r.	Presence in females of same family of atypical and minor choroidoretinal changes	Carrier females heterozygous for a gene for which the affected males are hemizygous; not all pedigrees manifest this relationship	2	Goedblad (1942), McCulloch and McCulloch (1948), Falls and Cotterman (1948)
7. Color blindness	s.l.r.	Females with minor impairment of color vision	Carrier females are heterozygous for a gene for which affected males are hemizygous	3	Wieland (1933), Schmidt (1934), Pickford (1949)
8. Congenital dislocation of hip	d.v.	Defective acetabular development	Heterozygous for same gene, which in carriers is so mildly expressed that dislocation of hip does not result	3	Faber (1938)
9. Diabetes mellitus	un.	Impaired glucose metabolism as shown by abnormal glucose tolerance curve	Not clear	3	Pincus and White (1934), Steiner (1936), Lemser (1938)
10. Dystrophia myotonica	d.v.	Cataract	Heterozygous for same gene	4	Thomasen (1948)

TABLE 6-1 — *Continued*

Disease	Mode of Inheritance	Characteristics of Carrier State May Be:	Genetic Relationship of Carrier to Manifest Disease	Reliability	Reference
11. Epilepsy	d.v. ?	Abnormal electroencephalogram	Heterozygous for same gene	3	Lennox, Gibbs, and Gibbs, (1940, 1942), Lennox (1946)
12. Friedreich's ataxia	a.r.	Pes cavus and absent tendon reflexes	As heterozygote to homozygote; findings in heterozygote inconstant	4	Davidenkov (1940), Spillane (1940)
13. Gout	d.v.	Hyperuricemia	Heterozygous for same gene, which produces gout in small fraction of the carriers	1	Smyth, Cotterman, and Freyberg (1948), Stecher, Hersh, and Solomon (1949)
14. Hemophilia	s.l.r.	Females with minor prolongation of coagulation time	Carrier females are heterozygous for a gene for which affected males are hemizygous	3	Günder (1938), Andreassen (1943), Fonio (1949)
15. Hereditary hemolytic jaundice	d.v.	Asymptomatic spherocytosis and increased erythrocyte fragility to hypotonic saline solutions	Heterozygous for same gene, with "subclinical" hematological effects in the carriers	1	Campbell and Warner (1926), Race (1942), Young, Izzo, and Platzer (1951)
16. Huntington's chorea	d.l.	Electroencephalographic abnormalities	Heterozygous for same gene, which has not yet reached the level of clinical expression in the younger carriers	2	Patterson, Bagchi, and Test (1948)
17. Hypertension (essential)	d.v. ?	Positive reaction to cold pressor test	Not clear; carriers may have same genetic constitution but do not show the disease because of youth or absence of specific eliciting factors	3	Hines (1937)
18. Juvenile amaurotic idiocy	a.r.	Increased incidence of vacuolated lymphocytes	As heterozygote to homozygote	3	Rayner (1952)
19. Keratosis follicularis spinulosa	s.l.r.	Mild keratosis follicularis	Carrier females are heterozygous for a gene for which males are hemizygous; only a single extensive pedigree of this disease known	1	Siemens (1925)
20. Laurence-Moon-Biedl syndrome	a.r.	Obesity, skeletal abnormalities, atypical retinal changes	As heterozygote to homozygote, but carrier state changes of a very nonspecific nature	4	Sorsby, Avery, and Cockayne (1939)
21. Morquio's disease	a.r.	Small stature and short mid-phalanges of hands and feet	As heterozygote to homozygote; relationship observed in only a single family	4	Grebe (1943)

81

TABLE 8-1—*Continued*

Disease	Mode of Inheritance	Characteristics of Carrier State May Be:	Genetic Relationship of Carrier to Manifest Disease	Reliability	Reference
22. Myopia (extreme)	a.r.	Mild myopia	As heterozygote to homozygote	4	Yamazaki (1927)
23. Ovalocytosis with hemolytic syndrome	d.v.	Asymptomatic ovalocytosis	Heterozygous for same gene	1	Lambrecht (1938), Mason (1938), Cooley (1942)
24. Pernicious anemia	d.v.	Achlorhydria; mild pernicious anemia-like blood changes	Heterozygous for same gene or genes	3	Hurst (1925), Connor (1930), Baggi and Romei (1949)
25. Peroneal atrophy (sex-linked type)	s.l.r.l.	Mild and nonprogressive peroneal atrophy	Carrier females are heterozygous for a gene for which males are hemizygous—not all s.l. pedigrees exhibit this relationship	3	Raffan (1907), Bell (1935)
26. Pick's disease of the brain	d.l.	Abnormal response to Rorschach test some years before onset of clincal disease	Heterozygous for same gene	4	Sanders (1939)
27. Schizophrenia	a.r.	Schizoid personality	As heterozygote to homozygote	4	Rudin (1916), Hoffmann (1921), Strohmayer (1925)
28. Sex-linked hypochromic anemia	s.l.r.	Females with increased numbers of pale, oval erythrocytes and/or splenomegaly	Carrier females are heterozygous for a gene for which males are hemizygous	1	Rundles and Falls (1946)
29. Sickle-cell anemia	a.r.	Sickle-cell trait	As heterozygote to homozygote	1	Neel (1949)
30. Spina bifida	d.v. ?	Spina bifida occulta	Both heterozygous for same gene	4	Schamburow and Stilbans (1932)
31. Thalassemia minor	a.r.	Thalassemia minor	As heterozygote to homozygote	1	Gatto (1942), Dameshek (1943), Valentine and Neel (1944)
32. Wilson's disease (hepatolenticular degeneration)	a.r.	Increased amounts of amino acids in urine	Homozygous for same gene, which has not yet reached level of clinical expression in carriers	3	Uzman and Hood (1952)
33. Xanthomatosis (extreme form)	a.r.	Hypercholesterolemia	As heterozygote to homozygote	1	Fliegelman, Wilkinson, and Hand (1949), Adlersberg, Parets, and Boas (1949)
34. Xeroderma pigmentosa	s.l.r. (incomplete)	Excessive freckling	As heterozygote to homozygote	4	Siemens and Kohn (1925), Cockayne (1933)

basis for positive action. While an accurate estimate is impossible, such is the frequency of inherited disease that it is quite likely that each of us is a carrier of not one but several undesirable genes. At the present time we can recognize the carriers of only a handful of those diseases, and this "recognition" is, for the most part, a very uncertain matter. It is the opinion of the authors that geneticists, faced with a problem of this complexity, would be wise to proceed with great caution in attempting to utilize the available information. We will return to this subject in chapter 20.

8.4. *Serological genetics.*—Thus far, under the chapter heading of "Genes neither Dominant nor Recessive," we have been considering, for the most part, cases in which an abnormal gene has a more extreme effect when homozygous than when heterozygous. We turn now to the consideration of a very important group of genes which may also be fitted under this chapter heading, although, as we shall see, for a somewhat different reason. These are the genes responsible for the serological differences between individuals. We shall first briefly review what is known concerning these genes and then consider the difficulties which arise when we attempt to classify these genes as either "dominant" or "recessive." In this presentation we shall draw heavily upon the books by Wiener (1943) and Race and Sanger (1950).

Human blood may readily be separated into two components—a cellular one primarily composed of erythrocytes, and a fluid component, which, depending upon its method of preparation, is termed "plasma" or "serum." Landsteiner in 1900 was the first to recognize that if erythrocytes from one person were mixed with the serum from another, the erythrocytes often tended to stick together in masses varying in size, a phenomenon known as "agglutination." The substance within the serum responsible for this clumping is called an "agglutinin" and is a particular type of antibody. The substance within the red cell which reacts with the agglutinin is called an "agglutinogen" and is a particular type of antigen.

The discovery of new blood antigens comes about in either of two general ways. They may be discovered (1) as a consequence of the natural occurrence of the specific agglutinins for this antigen in many people or (2) as a consequence of an immune reaction wherein the antibody-forming apparatus is stimulated to form an agglutinin by the presence of an incompatible antigen. There are three principal approaches to the recognition of immune reactions which are of particular importance to human serology: (1) The injection of human erythrocytes into an animal (rabbit, monkey) sometimes results in the production by the animal of an agglutinin which will be specific for that individual's cells and for the cells of certain other individuals, but which fails to agglutinate the cells of still other individuals who apparently do not possess the same agglutinogen. (2) Certain patients who have received one or more transfusions may develop antibodies against the blood cells of one of their donors, leading to the recognition of a new antigen. (3) Finally, it may happen that the developing fetus releases substances which find their way in minute quantities into the maternal circulation, there to stimulate the production of specific antibodies. An occasional sequel of this latter form of immunization is the presence of hemolytic disease, resulting in anemia, in infants born subsequent to the immunization.

8.5. *The A, B, O system.*—When one tests a whole series of bloods as to the ability of the serum of each to agglutinate the erythrocytes of others and the tendency of the erythrocytes to be agglutinated by the sera of others, one encounters striking differences between various bloods. On the basis of these differences, four distinct types of blood cells can be recognized. These are termed types AB, A, B, and O. Blood cells of type AB contain two antigens, A and B. Blood cells of type A contain the antigen A. Blood cells of type B contain the antigen B. Finally, blood cells of type O contain neither antigen A nor antigen B.

Individuals with type O blood cells contain in their serum two antibodies, usually designated as α and β, the α antibody capable of agglutinating type A and AB cells, the β antibody capable of agglutinating type B or AB cells. Individuals with type A blood cells contain in their serum only the β antibody (the presence of the α antibody would result in agglutination within the individual's own blood stream and hence would be incompatible with life). Individuals with type B cells contain in their serum only the α antibody (again, the presence of the β antibody would be incompatible with survival). Finally, where AB blood cells are present, neither antibody α nor antibody β is found. It is customary to designate an individual's blood type on the basis of the antigens present (AB, A, B, O). The relationship between the A, B, O antigens and antibodies in human blood and the cross-agglutinating properties of various bloods are summarized in Table 8-2.

The blood groups remain constant throughout life. Moreover, it was early observed that, with respect to the blood groups, certain types of marriages regularly yielded certain types of offspring. Thus, in marriages of AB × O, the children were either A or B; in marriages of A × B, the children might be AB, A, B, or O; and in marriages of O × O, the children were always O. The genetic explanation of these and other observations was for some time a matter of very active debate. Eventually, however, a general agreement was reached concerning the type of heredity involved. It is postulated that there are three different autosomal alleles involved in the determination of these blood groups. These may be termed I^A, I^B and I^O. These genes form a series of three multiple alleles. That is to say, the gene responsible for this particular agglutination reaction may occur in three different forms (alleles). Only one of these three alleles is contained in any particular germ cell, and only two in any fertilized egg. There are, then, six possible genotypes with respect to the A, B, O blood groups, namely, $I^A I^A$, $I^A I^B$, $I^B I^B$, $I^A I^O$, $I^B I^O$, and $I^O I^O$. In practice it is not feasible to distinguish between $I^A I^A$ and $I^A I^O$ and between $I^B I^B$ and $I^B I^O$—hence the recognition of only four blood types. The various possible matings and the results are shown in Table 8-3, only the superscripts A, B, and O being used for the sake of simplicity. Whenever the genes responsible for the A and B reactions are present in the same individual, neither is dominant to the other, and, in con-

TABLE 8-2

Cross-Agglutination Reactions between Individuals
of Various Blood Types

Type of Blood	Agglutinogens Present	Agglutinins Present	Agglutination Reaction of Serum with Red Blood Cells of Group			
			O	A	B	AB
O.........	O	α and β	−	+	+	+
A.........	A	β	−	−	+	+
B.........	B	α	−	+	−	+
AB.......	A, B	−	−	−	−	−

sequence, both A and B antigens are present. The gene responsible for group O, on the other hand, has usually been represented as producing no detectable antigen. It has been customary, therefore, to speak of the genes I^A and I^B as being dominant to I^O.

In 1911, Von Dungern and Hirszfeld definitely demonstrated that there are two forms of the A antigen, termed "A_1" and "A_2," and, by inference, at least two different genes responsible for the A reaction. Subsequently, the latter author has presented evidence of still further subdivisions in the A group, the new alleles being responsible for certain very weak A reactions.

In 1948, Boorman, Dodd, and Gilbey reported what is probably the first instance of a true anti-O serum. This serum was shown to agglutinate the red blood cells of individuals who were known to be heterozygous or homozygous for the gene I^O by virtue of the children which they had produced. On the basis of this work, it seems reasonable to conclude that the gene I^O is capable of producing an antigen O, which, in turn, can elicit the formation of an anti-O antibody in the blood of some individuals. This fact must lead to the rejection of the conventional idea of the dominance of the genes I^A and I^B over I^O, the apparent dominance being due to the fact that the antigen associated with the presence of gene I^O does not elicit agglutinin formation so readily as do the A and B antigens.

It has been accepted practice to designate the genes in the A, B, O system by using the superscript alone, thus I^A is denoted by A. This convention will be employed by the authors when there is no danger of confusing a gene with a group.

TABLE 8-3*

EXPECTATION FROM MARRIAGES CLASSIFIED ACCORD-
ING TO BLOOD GROUP OF PARENTS

No.	Type of Mating	Blood Groups of Children
1........	$OO \times OO$	OO
2........	$\begin{cases} OO \times AO \\ OO \times AA \end{cases}$	AO, OO AO
3........	$\begin{cases} OO \times BO \\ OO \times BB \end{cases}$	OO, BO BO
4........	$\begin{cases} AO \times AO \\ AA \times AO \\ AA \times AA \end{cases}$	AA, AO, OO AA, AO AA
5........	$\begin{cases} AO \times BO \\ AA \times BO \\ AO \times BB \\ AA \times BB \end{cases}$	AB, AO, BO, OO AB, AO AB, BO AB
6........	$\begin{cases} BO \times BO \\ BO \times BB \\ BB \times BB \end{cases}$	BB, BO, OO BB, BO BB
7........	$OO \times AB$	$AO \times BO$
8........	$\begin{cases} AO \times AB \\ AA \times AB \end{cases}$	AA, AB, AO, BO AA, AB
9........	$\begin{cases} BO \times AB \\ BB \times AB \end{cases}$	AB, BB, AO, BO AB, BB
10........	$AB \times AB$	AA, AB, BB

* The types shown in braces cannot in practice be distinguished from one another except on the basis of the kinds of children produced.

8.6. *The Rh system.*—In 1940, Landsteiner and Wiener reported that when rabbits were injected with the blood of rhesus monkeys, they produced antibodies which would agglutinate the blood of all rhesus monkeys. More surprising, these same antibodies would cause clumping of the erythrocytes of 85 per cent of the white population of the United States. There was thereby recognized a new antigen, termed "Rh" because of the role of the rhesus monkey in its discovery. Individuals whose erythrocytes clump with the antirhesus antibody are called "Rh-positive," while those whose erythrocytes fail to clump are "Rh-negative."

It had been recognized for some years that in certain instances where patients had received repeated blood transfusions, these patients exhibited a tendency to destroy the red blood cells of certain donors, even though they were compatible in terms of the A, B, O blood groups. Very shortly after the discovery of the Rh-reaction, Wiener and Peters reported that some of these reactions were due to the fact that the patients, who were Rh-negative, had earlier unwittingly received transfusions of Rh-positive blood and had developed antibodies toward such blood.

For many years physicians had recognized that about 1 in 500–1,000 newborn infants in the United States exhibited evidences of a very severe anemia which often resulted in death, sometimes even prior to the birth of the child. A larger number of newborns, perhaps 1 in 250, exhibited a similar but milder anemia. In 1941, a year after the Rh-factor was discovered, Levine, Katzin, and Burnham reported that the blood of the mothers of such children frequently contained an antibody for the cells of the child. It was then shown that the mother was usually Rh-negative, whereas the father of the child and the child itself were Rh-positive. The Rh-positive cells of the child, escaping in small quantities from the circulation of the developing fetus into the maternal circulation, stimulated the mother to produce antibodies, which, finding their way across the placenta, were responsible for the agglutination and destruction of the erythrocytes of the developing child. The entire chain of events is depicted in Figure 8-2.

It soon became evident that the tendency to be Rh-positive or Rh-negative was inherited, with Rh-negativeness recessive to Rh-positiveness, and that Rh-positive individuals differed significantly from one another in the type of antibodies which their erythrocytes elicited. Two explanations of this finding have developed. Wiener has suggested that the differences in the Rh-reaction are due to the existence of a series of multiple alleles, just as in the case of the A, B, O reactions. The number of alleles in the series is constantly being revised upward in consequence of new serological discoveries. The eight most common and most firmly established genes in the series are designated as r, r', r'', r^y, R^o, R^1, R^2, and R^z, certain of these genes having overlapping serological properties. Rh-negative individuals are rr; persons with any one of the seven other genes are Rh-positive to some degree. More recently, a group of English investigators has suggested an alternative explanation. They postulate that in the determination of Rh-positivity there is involved not one genetic locus but three very closely linked loci, each of which may be represented by at least two alleles—a "recessive" gene responsible for a tendency to Rh-negativity and one or more dominant genes responsible for Rh-positiveness. The three chief pairs of alleles are termed, respectively, *C* and *c*, *D* and *d*, *E* and *e;* the relationship of these factors to those postulated by Wiener are as given in the accompanying tabulation.

Wiener	Fisher, Race	Weiner	Fisher, Race
r.....................	*cde*	*R⁰*.....................	*cDe*
r'.....................	*Cde*	*R¹*.....................	*CDe*
r''.....................	*cdE*	*R²*.....................	*cDE*
rʸ.....................	*CdE*	*Rᶻ*.....................	*CDE*

An individual who is Rh-negative must therefore be *cde/cde*. Rh-positive individuals, on the other hand, may possess a wide variety of genotypes, the

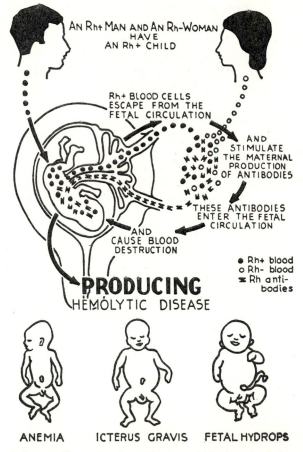

FIG. 8-2.—A diagram illustrating the mechanism which gives rise to hemolytic disease of the newborn. (By permission from: Rh . . . its relation to congenital hemolytic disease and to intragroup transfusion reactions, by Dr. Edith Potter, copyright 1947, Yearbook Publishers, Inc.)

more common being *CDe/cde*, *CDe/CDe*, *CDe/cDE*, and *cDE/cde*. An Rh-negative mother who has become sensitized by the presence of an Rh-positive fetus may, depending on the genotype of the fetus, produce anti-C, anti-D, or anti-E agglutinins, either singly or in various combinations. More rarely, an Rh-positive mother may produce an anti-c or an anti-e agglutinin. Thus far, an anti-d agglutinin has not been positively demonstrated to the satisfaction of all serologists.

A decision as to which of these two hypotheses provides the correct explanation of the observed relationship is extremely difficult. The English investigators postulate that linkage between the three genes is so close that crossing-over rarely or never occurs. If crossing-over never occurs, then, since we may define a gene as the unit of inheritance, which cannot be subdivided by crossing-over, there is no real difference between their scheme and Wiener's except as regards terminology. Under these circumstances, there is little "practical" difference between the two hypotheses; one uses whichever theory seems the easier to work with. It is important to realize, however, that under neither hypothesis concerning the genetics of the situation is true dominance of one gene over the other postulated.

In clinical medicine the conventional concept of Rh-positiveness as dominant to Rh-negativeness is a satisfactory working notion if this subdivision is based on the reactions of human beings to the anti-D serum. This latter serum is identical in its serological properties with the original antisera described by Landsteiner and Wiener. However, as has been pointed out by Mollison, though some 99 per cent of those individuals who become immunized by transfusion or pregnancy will be Rh-negative on the basis of this subdivision, this is not a satisfactory definition for the selection of Rh-negative blood donors.

8.7. *The MN system.*—In two brilliant papers published in 1927, Landsteiner and Levine described two new antigens, M and N. The antibodies for these two antigens rarely occur naturally and generally arise as a result of an immunization reaction after human blood has been injected into rabbits. These antigens were shown to be inherited, and it was postulated and accepted that they are due to a set of two alleles without dominance. It was possible, therefore, to specify serologically the three genotypes: *MM*, *MN*, and *NN*.

It is a well-established convention in animal genetics that the members of a series of multiple alleles all are designated by the same base letter, with differences indicated by appropriate sub- or superscripts. For this reason, it is somewhat confusing to speak of the *M*-gene and *N*-gene as members of an allelic series. Strandskov has suggested referring to the two genes as Ag^M and Ag^N. However, the terms *M* and *N* are so well intrenched that it seems unlikely that any change can be brought about now. Table 19-2 (p. 320) lists, using the usual terminology, the various possible matings for the MN series and their results.

In 1947, Walsh and Montgomery described a new blood cell antigen, which was termed "S." It was quickly observed that S-positiveness appeared to be dominant to S-negativeness and that the S antigen exhibited a peculiar relationship to the MN blood groups. Among the Englishmen examined, M-type individuals were more often S-positive than were N-type persons. Two possible explanations were advanced for this finding. It could, on the one hand, be postulated that the S-reaction still further subdivided the MN series, so that now we must recognize four genes which may be designated according to the serological reactions that they engender as MS, NS, M, and N, MS being more common than NS. With a series of four multiple alleles, we would expect $N(N + 1)/2 = 10$ genotypes. Alternatively, it could be postulated that the S-reaction was due to a gene separate from, but closely linked to, the alleles responsible for the MN reactions, with S-positiveness due to a gene, S, which was dominant to the gene responsible for the absence of the reaction, s.

Discoveries follow thick and fast in the field of serological genetics. The original anti-S serum was found in 1947. A second such serum was found in 1948. But in 1951, Levine and collaborators discovered an anti-s serum, capable of agglutinating ss and Ss bloods. Had this been discovered first, we would be designating just the opposite as regards the dominance relationships of these two genes. This discovery forces us to modify the first hypothesis described above regarding the relationship of MN to S; if there are four alleles, they must be designated MS, NS, Ms, and Ns.

With the description of the s antigen, it appeared that the $MNSs$ system had been completely determined. But, as is frequently the case in serological genetics, in very short order this system was plunged back into the confusion of increasing complexity. As early as 1934, Landsteiner, Strutton, and Chase had, by the injection of rabbits with blood from a Negro male, produced an antiserum which agglutinated the red cells of 7.3 per cent of a group of American Negroes tested, but only 0.5 per cent of a group of American Caucasians. The agglutinogen in question has come to be known as "Hunter" (Hu). The relationship of this antigen to the then known systems was not clear, although Wiener noted that all positive reactors were N or MN. In 1951, Ikin and Mourant, while studying blood samples from Nigerians, found two bloods which gave anomalous reactions with a rabbit antiserum thought to be specific for the M agglutinogen. Because none of the bloods from Caucasians which they tested proved positive for this antigen, Ikin and Mourant suggested that they might have found a second antiserum against the Hunter factor. Subsequently, Chalmers, Ikin, and Mourant (1953) proved that this was not true, and the new antigen which had thus been discovered was called "Henshaw" (He). These same investigators then found that all the Hunter- and Henshaw-positives whom they tested were also N-positive, from which it follows that the genes responsible for these antigens, Hunter and Henshaw, must be closely linked to the $MNSs$ system in a manner analogous to the C, D, and E loci (or gene complex) postulated for the Rh-system. It is not yet clear whether the genes responsible for the antigens, Hu and He, are alleles. Another interesting peculiarity of these antigens is their racial distribution. Up to the present time, the Henshaw antigen has not been found among Europeans, and the Hunter antigen appears to be about fourteen times more frequent among American Negroes than among non-Negroes living in the United States.

The problem as to the genetic relationship of the MNSs serological reactions is very similar to the problem of the genetic relationship of the Rh-reactions. Under either of the two interpretations which have been presented, however, the absence of dominance is the same.

8.8. *Other blood antigens.*—In addition to the three antigen systems just described, fourteen other systems have been recognized at the time of this writing. In the case of eight of these systems, the antigen by which the system is characterized has thus far been found in only one, two, or three families. Levine (1951; see also Wiener and Brancato, 1953) has aptly termed these antigens "private" blood factors. Presumably they are quite rare. It is of interest that, although the antibody responsible for the recognition of one of these "private" factors (Jay) has thus far been found in only four persons in three families, these four persons are distributed over three continents (Zoutendyk and Levine, 1952). The antigens characterizing the other six antigen systems (P, Lutheran, Kell, Lewis, Duffy, Kidd; cf. Table 8-4) are much more common, rendering these systems useful to the geneticist in a variety of ways to be discussed later.

Because of the gene relationships which have come to light in the field of serological genetics, serologists have developed an elastic system of terminology which has much to recommend it. When an antigen is discovered, it is identified in some appropriate fashion, usually by the initials of the person in whose blood the antisera was first recognized. The first gene which can be postulated is designated by these same initials with the superscript "a." When a second gene can be postulated, it is given the superscript "b"; and so on. This terminology is followed in Table 8-4, except where, as for the MN and A, B, O groups, current usage has already established an alternative nomenclature.

8.9. *Concluding generalizations.*—In the light of the information afforded by the ABO, MN, Rh, Kell, Kidd, and possibly the Lewis antigenic systems, it seems reasonable to conclude that the genes controlling the red blood cell antigens rarely, if ever, show dominance among members of the same allelic series. Within each allelic series, certain of the corresponding antigens are more prone to elicit antibody formation than are others. However, it now seems probable that, in time, the majority, if not all, of the genes which are

TABLE 8-4

INHERITANCE OF P, LUTHERAN, KELL, LEWIS, DUFFY,
AND KIDD ANTIGENIC SYSTEMS

Antigen System	Antibodies Demonstrated	Phenotypes (Serological Reactions)	Antigens Postulated	Presumed Genotypes
P..........	anti-P	$\begin{cases} P(+) \\ P(-) \end{cases}$	P	$\begin{cases} P+P+ \\ P+P- \\ P-P- \end{cases}$
Lutheran...	anti-Lu^a	$\begin{cases} Lu\ (a+) \\ Lu\ (a-) \end{cases}$	Lu (a+)	$\begin{cases} Lu^aLu^a \\ Lu^aLu^b \\ Lu^bLu^b \end{cases}$
Kell........	$\begin{cases} \text{anti-K} \\ \text{anti-k} \end{cases}$	$\begin{cases} K+k- \\ K+k+ \\ K-k+ \end{cases}$	$\begin{cases} \text{K (Kell)} \\ \text{k (Cellano)} \end{cases}$	$\begin{cases} KK \\ Kk \\ kk \end{cases}$
Lewis......	$\begin{cases} \text{anti-Le}^a \\ \text{anti-Le}^b \end{cases}$	$\begin{cases} Le\ (a+b-)^* \\ Le\ (a-b+) \\ Le\ (a-b-) \end{cases}$	$\begin{cases} \text{Le (a+)} \\ \text{Le (b+)} \end{cases}$	$\begin{cases} Le^aLe^a \\ Le^aLe^b,\ Le^bLe^b,\ Le^bLe^c \\ Le^aLe^c,\ Le^cLe^c \end{cases}$
Duffy......	anti-Fy^a	$\begin{cases} Fy(+) \\ Fy(-) \end{cases}$	Fy(a+)	$\begin{cases} Fy^aFy^a \\ Fy^aFy^b \\ Fy^bFy^b \end{cases}$
Kidd.......	$\begin{cases} \text{anti-Jk}^a \\ \text{anti-Jk}^b \end{cases}$	$\begin{cases} Jk\ (a+b-) \\ Jk\ (a+b+) \\ Jk\ (a-b+) \end{cases}$	Jk (a+)	$\begin{cases} Jk^aJk^a \\ Jk^aJk^b \\ Jk^bJk^b \end{cases}$

* The relationship of the serological reactions to the number of genes and hence genotypes is not at present clear. The explanation given here is that proposed by Race and Sanger (1950).

members of allelic series characterized by serological effects will be found to have some expression when heterozygous, as detected by the ability to elicit antibody formation.

Further evidence in support of the hypothesis that complete dominance is and will be rare with respect to genes responsible for serological reactions comes from a study of the quantitative aspects of the blood-group antigens. A number of blood-group antigens exhibit significantly stronger reactions when the responsible gene is homozygous than when heterozygous (dosage effect). These include the MN and Ss of the MNSs system, Rh, Kell, and Duffy antigens (literature in Race, Sanger, and Lehane, 1953). Here again are findings permitting differentiation in some cases between individuals homozygous and heterozygous for particular genes.

Because of the nature of serological investigation, it is especially feasible in this field of inquiry to identify an effect of a given gene when the latter is heterozygous. If one may argue by extension, then these findings further reinforce the opinion, expressed earlier in the chapter, that in time, as we sharpen our diagnostic tools, we will be able to recognize the effects when heterozygous of a very considerable number of human genes.

Bibliography

SPECIFIC REFERENCES

CHALMERS, J. N. M.; IKIN, E. W.; and MOURANT, A. E. 1953. A study of two unusual blood-group antigens in West Africans, Brit. M. J., 2:175–76.

FALLS, H. F., and COTTERMAN, C. W. 1948. Choroidoretinal degeneration, Arch. Ophth., 40:685–703.

IKIN, E. W., and MOURANT, A. E. 1951. A rare blood group antigen occurring in Negroes, Brit. M. J., 1:456–57.

LANDSTEINER, K.; STRUTTON, R. W.; and CHASE, M. 1934. An agglutination reaction observed with some human bloods, chiefly among Negroes, J. Immunol., 27:469–72.

LEVINE, P. 1951. A brief review of the newer blood factors. Tr. New York Acad. Sc., Ser. II, 13:205–9.

LEVINE, P.; KUHMICHEL, A. B.; WIGOD, M.; and KOCH, E. 1951. A new blood factor, s, allelic to S, Proc. Soc. Exper. Biol. & Med., 78:218–20.

MOHR, O. L., and WRIEDT, C. 1919. A new type of hereditary brachyphalangy in man, Carnegie Inst. Washington Pub., No. 295.

RACE, R. R.; SANGER, R.; and LEHANE, D. 1953. Quantitative aspects of the blood-group antigen Fy^a, Ann. Eugenics, 17:255–66.

SANGER, R., and RACE, R. R. 1951. The MNSs blood group system, Am. J. Human Genetics, 3:332–43.

SNYDER, L. H., and DOAN, C. A. 1944. Studies in human inheritance. XXV. Is the homozygous form of multiple telangiectasia lethal? J. Lab. & Clin. Med., 29:1211–16.

STERN, C.; CARSON, G.; KINST, M.; NOVITSKI, E.; and UPHOFF, D. 1952. The viability of heterozygotes for lethals, Genetics, 37:413–49.

STERN, C., and CHARLES, D. R. 1945. The rhesus gene and the effect of consanguinity, Science, 101:305–7.

WIENER, A. S., and BRANCATO, G. J. 1953. Severe erythroblastosis fetalis caused by sensitization to a rare human agglutinogen, Am. J. Human Genetics, 5:350–55.

ZOUTENDYK, A., and LEVINE, P. 1952. A second example of the rare serum anti-Jay (Tjᵃ), Am. J. Clin. Path., 22:630–33.

HUMAN HEREDITY

GENERAL REFERENCES

MULLER, H. J. 1950. Our load of mutations, Am. J. Human Genetics, 2:111–76.

NEEL, J. V. 1947. The clinical detection of the genetic carriers of inherited disease, Medicine, 26:115–53.

RACE, R. R., and SANGER, R. 1950. Blood groups in man. Oxford: Blackwell Scientific Publications.

WIENER, A. S. 1943. Blood groups and transfusions. 3d ed. Springfield: Charles C Thomas.

Problems

1. List the *impossible* paternal blood groups in each of the following cases:

Blood Group of Child	Blood Group of Mother		Blood Group of Child	Blood Group of Mother
O	O		N	MN
O	A		$Jk^a\ Jk^b$	$Jk^a\ Jk^a$
A	B		$r'r$	$R'r$
AB	AB		KK	Kk
MN	N		CDE/cde	cde/cde

2. In the examination of the uric acid levels among the relatives of gouty patients, it was found that males averaged higher uric acid levels (about 0.5 mg. per cent). Furthermore, the mean levels of uric acid in hyperuricemic persons were about 7–8 mg. per cent in the male and 5.5–6 mg. per cent in the female. Only about 10 per cent of hyperuricemics develop gout; of these gouty patients, 90 per cent are males. How can the differences between the sexes in uric acid levels be accounted for, and what criteria may be postulated as necessary before clinically recognizable gout appears?

3. Originally the ABO blood groups were assumed to be due to two independent pairs of genes, say A and a and B and b. It was believed that $aabb$ individuals were type O, $AAbb$ or $Aabb$ individuals were type A, $aaBB$ or $aaBb$ individuals were type B, and, lastly A–B– individuals were type AB. How could one disprove this hypothesis? (Hint: Investigate the outcome of various types of matings under this hypothesis.)

4. If for the moment we assume that hemolytic disease of the newborn is due solely to Rh-incompatibility, would the disease appear more frequently or less frequently among the children of first-cousin marriages? (Hint: Stern and Charles, 1945.)

5. If we think in terms of the simple dichotomy, Rh+ and Rh−, what effect would the elimination of some infants by hemolytic disease of the newborn have on the frequency of the gene responsible for Rh-negativity, say q, when $p = q$, $p < q$, and $p > q$?

CHAPTER 9

More Complex Genetic Situations

NUMEROUS examples of the simple ratios compatible with expectation on the basis of segregation for one pair of genes are available in human genetics; some of these examples have been discussed in previous chapters. We turn now to problems of identifying more complex modes of inheritance of human traits, problems such as occur when a trait depends on homozygosity or heterozygosity for two or more nonallelic genes.

9.1. *Two pairs of alleles.*—The logical point of departure in a discussion of the more complex genetic situations is the theoretical behavior to be expected in the segregation of two independent pairs of alleles. Consider the case of two independently assorting pairs of alleles, say, A and a and B and b. Three genotypes may be expected to exist for each pair of alleles; these genotypes are AA, Aa, and aa and BB, Bb, and bb. Since the pairs are assumed to assort independently, nine different combinations of the four types of genes, A, a, B, and b, may be expected. In the event of the mating of two individuals of genotype $AaBb$, the frequencies of these genotypes are given by $[p(A) + q(a)]^2[p(B) + q(b)]^2$, where $p(A)$ is the probability of gene A; $q(a)$ the probability of a; etc. Since in this particular mating $p(A) = q(a) = p(B) = q(b) = 1/2$, the frequencies of the various combinations are as follows:

1/16......$AABB$	2/16........$AaBB$	1/16.......$aaBB$
2/16.......$AABb$	4/16.......$AaBb$	2/16.......$aaBb$
1/16.......$AAbb$	2/16........$Aabb$	1/16.......$aabb$

Thus we see that the basic genotypic ratio is $1AABB:2AABb:2AaBB:4AaBb:1AAbb:2Aabb:1aaBB:2aaBb:1aabb$.

We may now ask: How is this ratio reflected in the phenotypes when modified by dominance within a pair or pairs of alleles and interactions between nonallelic pairs? The simplest conceivable case is that of no dominance in either pair of alleles. In such a case the phenotypic ratio is precisely the same as the genotypic, namely, $1:2:2:4:1:2:1:2:1$. In man such a ratio would be expected, for example, from the mating of a person of blood type $ABMN \times$ another $ABMN$. In order of increasing complexity, the next cases would involve dominance within one or the other, or both sets of alleles. If dominance exists in one set of alleles, say the A-a pair, then the genotypes AA and Aa are indistinguishable, and the phenotypic ratio to be expected from heterozygous parents would be $3A$-BB:$6A$-Bb:$3A$-bb:$1aaBB$:$2aaBb$:$1aabb$, where "-" indicates that the allele may be either A or a. Similarly, if dominance existed in both sets of alleles, the ratio to be expected would be $9A$-B-: $3aaB$-:$3A$-bb:$1aabb$. Here again situations arise in human heredity where this ratio would be expected. For example, as long as antisera against the postulated Fy^b and Lu^b antigens are unknown, such a ratio would be expected from the mating of two individuals of genotype $Fy^aFy^bLu^aLu^b$ (cf. Table 8-4, p. 92).

Now, if to interactions between members of a gene pair we add interactions between gene pairs, a whole new spectrum of possible phenotypic ratios is conceivable. If we consider only attributes of the individual, that is, characteristics which are classifiable in an "either-or" sense, then one interallelic interaction is readily apparent. This is the type of interaction in which a member of a pair of alleles masks the effects of a nonallelic gene or genes. This masking effect is known as "nonallelic dominance" or "epistasis." Snyder has proposed the terms "dominant" and "recessive" epistasis to differentiate between cases in which the gene which masks is dominant or recessive within its own set of alleles. This scheme of nomenclature has certain descriptive advantages in the simpler cases, but it is not widely used.

We shall consider fully only one specific type of epistasis—the case of dominant epistasis. It is understood that gene A is epistatic, that is, it masks the effects of genes B and b, and hence genes B and b find expression only in the absence of A. When A is epistatic in this manner, then the genotypes $AABB$, $AABb$, $AAbb$, $AaBB$, $AaBb$, and $Aabb$ are indistinguishable. The ratio of the recognizable phenotypes is $12A$---:$3aaB$-:$1aabb$ if B is dominant to b, or $12A$---:$1aaBB$:$2aaBb$:$1aabb$ if it is not.

The array of various phenotypic ratios resulting from a number of different forms of epistasis are given in Table 9-1. These various interallelic interactions are well recognized in plants and animals, and examples are to be found in any standard textbook on general genetics. In man, however, we recognize no clear-cut examples of such interactions, despite the certainty that such must exist. For example, we may logically expect masking when dealing with two pairs of genes, one pair of which is responsible for the elaboration of a metabolite or a precursor of a metabolite utilized by the other pair. The reasons for this lack of identification of such interactions undoubtedly find their bases in the problems posed in the analysis of family data, some of which have been discussed in the preceding chapters and others of which will be presented in the chapters to follow.

9.2. *Three pairs of alleles.*—We need not explore at any length the problems presented by three pairs of alleles. It is apparent that an even greater number of modifications of the basic genotypic ratio are possible. The reader may verify for himself that an individual whose genotype is $AaBbCc$ will produce the following 2^3 gametes, ABC, ABc, AbC, abC, Abc, aBc, abC, and abc and that if that individual was married to a similar individual, the genotypic

TABLE 9-1*

PHENOTYPIC RATIOS TO BE EXPECTED FROM
VARIOUS TYPES OF GENE INTERACTION

GENIC RELATIONSHIP			PHENOTYPIC RATIO								
Within A-a Pair	Within B-b Pair	Between Pairs	AABB	AABb	AaBB	AaBb	AAbb	Aabb	aaBB	aaBb	aabb
No dominance....	No dominance	None	1	2	2	4	1	2	1	2	1
No dominance	Dominance	None	3		6		1	2	3		1
Dominance.......	Dominance	None	9				3		3		1
Dominance.......	Dominance	aa masks B, b	9				3		4		
Dominance.......	Dominance	A masks B, b	12						3		1
Dominance.......	Dominance	$\{A$ masks B, b / $\{bb$ masks A, a	12						3		1
					←————(13)————→						
Dominance.......	Dominance	$\{aa$ masks B, b / $\{bb$ masks A, a	9						7		
Dominance.......	Dominance	$\{A$ masks B, b / $\{B$ masks A, a	15								1

* Reproduced by permission from The principles of heredity, by Dr. L. H. Snyder, copyright 1946, D. C. Heath and Co.

ratio to be expected among a large number of offspring would be as shown in the accompanying tabulation.

1$AABBCC$	2$AaBBCC$	1$aaBBCC$
2$AABBCc$	4$AaBBCc$	2$aaBBCc$
1$AABBcc$	2$AaBBcc$	1$aaBBcc$
2$AABbCC$	4$AaBbCC$	2$aaBbCC$
4$AABbCc$	8$AaBbCc$	4$aaBbCc$
2$AABbcc$	4$AaBbcc$	2$aaBbcc$
1$AAbbCC$	2$AabbCC$	1$aabbCC$
2$AAbbCc$	4$AabbCc$	2$aabbCc$
1$AAbbcc$	2$Aabbcc$	1$aabbcc$

9.3. *Quantitative characteristics and continuous variation.*—We have seen that the distribution of a given attribute, that is, a characteristic which is either present or not present within a family or a group of families, is explicable on the basis of postulating a discontinuity in genotype. Not all human characteristics, however, are discontinuous attributes. Some characteristics we all possess but in varying degrees, e.g., stature or intelligence. These are the characteristics which give rise to "measurement" data. It behooves us, therefore, to demonstrate that differences between individuals with respect to a trait which exhibits a continuous spectrum of variation from the highest to the lowest value may, like qualitative differences, also be attributed to a discontinuity in genotype.

The notion of continuous variation may be briefly characterized by visualizing a variable capable of assuming an "infinite" number of values, that is, different measurements. In practice, this is rarely realized, both because our means of obtaining the measurements are not sufficiently refined and because there is a practical limit to the extent to which further subdivision is useful; however, we may still conceive of values intermediate between those which are obtained by the standard of measure at our disposal. Any given measurement with respect to this variable may be regarded as a sample of one observation from an infinite population of possible measurements. This population may be mathematically defined in many ways; two of the most convenient measurements in this respect are the average value, or mean, and the variance, a measure of the amount by which the individual observations depart from the mean (cf. sec. 5.4, p. 38). When we have a variable capable of assuming an "infinite" number of values, it is obviously impossible to record every value, and we are forced to estimate the population mean and variance from a sample of measurements chosen at random from the population in question. Since numerous samples may be chosen or drawn, the sample is itself a random variable having a sampling distribution characterized by a mean and a variance. The sample mean, $\bar{x}$, and the sample variance, s^2, which characterize a particular sample are determinable. These values, when considered as estimates of the population values, are defined as follows:

$$\bar{x} = \frac{1}{n} \sum_i x_i, \qquad s^2 = \frac{1}{n-1} \sum_i (x_i - \bar{x})^2,$$

where x_i is the ith observation in the sample and summation is over all observations. Now these sample values, unlike their population counterparts, are also variables, since the sample mean and variance may be expected to vary with different samples. Intuitively, we appreciate that these sample values will converge on the population values as the sample size increases. This belief may be formulated in terms of a precise mathematical argument.

Such theories, however, fall within the domain of statistics and need not concern us here. Suffice it to say that these theories permit us to argue from the sample to the population with some predictable degree of accuracy.

If, now, a large sample of individuals was drawn and each was measured with respect to some quantitatively variable characteristic, such as height or weight, we would have a large number of measurements which could be graphically represented in either of two ways. We might elect to plot (1) each measurement by the frequency of its occurrence or (2) each measurement by the frequency with which measurements equal to, or smaller than, the measurement occur. If the measurements were plotted by the frequency of their occurrence (method 1), we should find in many instances that, as the sample increased, these frequencies would exhibit an empirical distribution which is similar to a specific smooth curve. This smooth curve is the so-called "normal frequency" or "normal density" function and is a characteristically symmetrical bell-shaped curve. Figure 9-1 indicates the approximation to the normal frequency distribution observed in a specific set of data concerned with the height of American Indians. If, on the other hand, the measurements were plotted by the frequency with which measurements equal to, or smaller than, a given measurement are observed, that is, by method 2, we should find, again, that these frequencies would exhibit an empirical distribution similar to a smooth curve. In this case, the curve is the S-shaped curve of the normal cumulative distribution function. Figure 9-2 may serve as an illustration of the approach of an empirical distribution (again, the height of American Indians) to the normal cumulative distribution.

In applying the normal distribution, it is often convenient to speak of the probability of a given measurement's departing from the mean by some specified unit. The unit most frequently chosen is the standard deviation, σ, the square root of the variance. If a continuous variable is normally distributed with mean, m, and standard deviation, σ, then the intervals $m \pm 1\sigma$, $m \pm 2\sigma$, and $m \pm 3\sigma$ will contain approximately 68.3, 95.5, and 99.7 per cent of the measurements, respectively. It is thus possible to formulate a probability for the likelihood that a given measurement will not differ from the mean by more than one standard deviation, etc.

9.4. *The genetic theory of continuous variation.*—"The idea that the occurrence of numerous heritable grades of a single character might be due to multiple independent factors was suggested by Mendel himself in 1865" (Wright, 1952). This fact notwithstanding, it was not until 1908 that the experimental and logical bases of quantitative variation began to attract the interest of geneticists. In this year Nilsson-Ehle demonstrated the genic nature of seemingly continuous variability in the color of wheat; and in 1909

Johannsen, in a classic study of the causes of variability in the size of beans, clearly differentiated between genetic and nongenetic causes of variation in the same character.

Let us turn, now, to a small demonstration that differences between individuals with respect to a continuously distributed variable may be attributed to the inheritance by one individual of discrete potentialities not present in another. When we have a collection of individuals between whom matings may occur, that is, a population in the genetic sense, we may refer to the sum total of their genes as a "gene pool." We may also, in a more restricted sense,

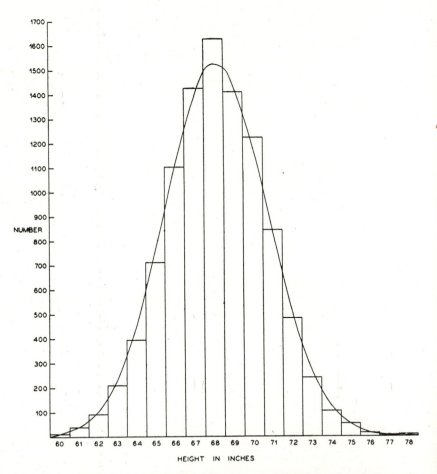

FIG. 9-1.—The observed distribution of the statures of 10,004 American Indian males. The normal distribution having the same mean and variance is superimposed. (Data of Davenport and Love, 1921.)

refer to the gene pool with respect to a specific gene or group of genes. Imagine, now, a pool of genes which have in common the fact that they influence a particular physiological reaction. Let us assume, for simplicity, that basically there are only two types of genes, say $g_{.1}$ and $g_{.2}$, in this pool. From this pool we propose to make a series of draws, each of two genes, and we shall say that the sum total of the draws represents the genotype of an individual with respect to n loci and that the outcome of one draw of two genes determines the genotype of the individual with respect to a specific locus. Each individual will, then, possess n sets of two genes, or $2n = N$ total genes.

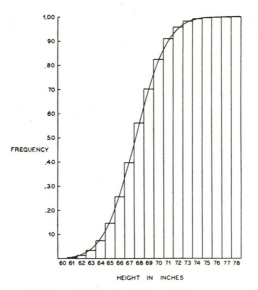

HEIGHT IN INCHES

FIG. 9-2.—The empirical cumulative distribution of the statures of 10,004 American Indian males. The normal cumulative distribution is superimposed. (Data of Davenport and Love, 1921.)

At the ith locus, an individual's genotype may be either $g_{.1}g_{.1}$ or $g_{.1}g_{.2}$ or $g_{.2}g_{.2}$. If p_i is the probability that the gene at the ith locus will be $g_{.1}$ and $q_i = 1 - p_i$ is the probability that the gene at the ith locus will be $g_{.2}$, then the probabilities of the various genotypes are p_i^2, $2p_iq_i$, and q_i^2, respectively. Now if the initial gene pool is very large, then we may assume that p is constant for all loci; and the same obviously holds true for q. If this obtains, then it follows that the probability of an individual's being homozygous $g_{.1}g_{.1}$ at all n loci, assuming the loci are independent, will be $(p^2)^n$; similarly the probability that an individual will be homozygous $g_{.1}g_{.1}$ at $n - 1$ loci and heterozygous at the other is $n(p^2)^{n-1}(2pq)$, since there are n ways in which heterozygosity could occur. It may be seen, then, that the array of possible genotypes with their frequencies are given by the successive terms in $(p^2 + 2pq + q^2)^n$. In terms of the frequency of different numbers of genes of type $g_{.1}$ or $g_{.2}$, irrespective of locus, the distribution is merely $(p + q)^N$, since the distribution of genotypes is equal to $[(p + q)^2]^n$ or $(p + q)^{2n}$. Consequently, the mean number of genes of, say, type $g_{.1}$ to be expected among N is Np. The variance of this distribution of number of genes of type $g_{.1}$ is Npq. It should be noted that in the event that p is not constant for all loci, as would very likely be the case in many instances, the frequencies of the various genotypes are given by $(p_1 + q_1)^2(p_2 + q_2)^2, \ldots, (p_n + q_n)^2$ if all n loci are assumed to be independent. When this obtains, the mean may or may not be altered over the preceding situation, that is, when p is constant. In either event, that is, whether the mean is or is not altered, the variance will be less when p varies from locus to locus.

Now let us assume that each of these two types of genes makes a contribution, differing in amount, to the measurement obtained on an individual and that these contributions are additive. Since the genotype of an individual with respect to any locus is a random event, the contributions of a pair of genes at this locus is a random variable, say m_i, capable of assuming three values with probabilities specified by $(p + q)^2$ corresponding to the three possible genotypes at this locus. If there are two such independent loci, there would exist nine genotypes but only five different measurements corresponding to $0, 1, \ldots, 4$, of the genes making the greater contribution, say $g_{.2}$, since the specific location of the latter gene or genes is immaterial. If there are n such independent loci, then the measurement of an individual will be the cumulative effect of many random variables. Moreover, since we have assumed the genes to be additive in effect, the measurement will be the sum of these random variables, say M. Since the value of M is dependent upon the number rather than the location of the $g_{.2}$ genes, there would exist only $2n + 1$ different values of M, despite the fact that 3^n different genotypes may exist. As we have already seen, the number of $g_{.2}$ genes is binomially distributed; consequently, the intermediate values of M would be most common. Furthermore, the frequency of any one value of M would be determined by the frequency with which the appropriate number of $g_{.2}$ genes occurred. For example, the probability that M would be equal to x, that is, that there would be x $g_{.2}$ genes present, would be

$$P\,(M = x) \,=\, \binom{2n}{x}\, p^x q^{2n-x}.$$

Now, as the number of loci increases, the number of $g_{.2}$ genes possible increases, as does the possible number of $g_{.1}$ genes, that is, as n and x increase, so may $2n - x$. Under these circumstances it may be shown that the distribution of the number of genes of type $g_{\cdot2}$ will tend toward the normal distribution. It follows, then, that the distribution of measurements would be almost continuous. This continuity would then be completed if environmental variables led to an overlapping of the phenotypes resulting from differing numbers of $g_{\cdot2}$ genes. In effect, then, if we have a "large" number of sets of alleles acting on a single characteristic in a uniform manner, we may expect that the empirical distribution of this characteristic will approximate the normal distribution as the number of gene pairs increases.

We have assumed in the foregoing model (1) that the genotype of an individual is a random event, (2) two alleles at any given locus, (3) n independent loci, and (4) additive action of the genes. Let us now examine these assumptions. (1) The first assumption is not unique to this problem, and its legitimacy is amply attested to by the very nature of gene segregation. (2) On the other hand, it is quite conceivable that some loci may have more than two alleles. This does not materially alter the argument, since the primary effect of more than two alleles at any one locus would be to increase the number of different values of M. (3) The third assumption, however, does not hold if n is sufficiently large, since some of the sets of alleles will be borne on the same chromosome and the segregation of one may not be independent of the segregation of another within a given family. Fortunately, this does not necessarily invalidate the argument, since it can be shown that some non-independent variables (e.g., linked genes) also approach the normal distribution, and, furthermore, we may legitimately expect that the majority of the n sets of alleles will be, in fact, independent of one another. (4) In the example we have assumed an additive effect of the genes involved, since this is the simplest conceivable hypothesis of cumulative gene action. Many other types of gene interaction are possible. For example, a few or many of the genes may exhibit dominance, in which case their effects are not independently additive; some of them may be multiplicative in their action; etc. If dominance were exhibited within the gene pairs, we should expect a skewing, that is, an asymmetry in the empirical distribution of the measurements. This skewness would be more difficult to demonstrate as the number of gene pairs increased, even though dominance existed within each pair. If the genes were multiplicative in their action, i.e., if, instead of an absolute amount, their contributions were exponential, so that the absolute contribution of any gene to the character in question was a function of the number of similar genes present, an asymmetry in the distribution would again result when the frequencies of arithmetically equal classes were plotted. However, for the purpose of analysis of the genetic and nongenetic components of variation, the choice of the scale of measurement is in the hands of the investigator; frequently it is possible to select a scale on which the effects of factors, both genetic and environmental, are additive, even though in an absolute sense their contribution to variation may not be so.

The mere fact that a genetic argument can be constructed which will account for continuous variation does not in any sense mean that the underlying cause of all variation or even variation with respect to a specific characteristic is genetic. In point of fact we recognize that, given a uniform genetic population, differences in the environment may lead to variation among individuals and that such variation may be normally distributed. Since neither heredity nor environment operates independently of each other, demonstrable variation is quite likely to be the result of the interplay of the two. Fortunately, this fact does not vitiate attempts to detect the role of nature and of nurture in quantitative variation. It does, however, necessitate the use of rather special analytic techniques, one of which, and the one which we shall now consider, being known as "correlation analysis."

9.5. *The role of correlation in the analysis of continuous variation.*—We recognize the existence of resemblances between parent and child. Such resemblances may be due to heredity or to environment. They may be due to heredity, because the genotype of an offspring is dependent upon the parental genotypes; they may be due to environment, because the environment in which an individual is reared is more likely to be comparable to that in which the parent or parents were reared than to that of an unrelated individual. To dissociate these factors of heredity and environment, it is essential that the degree of resemblance be quantitated under varying degrees of relationship or environments.

A means of determining whether a dependent relationship exists between parent and child measurements and the subsequent measure of the "strength" of this relationship may be developed as follows: Assume that we are given a series of measurements obtained from parents and children. Let us designate the measurement on the ith parent by x_i and the measurement on the offspring of the ith parent by y_i. Let us assume that both x and y are capable of assuming n different values. It is now possible to construct a table in terms of the two variables, as, for example, Table 9-3. This bivariate table would have n columns corresponding to the n values of x (parental measurements) and n rows corresponding to the n values of y (child measurements), or n^2 cells in all. Each pair of measurements, say x_i and y_i, could be entered in one and only one cell, namely, that cell corresponding columnwise to the parental measurement and rowwise to the child measurement. For each row and each column there will exist a mean and a variance. If x and y are independent, then all row means will be equal and all column means will be equal; of course, within any specific set of data they may vary because of sampling error. On the other hand, if x and y are not independent, then the row and column means and variances will differ among themselves.

Let us now visualize a process wherein, for each successive value of x (that is, for each column) in the population of x's, we determine the mean of the y's in the population of y's corresponding to the fixed x. These means could be plotted against their appropriate x's and a line drawn such that it passed through each of the points so determined. Such a line is called the "regression line," and we shall assume that in this case it is a straight line, though this need not be true. Now this regression line, if it were known, would indicate whether or not prediction of y for a given x was possible. The predictive value of the regression line may vary from situation to situation, and so we desire a measure of the usefulness of a given regression line for predictive purposes. Such a measure is the correlation coefficient, sometimes termed the "coefficient of product-moment correlation." The correlation coefficient is so constructed that it will have a value close to zero for a regression line useless for prediction, and close to ± 1 for a line capable of perfect prediction.

In general, the "true" correlation coefficient between two variables is unknown. It may be estimated, however; and we take as our estimate of the correlation coefficient, r, the ratio of the covariance to the geometric mean of the two variances, that is, of the variances of x and y. The covariance is merely a means of measuring the extent to which the values of x and y vary dependently. Mathematically, the covariance is defined as the mean product of two variables measured from their means, that is, $\sum_i (x_i - \bar{x})(y_i - \bar{y})/n$.

From the formula it is apparent that the covariance will be large if positive deviations in x are associated with positive deviations in y, or negative with negative. On the other hand, if x and y do not vary dependently, then we should expect the association of minus with minus and plus with plus deviations to be no more frequent than minus with plus or plus with minus deviations. In this case the covariance would tend toward zero. Now the geometric mean of the variances is merely the square root of the product of the variances of x and y. In terms of the simplest computational formula, this definition of the estimate of the correlation coefficient is equivalent to

$$r = \frac{n\Sigma xy - \Sigma x \Sigma y}{\sqrt{[n\Sigma x^2 - (\Sigma x)^2][n\Sigma y^2 - (\Sigma y)^2]}}, \qquad (9.5.1)$$

where n is the number of pairs of measurements.

The estimate of the correlation coefficient derived from a sample will rarely be zero, even when it is known that such is the value of the "true" correlation coefficient. It is necessary, therefore, to determine the statistical significance of a correlation coefficient derived from a sample. In human biology a statistically significant correlation coefficient generally implies a correlation coefficient of magnitude wherein equal or larger values would be expected among only 5 per cent (or 1 per cent) of an infinite number of comparable estimates if the true correlation coefficient in the population from which these estimates were derived was zero.

It must be recognized that a correlation coefficient, per se, is purely a mathematical interpretation of the relationship of two variables and is completely devoid of any cause or effect implications. The mere fact that two variables tend to increase or decrease together does not necessarily imply a direct or indirect effect of one on the other. It may merely mean that they both respond similarly to some third factor. How, then, are we to utilize the correlation technique in detecting differences due to environment and heredity, since either or both may lead to an association of parental with child measurements? In the absence of ancillary evidence, a single correlation coefficient will not permit the dissociation of these two causes of variation if both are known to be operative. On the other hand, a comparison of an array of correlation coefficients derived from material wherein the genetic dependency varies will be informative. For example, if genetic factors are paramount as the cause of variation, we should expect the correlation of a parent with child to be greater than the correlation of parent with foster-child, since in the latter case only the environments are comparable. On the other hand, if the role of heredity is minimal in the observed variation, we should expect these correlations to be similar. It should be noted, however, that, in the event of dissimilar correlation coefficients between parent and child and between parent and foster-child, we are not fully justified in attributing, categorically, the difference between these coefficients to heredity. We are not justified in using these correlation coefficients to determine numerical estimates of the contributions of heredity and environment to the total variation, unless we assume that the whole set of factors which tend to make related individuals alike is their heredity. This, however, amounts to the assumption that types of environments are randomly distributed; that is to say, that all genotypes have equal probability of experiencing the various sets of nongenetic intrinsic and extrinsic factors which are termed "environments." To accept this assumption as legitimate is to overestimate the effects of heredity by ignoring those factors which lead to a nonrandom distribution of environments, such as "the viviparous habit of the human species, the existence of the family as a social institution, and the stratification of human society in widely different social levels" (Hogben, 1933). It is precisely this fact—the existence of a correlation between hereditary and environmental factors—which has vitiated all attempts to define sharply, in terms of components of variation, the roles of heredity and environment in quantitative human variability.

9.6. *Quantitative genetic characteristics.*—Stature and intelligence are two human characteristics which exhibit a wide spectrum of variation. There are recognized certain single-gene substitutions which may result in an individual's differing very appreciably from the mean in stature or intelligence. Thus individuals heterozygous for the dominant gene responsible for chondrodystrophy are invariably much shorter than their brothers or sisters who do not have the gene. Likewise, there is known a recessive gene which, when homozygous, results in a severe type of mental retardation known as "phenylketonuria" because of the excretion of large amounts of phenylketones in the urine. When, however, one considers the range of stature and intelligence in any population, there is little evidence that single genes with large effects contribute to any great extent to the total variability. What, now, is the evidence that stature and intelligence conform to the behavior expected from genetic characteristics controlled by a larger number of genes whose effects are equal and additive?

1. STATURE

Historically, stature was the first of the continuously distributed characteristics to be investigated genetically. The pioneer work of Galton and Pearson on this characteristic laid the foundation for the genetic theory of continuous variation and, what is more, for the application of statistical methods to the analysis of biological data.

In the simplest terms we may conceive of stature as being the cumulative effect of a number of genes whose actions are similar and whose effects are additive. Under this hypothesis, then, as we have seen, we may expect that the distribution of statures will be reasonably well approximated by the normal distribution and that a correlation will exist between parent and child stature. Does this obtain?

In Table 9-2 are given the heights of 10,004 American Indian males obtained at selective service examinations during World War I (Davenport and Love, 1921). The sample mean and standard deviation were computed and found to be 68.12 and 2.61 inches, respectively. The observed frequencies are plotted graphically in Figure 9-1, as are the theoretical frequencies expected on the basis of a normally distributed variable with the same mean and variance as the sample values. The correspondence between the observed and expected values is fairly good. This would suggest, then, that the normal distribution is a reasonable approximation to the empirical distribution of statures.

Does there exist a parent-child correlation in stature? The distribution of statures of 1,078 sons according to the stature of the fathers is presented in Table 9-3 (Pearson and Lee, 1903). The use of fractional entries in this table is designed to minimize error and need cause no confusion. Whenever a measurement fell on the border line of two intervals, half a unit was assigned to each of the contiguous intervals. It was thus possible when both of a pair of measurements fell on border lines to record the pair as four quarter-entries, one to each of the four cells surrounding the boundary at which the pair of measurements occurred.

Certain features of these data are apparent merely on inspection of this table. We note (1) that the upper right-hand and lower left-hand quadrants

TABLE 9-2*

FREQUENCY DISTRIBUTIONS OF STATURES FOR ADULT INDIAN MALES
OBTAINED AT TIME OF PHYSICAL EXAMINATION FOR SELEC-
TIVE SERVICE DRAFT, 1917 AND 1918

Height (Inches)	Frequency Observed	Frequency Expected	Height (Inches)	Frequency Observed	Frequency Expected
60 −	11	13	71 −	845	833
61 −	40	38	72 −	482	511
62 −	92	101	73 −	239	269
63 −	210	228	74 −	104	123
64 −	396	442	75 −	52	50
65 −	712	750	76 −	16	16
66 −	1,103	1,096	77 −	3	5
67 −	1,427	1,389	78 −	7	2
68 −	1,629	1,524			
69 −	1,411	1,437	Total..	10,004	10,003
70 −	1,225	1,176			

$\bar{x} = 68.12, s = 2.61$

* Data from Davenport and Love (1921).

of the table contain relatively few entries; that is to say, there are relatively few combinations of a tall or a very tall parent with a short or very short child and vice versa, and (2) that the entries tend to cluster along the diagonal leading from the upper left-hand to the lower right-hand corner of the table. These observations would suggest the presence of a positive association of the two measurements. This association may be measured by the correlation coefficient, as previously indicated. In this case, the correlation coefficient estimated from the sample is 0.51.

It has been previously pointed out that genetic theory demands a positive correlation if stature is the result of the cumulative effect of a large number of genes; but we may also conceive of such factors as diet, which is more likely to be similar for cultural and economic reasons, operating to make related

TABLE 9.3*

DISTRIBUTION OF STATURES OF 1,078 SONS ACCORDING TO STATURE OF FATHER

STATURE OF SON	\[STATURE OF FATHER\] 58.5–59.5	59.5–60.5	60.5–61.5	61.5–62.5	62.5–63.5	63.5–64.5	64.5–65.5	65.5–66.5	66.5–67.5	67.5–68.5	68.5–69.5	69.5–70.5	70.5–71.5	71.5–72.5	72.5–73.5	73.5–74.5	74.5–75.5	TOTAL
59.5–60.5	1.00						1.00											2.0
60.5–61.5	2.00				0.50													1.5
61.5–62.5		0.25	0.25		0.50	0.50												3.5
62.5–63.5		0.25	0.25	2.25	0.50	1.00	0.25		1.00									20.5
63.5–64.5			1.50	3.75	2.25	2.00	4.00	0.25	0.50	0.50								38.5
64.5–65.5		1.00	0.50	2.00	3.00	4.25	8.00	5.00	2.75	1.25	1.50		0.25					61.5
65.5–66.5		0.50	2.00	2.25	3.25	9.50	13.50	9.25	3.00	1.25	3.50	0.25	1.25		1.00			89.5
66.5–67.5		1.50	1.50	4.75	5.25	9.50	10.00	10.75	7.50	5.50	5.25	0.75	2.50	1.00				148.0
67.5–68.5				2.00	3.50	13.75	19.75	16.75	17.50	16.00	12.50	2.50	3.25	0.50				173.5
68.5–69.5			1.00		7.50	10.00	10.25	26.50	25.75	19.50	29.50	2.00	8.50	9.50	2.25		1.00	149.5
69.5–70.5					5.25	5.00	12.75	24.25	31.50	23.50	29.00	13.75	10.00	6.25	2.25			128.0
70.5–71.5					1.00	2.50	5.75	18.25	16.00	24.00	22.50	13.25	14.50	8.00	3.50		1.00	108.0
71.5–72.5						3.25	5.00	18.75	11.75	19.50	14.75	21.50	10.75	6.25	5.00	1.50	1.00	63.0
72.5–73.5					1.00	0.25	3.00	8.75	10.75	19.00	10.75	19.50	10.00	3.25	2.75	1.00	0.50	42.0
73.5–74.5							0.75	1.25	7.00	7.75	6.50	20.75	7.50	0.75	3.25	0.50	2.00	29.0
74.5–75.5							1.50	0.75	2.50	7.50	2.25	10.75	6.50	1.00	3.25	0.50		8.5
75.5–76.5								1.50		5.25	2.00	6.00	2.50		1.75	0.50		4.0
76.5–77.5										1.00	0.25	2.50	0.50	0.25	1.50			4.0
77.5–78.5										1.25	0.25	1.00		0.25	0.75			3.0
78.5–79.5										1.25	1.00	1.00			0.25			0.5
Total	3	3.5	8	17	33.5	61.5	95.5	142	137.5	154	141.5	116	78	49	28.5	4	5.5	1078

* Data of Pearson and Lee, 1903, from The advanced theory of statistics, Vol. 1, M. G. Kendall, Griffin and Co., London, England, by permission of author and publishers. Note that when a height falls on the border line of an interval, one-half of the individual is assigned to each of the adjacent intervals.

individuals more alike than unrelated (cf. sec. 4.2, p. 18). It is logical, therefore, to ask whether or not environmental similarities could account for, or markedly influence, this correlation of parental with child stature. One method of detecting differences due to environment has been suggested, namely, to compare the correlation of parent and child with parent and foster-child. The requirements of an adequate foster-child study will be discussed in connection with such studies on intelligence. A second method which may be used is to compare the effects of different environments on a uniform genetic background such as would occur in the case of identical twins reared apart. If the latter approach is used, we find that the correlation between statures of identical twins reared apart ($r = .969$) is not significantly different from that of identical twins reared together ($r = .932$) (cf. Table 16-4, p. 276). We may surmise, therefore, that, under the environmental conditions in which this study was conducted, the primary cause of variation in stature is genetic. The limitations of extrapolating from twin data will be discussed in a later chapter.

2. INTELLIGENCE

Probably in no other single way do human beings differ as much as in their conscious response to external stimuli. This voluntary or wilful response to a new situation, or to the repetition of an old situation, may be termed "intelligence." Inherent in this definition is the ability to ascertain new courses of action through the interpretation of previous experiences.

We may divide intelligence into general and special abilities. This division is somewhat arbitrary and artificial, but it will serve to preclude from the following discussion those "special intelligences," such as may be exemplified by high levels of achievement in the arts, handicrafts, etc. We are concerned primarily, then, with the innate ability to cope with the complexities of human society as they affect the individual member.

An evaluation of the role of heredity in determining general intelligence is considerably more difficult than the comparable evaluation of the role played by heredity in determining stature. Before the relative contributions of heredity and environment to variation may be evaluated, one must define (1) the characteristic to be measured and (2) the standard or unit of measure to be used. In the discussion of stature this posed no problem; but the characteristic, intelligence, is not so amenable to a precise, clearly delineated definition as is possible in stature, nor is an objective standard of measure of the degree of intelligence readily apparent. The definition of general intelligence previously given suffers from the same vagueness as that found in all definitions of intelligence. Under such circumstances we cannot be sure that any unit of measure which may be elected is actually measuring the characteristic of interest. Currently the most widely recognized measure of intelligence, in the general sense, is the intelligence quotient (I.Q.), that is, the ratio of the mental age, as measured by a series of tests, to the chronological age. As a unit of measure of intelligence, I.Q. is only as good as the tests by which mental age is determined. The ideal test of mental age would possess the following minimal characteristics: (1) the mean mental-age score for un-selected individuals of any chronological age would agree closely with the mean chronological age; (2) the score should be duplicable within any age

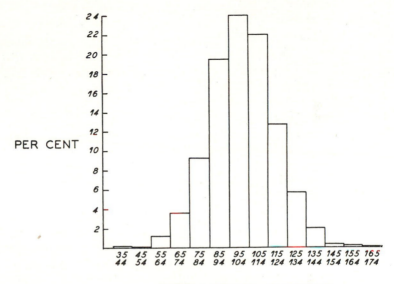

PER CENT

INTELLIGENCE QUOTIENT

Fig. 9-3.—The distribution of intelligence quotients in the United States. (Modified, by permission, from Measuring intelligence, by Drs. L. M. Terman and M. A. Merrill, copyright 1937, Houghton Mifflin Co.)

and should not vary between ages for the same subject; and (3) the score should be an equally reliable index of intelligence over all ranges of mental age. None of the existing tests fully satisfies these minimal requirements. Furthermore, the tests are reliable only in terms of the general set of experiences on which the test was standardized. Thus a test standardized on urban experiences is hardly applicable to a group whose experiences have been primarily rural.

Cognizant of these difficulties, we may still pose the question, "Are there detectable differences in I.Q. ascribable to heredity?" The distribution of the I.Q.'s of 2,904 individuals of varying ages is given in Figure 9-3. The shape of the distribution is reasonably close to that of the normal curve. It is a matter of custom to divide this distribution into three segments, namely, those of inferior mental ability (I.Q. below 70), those of average ability, and those of superior ability (I.Q. over 140). Within each of these three groups we may seek evidence of the role of heredity.

Terman (1925) and Hollingsworth (1940) have studied extensively the group of individuals with superior ability. Terman selected a group of superior children, all with I.Q.'s over 140, and conducted a long-term study of their physical and mental traits. He found that their initial superiority was maintained at all successive stages in the study. Furthermore, he noted that the offspring of these superior individuals were, in turn, appreciably above the mean in intelligence. The average I.Q. of these offspring was approximately 128. From such data one would be tempted to surmise that the superior beget the superior; but one cannot reach conclusions as to what extent this is a function of heredity and to what extent a function of environment.

In the other class of extreme deviates, those of inferior ability, it has long been suspected that they tend to produce more individuals of the same inferior mental capabilities. A number of studies bear out this suspicion; of these studies two are worthy of special mention. These are the Colchester (England) survey of Penrose (1938) and the study of mental defectives in Ohio by Halperin (1945). Each of these studies had as its starting point institutionalized mental defectives, and, in this connection, it should be borne in mind that social irresponsibility is a major factor in dictating institutionalization. The groups within these studies which are of special interest are those defectives who were free of physical and psychiatric disease, the group variously called "aclinical amentia," "simple primary amentia," "subcultural," or "residual." Penrose found that 37.1 per cent of the parents of aclinical aments were of borderline or lower levels of intelligence and that 15.1 per cent of the parents of aclinical aments were clearly defectives themselves. Halperin has summarized his data on aclinical amentia as follows: "In examining the mental status of the parents who contribute to the aclinical population [of a mental institution] we cannot escape the fact that 31 per cent of the parents are themselves defective and that 59 per cent of the children from these matings are defective and inferior." Obviously, the inferences which may be drawn from these data are analogous to Terman's and Hollingsworth's studies on the superior deviates.

The most pertinent data regarding the relative importance of genetic and environmental factors in intelligence come, not from a study of the extremes, but from a study of the intermediate group, utilizing the "foster-child" technique. Of the various studies of this type which have been carried out, those of Burks (1928) and Leahy (1935) appear most satisfactory. The results of these studies are reproduced in Table 9-4. Before interpreting these results, it would be wise to consider the requirements which must be met by foster-child studies if they are to be deemed adequate. These requirements have been summarized by Osborne (1951) as follows: (1) foster-children must be placed in the adoptive home sufficiently early to be relatively uninfluenced by the environment of the original home; (2) there must be little or no selective placement of the children; (3) an adequate sample of adoptive homes at the various social levels must be included in the survey; and (4) the foster-children should be of one race and nationality, to eliminate

TABLE 9-4

CORRELATION OF INTELLIGENCE BETWEEN PARENT AND OWN
CHILDREN AND PARENT WITH FOSTER-CHILDREN

	FOSTER-GROUP: CHILDREN AND FOSTER-PARENTS		CONTROL GROUP: CHILDREN AND TRUE PARENTS	
	r	N	r	N
a) Data of Burks (1928)				
Father's MA............	.07	178	.45	100
Mother's MA..........	.19	204	.46	105
b) Data of Leahy (1935)				
Father's Otis score......	.15	178	.51	175
Mother's Otis score......	.20	186	.51	191

racial sources of variation. Burks's and Leahy's studies most nearly approximate these requirements of all available studies of foster-children.

On inspection of Table 9-4, it is immediately apparent that the correlation of parent with child is consistently higher within the control group than within the group of parents and foster-children. The control group was so selected as to yield comparability in race, age of child, cultural level, occupation, and education. This consistently higher association would seem to indicate that, within the limits of this group, heredity is an important determiner of parent-child similarities in intelligence. Data from foster-child studies are not alone in indicating the importance of hereditary endowment in intelligence. Additional data bearing on this problem of the role of heredity and environment in intelligence may be gleaned from twin studies. **The re-**

sults of such studies will be discussed in the chapter on twins. Suffice it to say at this time that twin studies lead the investigator to essentially the same conclusions, namely, that a major source of variation in intelligence is ascribable to differences in heredity.

9.7. The present status of the analysis of continuous variation in man.—The present status of our knowledge of quantitative variability has been excellently summarized by Wright (1952) as follows: "The evidence that the genetic component of quantitative variability depends in general on multiple loci, distributed through the chromosomes, seems at present adequate. Satisfaction with this theory has, however, fallen short of completeness. The reason is the extraordinary difficulty in reaching conclusions on the details of the genetic situation in particular cases, even after very great effort. This difficulty rests in the large number of variables that must be contended with." In brief, while we recognize that quantitative variation in man undoubtedly frequently has a genetic basis, we have made little progress toward sharply defining the genetic and nongenetic components of variation in particular cases.

Bibliography

SPECIFIC REFERENCES

BURKS, B. S. 1928. The relative influence of nature and nurture upon mental development, 27th Yearbook Nat. Soc. Stud. Educ., Part I, pp. 219–318 .

DAVENPORT, C. B., and LOVE, A. G. 1921. The Medical Department of the U.S. Army in the World War. Vol. **15**: Statistics. Part I: Army anthropology. Washington: Government Printing Office.

HALPERIN, S. L. 1945. A clinico-genetical study of mental defect, Am. J. Ment. Deficiency, **50**:8–26.

HOLLINGSWORTH, L. S. 1940. The significance of deviates. *In:* Intelligence: its nature and nurture: 39th Yearbook Nat. Soc. Stud. Educ., Part I, pp. 43–66.

KENDALL, M. G. 1947. The advanced theory of statistics, Vol. **1**. 3d ed. London: Griffin & Co., Ltd.

LEAHY, A. M. 1935. Nature-nurture and intelligence, Genet. Psychol. Monogr., **4**:236–308.

OSBORNE, F. 1951. Preface to eugenics. Rev. ed. New York: Harper & Bros.

PEARSON, K., and LEE, A. 1903. Inheritance of physical characters, Biometrika, **2**:357–462.

PENROSE, L. S. 1938. A clinical and genetic study of 1,280 cases of mental defect (Coldchester survey). Spec. Rep. Ser. Med. Res. Council, No. 229. London: H.M. Stationery Office.

TERMAN, L. M. (ed.). 1925–47. Genetic studies of genius, Vols. 1–4. Palo Alto, Calif.: Stanford University Press.

TERMAN, L. M., and MERRILL, M. 1937. Measuring intelligence. Boston: Houghton Mifflin Co.

GENERAL REFERENCES

FISHER, R. A. 1950. Statistical methods for research workers. 11th ed. New York: Hafner Publishing Co.

HOEL, P. G. 1947. Introduction to mathematical statistics. New York: John Wiley & Sons, Inc.

HOGBEN, L. 1939. Nature and nurture. Rev. ed. London: Allen & Unwin, Ltd.

MATHER, K. 1947. Statistical analysis in biology. 2d ed. New York: Interscience Publishing Co.

————. 1949. Biometrical genetics. New York: Dover Publications, Inc.

SNEDECOR, G. W. 1946. Statistical methods. 4th ed. Ames: Iowa State College Press.

WRIGHT, S. 1952. The genetics of quantitative variability. *In:* REEVE, E. C. R., and WADDINGTON, C. H. (eds.), Quantitative inheritance, pp. 5–41. London: H.M. Stationery Office.

Problems

1. Assume that a given inherited disease is the result of two rare dominant genes. Assume that these two genes, say A and B, are equally common in the population and that affected individuals, $A-B-$, are incapable of reproducing. What would be the expected ratio of affected offspring to normal offspring in families exhibiting this pathology? How could this mode of inheritance be differentiated from the other common mode of inheritance that it would simulate?

2. What would the proportion of affected individuals be, assuming a trait to be due to homozygosis for a rare autosomal recessive gene and a sex-linked recessive gene, when both parents were normal, but carriers? When only the mother was normal? When only the father was normal? In the latter instances the normal parent is assumed to be a carrier.

3. Over the past several generations there has been a gradual increase in mean stature. What effect, if any, would this have on parent-offspring correlation in height?

4. What effect would assortative mating, i.e., the mating of "like with like," have on the parent-offspring correlation?

5. It has long been observed that when progeny from two inbred lines differing in some quantitative characteristic are crossed, the resulting hybrid is not infrequently more vigorous than either parent. This phenomenon has been termed "heterosis," or "hybrid vigor." Comment on the implications of this phenomenon with respect to man.

CHAPTER 10

Linkage

IN THE previous chapter, we considered the ratios to be expected when two or more genes were segregating independently. We recognize, however, that the number of genes in man is obviously far greater than the number of chromosomes, and hence not all genes can be independent. Otherwise stated, a given chromosome must contain numerous genes. Genes located on the same chromosome are said to be "linked." It is obvious that two genes located on the same chromosome have certain physical restrictions on their behavior with reference to one another which are not shared by two genes on different chromosomes. The genetic ratios observed in matings involving two linked genes may differ greatly, therefore, from those seen when the genes are free to segregate independently of one another.

Two genes may be closely linked, that is, the initial relationship between them is almost always maintained in transmission, or they may be loosely linked, the initial relationship between them being subject to frequent alteration by a segmental interchange between nonsister-chromatids of the homologous pair of chromosomes. When this segmental interchange occurs, the genes are said to have "crossed over." The frequency of crossing-over between two genes—the recombination value—is used as a measure of their spatial relationship on the chromosome.

Linked genes may occur in either of two phases. If the two genes whose linkage relationship is being determined are on the same chromosome of the homologous pair of chromosomes, they are said to be "coupled"; if they are located on opposite members of the same pair of homologous chromosomes, they are said to be "repulsed." Linkage can be detected irrespective of its phase.

It will be our purpose in this chapter to indicate (1) the general problems posed in the detection of linkage in man, (2) the methods of linkage analysis currently available, (3) the status of our knowledge of linkage relationships in man, and (4) the application of linkage information to other problems of genetical concern.

10.1. General problems in the detection of linkage in man.—The study of linkage relationships in any organism poses two different, albeit related, problems.

116

lems: (1) the detection of linkage and (2) the estimation of the recombination frequency. In man these problems are rendered more difficult as a consequence of (1) the small number of offspring from any one mating and the consequent necessity for the use of pooled data; (2) our inability to determine, in most instances, whether or not the genes in question are in the coupled or repulsed phase in a specific mating; and (3) the biases inherent in the special manner in which human data are frequently collected. Methods applicable to the study of linkage relationships in man, therefore, not only must be logically deduced from the general behavior to be expected of linked genes but must compensate for these latter difficulties.

In general, linkage can be demonstrated only in those matings wherein opportunity is afforded for new phenotypes to arise by crossing-over. This means that at least one of the parents must be a double heterozygote, that is, one parent must be heterozygous for each of the two sets of alleles whose linkage relationships are to be determined. Furthermore, if we are dealing with sets of alleles in which one of the alleles is dominant to the other, not all the types of matings involving double heterozygotes for these two loci will yield linkage information. More precisely, only those matings involving parents (1) both of whom are doubly heterozygous, (2) one of whom is doubly heterozygous, while the other is a single heterozygote but homozygous for one or the other of the recessive alleles, or (3) one of whom is the double heterozygote, while the other is homozygous for both the recessive genes, will give rise to demonstrable linkage.

The initial step in the argument of all the methods of detecting loose linkage in man consists of determining how the expected phenotypic proportions from a given type of mating when the genes are postulated to be linked with recombination frequency, χ, will differ from those to be found in the absence of linkage. The change in the phenotypic proportions as a consequence of linkage will vary with (1) the type of mating, (2) the phase of linkage and (3) the presence of dominance within the gene pairs. This dependence of the expected proportions upon the latter three variables may be illustrated as follows: Consider the mating $TtWw \times ttww$, where T is dominant to t and W is dominant to w. Assume that the genes T and W are coupled in the double heterozygote and that the frequency of recombination is χ ($1 - \chi$ is the frequency of noncross-over gametes). Now in this particular mating, since the $ttww$ individual produces only one type of gamete, tw, the frequency of the various possible offspring phenotypes is dependent solely upon the frequency of the four classes of gametes produced by the double heterozygote. The double heterozygote produces, as noncross-over gametes, the gametes TW and tw, whose combined frequency is $(1 - \chi)$, and, as crossover gametes, the gametes Tw and tW, whose combined frequency is χ. If we were to consider this same type of mating and recombination frequency but were to assume that T and W were repulsed rather than coupled, we should find that a reciprocal relationship exists wherein the cross-over categories under coupling become the noncross-over categories under repulsion. The frequencies of the various phenotypes to be expected from this mating are shown in the accompanying tabulation.

PHASE	FREQUENCY OF PHENOTYPE				TOTAL
	TW	Tw	tW	tw	
Coupling.......	$(1-\chi)/2$	$\chi/2$	$\chi/2$	$(1-\chi)/2$	1
Repulsion......	$\chi/2$	$(1-\chi)/2$	$(1-\chi)/2$	$\chi/2$	1

As a second illustration of the effect of type of mating, phase of linkage, and the presence of dominance on the expected phenotypic proportions, let us consider the mating $TtWw \times Ttww$. Assume, again, that the genes T and W are dominant to their alleles, that they are coupled in the double heterozygote, and that the frequency of recombination is χ. In this case the types and relative frequencies of the gametes produced by the double heterozygote remain the same as in the preceding example, but now the other parent, $Ttww$, is capable of producing two equally frequent types of gametes, Tw and tw. This will lead to the more frequent occurrence of the TW and Tw phenotypes, as in the accompanying tabulation. It should be noted that, in both

PHASE	FREQUENCY OF PHENOTYPE				TOTAL
	TW	Tw	tW	tw	
Coupling.......	$(2-\chi)/4$	$(1+\chi)/4$	$\chi/4$	$(1-\chi)/4$	1
Repulsion......	$(1+\chi)/4$	$(2-\chi)/4$	$(1-\chi)/4$	$\chi/4$	1

the examples, if $\chi = \frac{1}{2}$, that is, if cross-over gametes are as frequent as non-cross-over gametes, the distribution of phenotypes is that to be expected of independently segregating genes. Thus, if chance linkage between two genes did exist and if the cross-over value was slightly less, equal to, or greater than 50 per cent, only in the rarest of circumstances could such linkage be demonstrated.

These brief considerations will permit a listing of the minimal requirements of a satisfactory test of linkage in man. Such a test must (1) be independent of phase of linkage; (2) account for differing proportions of phenotypes as a consequence of dominance; (3) reflect the effect of differing mating types on the frequency of those phenotypes of interest; (4) permit the pooling of observations from a number of families, since rarely will one family possess all the phenotypes and, even when such is the case, it would be impossible to determine whether independent or loosely linked genes are involved; and (5) account for biases in the data which may result from the manner in which these data are selected.

10.2. *The methods of linkage analysis.*—The first mathematically sound method for the detection of linkage was suggested by Bernstein (1931). This method, as did those subsequently suggested by Wiener (1932) and Hogben (1934), required knowledge of two generations of individuals. In 1934 Haldane extended and improved the Bernstein method, without, however, obviating the necessity of two generations of data. The next step in the development of an efficient test was Fisher's (1935 ff.) introduction of the so-called "*u*-statistics." Fisher's approach was extended, in turn, by Finney (1940 ff.) to cover, among other things, those situations in which neither or only one parent was tested. More recently, Bailey (1951) has developed a generalized method of determining the appropriate multiple of the *u*-score to be used in linkage testing. Concomitant with this development was the origin of a test devised by Penrose (1935 ff.), the avowed purpose of which was to eliminate the necessity of obtaining data from two generations and to minimize the possible complicating factor of age variation in the manifestation of the traits under study. At present, then, there are two methods of linkage analysis, the method of paired sibs of Penrose and the Finney modification of Fisher's method. Of these two methods, in those situations where both are applicable, the method of Fisher-Finney is preferable because of its greater efficiency. Neither method, however, gives rise to an efficient estimate of the cross-over value; consequently, the problem of the estimation of the frequency of recombination will not be considered. Furthermore, since an understanding of the origin of the Fisher-Finney scores requires a mathematical background not usually possessed by students of human biology, only the paired-sib method will be presented.

The principle on which the method of paired sibs is based has been stated by Penrose (1935) as follows: "When pairs of sibs are taken at random from a series of families, certain types of sibling pairs will be more frequent if there is linkage than if there is free assortment of the characters studied." Briefly, and in its simplest form, the method is as follows: Assume we have two pairs of genes, say T and t and W and w, producing the phenotypes T (or t) and W (or w). In a series of sibships, we can compare all possible pairs of siblings within each sibship with respect to whether they are alike or unlike in these characteristics. If this is done, then every pair of siblings will occur once, and only once, in a 2 × 2 table, where the four possible alternatives are as indicated in the accompanying tabulation. If the genes in question are

Type of Sib-Pair	TW (or tw)	tW (or Tw)	Total
TW (or tw)......	n_1	n_2	n_1+n_2
Tw (or tW)......	n_3	n_4	n_3+n_4
Total.......	n_1+n_3	n_2+n_4	n

segregating independently, then a simple proportionality exists, that is to say, the probability of a sib-pair being included among n_1 or n_4 will be equal to the probability of a sib-pair being included among n_2 or n_3. If, on the other hand, the genes in question are linked, then these alternatives are not equally probable, and we would expect a concentration of the sib-pairs among n_1 and n_4. That this is true may be shown as follows: Assume that the genes T and W are linked with recombination frequency χ. Consider, now, the mating $Ttww \times TtWw$. If the genes are coupled in the double heterozygote, then the frequencies of the four phenotypes TW, tw, Tw, and tW are given by $(2 - \chi)/4$, $(1 - \chi)/4$, $(1 + \chi)/4$ and $\chi/4$ (cf. p. 118). It follows that from this mating the probability that a sib-pair will fall in cell n_1 is the sum of the frequencies of TW-TW, tw-tw, TW-tw, and tw-TW pairs or $(3 - 2\chi)^2/16$. Similarly, the probabilities for the other cells are: cell n_2, $(3 - 2\chi)$ $\times (2\chi + 1)/16$; cell n_3, $(3 - 2\chi)(2\chi + 1)/16$; and cell n_4, $(2\chi + 1)^2/16$. If we had assumed that the genes were repulsed in the double heterozygote, then the probabilities for cells n_1 and n_4 would be interchanged, while cells n_2 and n_3 would remain the same. Now the excess of sib-pairs occurring in cells n_1 and n_4 over those occurring in n_2 and n_3 would be, say, D, where

$$D = \tfrac{1}{16} \left[(3 - 2\chi)^2 + (2\chi + 1)^2 - 2(3 - 2\chi)(2\chi + 1) \right] = (\chi - \tfrac{1}{2})^2.$$

It may be similarly shown that the three other types of matings which give rise to demonstrable linkage effects, namely, $ttww \times TtWw$, $ttWw \times TtWw$, and $TtWw \times TtWw$, also give rise to an excess of sib-pairs in cells n_1 and n_4. It should be noted that in the event $\chi = \tfrac{1}{2}$, that is, if the genes are segregating independently, then $D = 0$.

The test of linkage consists in merely determining whether the observed distribution of sib-pairs among the four cells differs significantly from that expected on the basis of independent genes. Otherwise stated, the test of linkage consists in determining whether the observed frequencies are compatible with those expected of independent genes, namely, whether the cell frequencies are proportional to the marginal totals. If the total number of sib-pairs is sufficiently large, say of the order of 100, then the significance of the observed distribution may be judged by χ^2 (cf. sec. 13.6, p. 197), where

$$\chi^2 = \frac{n\,(n_1 n_4 - n_2 n_3)^2}{(n_1 + n_3)\,(n_1 + n_2)\,(n_2 + n_4)\,(n_3 + n_4)}.$$

Chi-square is adjudged significant if it exceeds 3.841 (5 per cent level) or 6.635 (1 per cent level). The choice of the level depends upon the frequency of error which the individual worker elects to make. Convention favors the use of the 5 per cent level. The chi-square method is inapplicable if the ex-

TABLE 10-1a*

TEST OF THE LINKAGE OF THE MN AND ABO LOCI

Sibship No.	Antigen A	Antigen M	Sibship No.	Antigen A	Antigen M	Sibship No.	Antigen A	Antigen M
3......	+	−	5.....	+	+	11....	−	+
	+	−		−	+		−	−
	+	−		−	−		+	+
4......	−	+		−	+	17.....	+	+
	−	−		−	+		+	−
	+	+		+	+		+	+
	+	+		+	+		+	+
	−	+		−	+		+	−
	−	+	11.....	−	−		+	−
				+	−		+	−
				+	−		+	+
				+	−		+	+

* Data from Kloepfer (1946). The presence of an antigen is indicated by +, its absence by −; "L" and "U" designate like and unlike, respectively.

pected number in any cell is less than five. Under such circumstances, the appropriate test is that of exact probability (see Fisher, 1950, sec. 21.02).

As an illustration of this method, consider the data in Table 10-1a. These thirty-eight individuals distributed among five sibships were tested by Kloepfer (1946) to determine their antigenic makeup with respect to the MN and ABO antigens. It is of interest to know whether these loci may be linked. To apply the sib-pair method, we need merely classify each individual as to the presence or absence of the M and A antigens. These findings have been summarized in Table 10-1b. Now, in proceeding to the test, one must make all possible sib-comparisons within each sibship. The number of possible comparisons within a sibship of size s is $s(s-1)/2$. A total of 143 comparisons is possible for these 38 individuals; these 143 sibling pairs are presented in Table 10-1c, where they have been classified as to whether they are alike with respect to both antigens, alike with one and unlike with the other, or unlike with both. Chi-square is found to be

$$\chi^2 = \frac{143\,[\,(51)\,(15) - (47)\,(30)\,]^2}{(98)\,(81)\,(62)\,(45)} = 2.686.$$

TABLE 10-1b

Sibship No.	Phenotype of Offspring				Total
	++	+−	−+	−−	
3........	0	3	0	0	3
4........	2	0	4	1	7
5........	3	1	4	1	9
11........	1	4	1	2	8
17........	6	5	0	0	11
Total...	12	13	9	4	38

TABLE 10-1c

Sibship No.	Sibling Pair				Total
	LL	LU	UL	UU	
3........	3	0	0	0	3
4........	7	4	8	2	21
5........	9	7	13	7	36
11........	7	6	9	6	28
17........	25	30	0	0	55
Total...	51	47	30	15	143

$$\chi^2 = 2.686;\ df = 1;\ 0.20 > P > 0.10$$

The probability of obtaining a χ^2 larger than this by chance alone, when the genes are, in fact, independent, is greater than 10 chances in 100 (cf. Table 13-9, p. 199). We conclude, therefore, that these data do not support an assertion of linkage between the MN and ABO loci.

Penrose (1946) later suggested an alternative to the 2×2 method just described and has shown that this alternative approach may be as much as six times more efficient in the detection of linkage than the 2×2 method.

Briefly, this revised method consists of distributing the sib-pairs in a 3 × 3 table, where the nine alternatives, using the notation of the preceding section, are as shown in the accompanying tabulation. If the genes in question

Type of Pair	T-T	T-t	t-t	Total
W-W..........	n_{11}	n_{12}	n_{13}	$n_{1.}$
W-w..........	n_{21}	n_{22}	n_{23}	$n_{2.}$
w-w..........	n_{31}	n_{32}	n_{33}	$n_{3.}$
Total......	$n_{.1}$	$n_{.2}$	$n_{.3}$	n

are linked, then there will be a concentration of sib-pairs in those cells which constitute the diagonals of the table, whereas, in the absence of linkage, the cell entries should be proportional to the marginal totals. Again, the significance of an observed distribution may be determined by χ^2; now χ^2 is

$$\chi^2 = \sum_{i,j} \frac{(o_{ij} - e_{ij})^2}{e_{ij}},$$

where o_{ij} and e_{ij} are the observed and expected numbers in the ith row and jth column. The expected numbers are determined from

$$e_{ij} = \frac{(n_{i.})(n_{.j})}{n}.$$

Chi-square is judged to be significant, in this instance, if it exceeds 9.488 (5 per cent level) or 11.668 (1 per cent level). Chi-square is applicable in all those cases in which some of the expected numbers exceed five; if all expected numbers are less than five, then Fisher's exact test is appropriate. The latter situation rarely obtains if a sufficient body of data has been gathered. In Table 10-2 the data of Table 10-1a have been classified in this revised manner. The value of χ^2 was found to be 10.889, which slightly exceeds the 5 per cent level of significance. The deviation, however, is in the direction opposite to that expected on linkage, that is, there is a deficiency rather than an excess of sib-pairs in the diagonals of the table. Again, we are led to the conclusion that there is no linkage of the MN and ABO loci.

Thus far, in the application of the paired-sib method the traits whose linkage relationship was the subject of testing were assumed to be simply inherited. Penrose (1938) has shown that this method is equally applicable to metrical or graded characteristics having a somewhat obscure but nonetheless real mode of inheritance. This may be illustrated by considering two graded characteristics. Assume that the variation in each trait is primarily genetic in origin and that these variations are scorable in some unit scheme of measure. Let g_1 and g_2 and h_1 and h_2 be the measurements of these traits on a pair of siblings, and let D_g and D_h be the intra-pair difference in g and h measurements. If the deviations between all sib-pairs for these two sets of measurements were computed and scored in a symmetrical table whose axes were the differences between sibs, that is, D_g and D_h, we should find that, in the absence of linkage, a simple proportionality between the entries and the marginal frequencies would exist. On the other hand, if linkage existed, this simple proportionality would be distorted, with an accumulation of entries

TABLE 10-2

DATA OF TABLE 10-1a TESTED FOR LINKAGE OF MN
AND ABO LOCI BY 3 × 3 METHOD OF PAIRED SIBS

ANTIGEN M	ANTIGEN A			TOTAL
	Present Both	Absent One	Absent Both	
Type of pair:				
Present both.......	19	21	12	52
Absent one.........	37	15	10	62
Absent both........	19	9	1	29
Total............	75	45	23	143

$\chi^2 = 10.889; df = 4; 0.05 > P > 0.02$

along the diagonals of the table. The statistic proposed to measure this distortion is the ϕ statistic, where

$$\phi = \left\{ \frac{n \left[\Sigma f_{gh} \left(D_g^2 \cdot D_h^2 \right) \right]}{\Sigma f_g \left(D_g \right)^2 \cdot \Sigma f_h \left(D_h \right)^2} \right\} - 1$$

and f_g, f_h, and f_{gh} are the frequencies of D_g, D_h, and $D_g \cdot D_h$, respectively. In the absence of linkage, ϕ is equal to zero, whereas in the event of complete linkage $\phi = \frac{1}{2}$. It should be noted, however, that if there is an association of the characteristics in the population, then a natural excess of entries in the diagonals will result. Under these circumstances, to determine whether linkage exists, it is necessary to correct ϕ by that amount of distortion due to the fact that the characteristics are correlated in the general population. The suggested correction is $\phi - r^2$, where

$$r^2 = \frac{\Sigma f_{gh} \left(D_g \cdot D_h \right)^2}{\Sigma f_g \left(D_g \right)^2 \cdot \Sigma f_h \left(D_h \right)^2}.$$

The test of significance consists of forming the ratio $\phi/\sqrt{V_\phi/n-1}$ or $(\phi - r^2)/\sqrt{V_\phi/n-1}$, where V_ϕ, the variance of ϕ, is

$$V_\phi = \frac{n^2 \left[\Sigma f_g \left(D_g\right)^4 \cdot \Sigma f_h \left(D_h\right)^4\right]}{\left[\Sigma f_g \left(D_g\right)^2 \cdot \Sigma f_h \left(D_h\right)^2\right]^2} - 1 .$$

If this exceeds approximately 2, that is, if ϕ is twice as large as its standard error, one may assert the existence of linkage.

As an illustration of this method, Kloepfer's (1946) observations on ear size and the attachment of the ear lobe in the sibships used in the preceding

TABLE 10-3a*

TEST OF LINKAGE OF MAJOR GENES RESPONSIBLE FOR
EAR SIZE AND ATTACHMENT OF EAR LOBE

Sibship No.	Ear Lobes	Ear Size	Sibship No.	Ear Lobes	Ear Size
3.......	3	2	11......	2	3
	1	1		3	2
	2	2		3	1
				1	3
4.......	1	1		1	1
	1	3		1	2
	2	2		2	2
	3	2		2	3
	1	2			
	1	1	17......	1	2
	2	2		1	3
				1	1
5.......	1	2		1	3
	1	2		1	2
	1	1		1	1
	1	2		1	1
	2	1		1	3
	2	2		1	1
	1	2		1	1
	1	2			
	1	3			

* Data from Kloepfer (1946). The method of scoring size and attachment is described in the text.

examples are presented in Table 10-3a. Ear size and ear lobes are scored as 1, 2, and 3 corresponding to small (or free), intermediate (or indeterminate), and large (or attached). Thus both D_g and D_h may assume any integral value between 2 and -2. Our problem consists, then, of determining whether the observed distribution given in Table 10-3b is compatible with a hypothesis of no linkage. We find that $\phi = -0.192$, that r^2 is negligible, and that $V_\phi = 9.057$. The ratio of ϕ to its standard error is thus equal to -0.759. Clearly, no evidence for linkage exists, and we may assert that the major genes determining these two characteristics are not linked.

10.3. *Linkage relationships in man.*—Two or more linked genes are said to constitute a "linkage group." In those organisms, such as Drosophila and corn, which have been thoroughly investigated, it has been shown that the number of linkage groups corresponds to the haploid number of chromosomes. Furthermore, it has been shown that the linear order of the genes within a linkage group could be specified by a study of the frequency with which genes within this group crossed over with one another. Thus, for example, if gene A recombined with gene B with a frequency of 10 per cent, B recombined with gene C with a frequency of 5 per cent, and A recombined with C with a frequency of 15 per cent, then the order would be either A-B-C or C-B-A. The distances between these genes are specified in terms of their recombination frequencies. By inference, then, since man has a haploid num-

TABLE 10-3*b*

EAR SIZE (D_h)	EAR LOBES (D_g)					TOTAL
	−2	−1	0	1	2	
2.........	0	1	15	1	0	17
1.........	1	7	20	2	3	33
0.........	0	8	32	7	3	50
−1.........	1	7	17	5	2	32
−2.........	0	1	7	2	1	11
Total....	2	24	91	17	9	143

$$\phi_{gh} = -0.192; \; V_\phi = 9.057$$

ber of twenty-four chromosomes, we may expect twenty-four independent linkage groups and ultimately twenty-four maps. Twenty-three of these linkage groups would correspond to the twenty-three autosomal chromosomes, and one linkage group would correspond to the sex chromosome in the haploid set.

At present, only one linkage group has been reasonably well demonstrated in man, namely, the linkage group which corresponds to the sex chromosome. This linkage group may be considered to be tripartite: one part identified with the nonhomologous portion of the X-chromosome, the second part with the nonhomologous portion of the Y-chromosome, and the third part with the homologous portions of the X- and Y-chromosome. Convention has conferred upon each of these three alternatives a specific name. Genes which are borne on the nonhomologous segments of the X-chromosome are generally called "sex-linked"; genes borne on the nonhomologous segment of the Y-chromosome are termed "Y-borne" or "holandric" genes; and, lastly, those genes borne on the homologous segments of the X- and Y-chromosome are called "incompletely or partially sex-linked." In chapters 6 and 7 attention was called to the methods of demonstrating such heredities, and specific examples were given.

Let us turn now to the estimation of crossing-over and of the linear order of the genes associated with these three segments. We may dispense with the holandric genes at the outset. Estimation of the order of these genes through linkage studies is impossible, since no opportunity is afforded for crossing-over to occur. If we consider only the sex-linked genes, it is obvious that if all these genes are, in fact, sex-linked, then they must be linked to one another. However, many of the sex-linked pathologic traits are rare or, at best, uncommon in their occurrence in human populations. As a consequence, families exhibiting two or more of these sex-linked genes are extremely rare, and, with but one exception, the opportunity to estimate the frequency of crossing-over and hence of a more precise spatial relationship between any two of these genes has not arisen. The single exception concerns the frequency of crossing-over between the genes responsible for hemophilia and color blindness. Haldane and Smith (1947) have estimated that the recombination value between these two genes is 9.8 per cent and that the true value probably lies between 5 and 20 per cent.

We recognize some nine genes which may be incompletely sex-linked genes in man (Haldane, 1936, 1941; Kalliss and Schweitzer, 1943; Snyder and Palmer, 1943); but, again, these genes are sufficiently uncommon in the population that families exhibiting the segregation of two or more of these genes are extremely unlikely. The fact, however, that the demonstration of an incompletely sex-linked gene is to some extent a function of the frequency with which the gene crosses over with sex permits us to infer indirectly the spatial relationship between these nine-odd genes. It is possible to reconstruct directly or indirectly the number of cross-overs necessary to produce a given array of offspring, and hence to estimate the frequency of recombination between the gene in question and sex. If, then, the frequencies of recombination of all the incompletely sex-linked genes are compared, using the differential sex segment as the common point of reference, we obtain estimates of (1) the linear order of these genes and (2) the frequency of recombination to be expected between any two given genes. A map of the homologous segments of the sex chromosomes of man constructed in this fashion is presented in Figure 10-1. There exist at least two reasons for doubt, however, as to the reliability of this map: (1) Macklin (1952) finds certain inconsistencies in the xeroderma pigmentosum data. She notes that in incomplete sex-linkage one should find (a) an excess of affected males when the inheritance is through the paternal side of the father of the affected individual;

(*b*) an excess of affected females when the inheritance is through the father's maternal side; and (*c*), collectively, more affected females than males. Macklin finds 16 affected males among 24 affected children when the relationship is like *a* above; 9 affected females among 14 affected children when the relationship is like *b* above; but, collectively, only 17 of 39 affected children are females. Thus the data do not fully bear out expectation. (2) A second reason for reservations in accepting too literally the spatial relationships of Figure

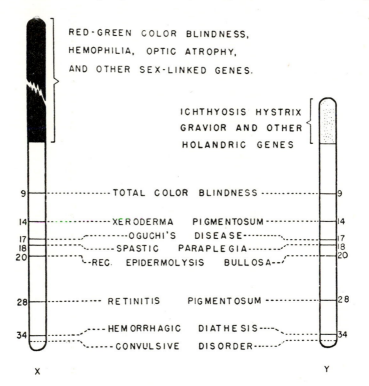

Fig. 10-1.—The most probable spatial relationships with the homologous segment of the X-chromosome of the genes thought to be incompletely sex-linked. (By permission from The principles of heredity, by Dr. L. H. Snyder, 3d ed., copyright 1946, D. C. Heath and Co.)

10-1 is Fisher's (1936) demonstration of the degree to which the calculations are influenced by assumptions regarding ascertainment (cf. chap. 14).

In so far as the legitimacy of this approach to the estimation of the linear order of incompletely sex-linked genes is concerned, the accuracy or inaccuracy of the provisional map is immaterial. The major question, which can be solved only cytologically, is whether or not incomplete sex-linkage is a tenable hypothesis. The existence of homologous segments of the X- and Y-chromosome with the spatial relationships depicted in Figure 10-1 has not been proved to the satisfaction of all human geneticists. The assignment of any specific autosomal gene to the proper autosomal linkage group presents a number of difficulties not found in the study of sex-linkage, Y-borne, or incompletely sex-linked genes. Not the least of these are (1) the number of expected groups, twenty-three; (2) the small number of simply inherited, common traits in man suitable for use in linkage studies; and (3) the fact that if two genes are more than twenty cross-over units apart, then it is difficult to demonstrate with a high degree of reliability that they are linked and not assorting independently, unless there is available a large collection of pertinent data. Moreover, the a priori probability that two

TABLE 10-4

SOME POSSIBLE AUTOSOMAL LINKAGES IN MAN

Traits	Method	P-Value	Investigator
Ear flare—MN...............	Paired sibs	0.035	Kloepfer (1946)
Finger length—ear lobe.........	Paired sibs	.01–0.001	Kloepfer (1946)
Ear size—PTC...............	Paired sibs	.001	Kloepfer (1946)
Red hair—strabismus...........	Paired sibs	.001	Kloepfer (1946)
Dominant eye—hair form.......	Paired sibs	.01– .001	Taillard (1951)
Dominant eye—hair whorl......	Paired sibs	.01– .001	Taillard (1951)
Eye color—myopia............	Paired sibs	....	Burks (1937)
Ear lobe—handedness.........	Paired sibs	.05– .02	Taillard (1951)
Ear lobe—PTC...............	Paired sibs	{ .01– .001 / .05–0.02 }	Kloepfer (1946), Taillard (1951)
Dermatoglyphics-handedness....	Piared sibs	.001	Rife (1941)
Hair color—tooth deficiency.....	Paired sibs	.025	Burks (1938)
ABO—allergy.................	Fisher-Finney	.045	Finney (1940)
ABO—phenylketonuria.........	Paired sibs	.046	Penrose (1946)
ABO—red hair................	Paired sibs	.033	Penrose (1935)
Lutheran—Lewis..............	Fisher-Finney	0.000001	Mohr (1951)

sets of alleles selected at random will be autosomally linked in man is only about one twenty-third, as opposed to approximately one chance in two in *Drosophila melanogaster* (ignoring the chance that the alleles will be located in the small fourth chromosome).

A considerable number of pairs of inherited traits have been studied for the existence of linkage between the genes responsible for them (summary in Mohr, 1953). The majority of such studies have, however, yielded negative information, that is, segregation has been shown to be compatible with independent sets of alleles (see, for example, Holt *et al.*, 1952). Some of the possible "linkages" discovered to date are summarized in Table 10-4; of these, the most likely prospect and that most closely approximating the ideal is the linkage of the genes responsible for the Lutheran and Lewis antigens reported by Mohr (1951). These possible "linkages" include only those studies based on a number of families. In cases where the findings of two authors disagree, the results have not been included in the table. It should be noted that the majority of these possible "linkages" include traits which are not sharply defined and whose genetic backgrounds are obscure. The interpretation of such linkages is complicated by our present lack of knowledge of gene interaction in man. Furthermore, many of these linkages are the result of testing a single group of individuals for a large number of traits, and we should remember that, quite by chance, a number of such comparisons should be "significant" at the 5 per cent level. It follows that independent confirmation of any reported linkage is absolutely essential before a linkage can be considered to be established.

10.4 *The distinction between independently segregating, linked, and allelic genes in man.*—On a priori grounds, two autosomal genes which appear to have somewhat similar effects may be located on the same or different chromosomes and, if on the same chromosome, may be alleles or not. In Drosophila and the mouse the question of the genetic relationship of two genes may be speedily resolved by the appropriate crosses. In man the dispatch with which the question is answered depends on whether the genes in question behave as dominants or recessives and how common the two traits determined by the genes in question are in the population.

Where the two genes in question behave as dominants, the simplest approach to a decision consists in studying the results of marriages of persons known to be heterozygous for both genes with normal individuals. Expectation under the three possible relationships mentioned above is shown in Figure 10-2. Obviously, loose linkage, permitting 40 per cent or more crossing-over, will be difficult or impossible to detect.

The solution of this question is more difficult where recessive genes are involved. Where two recessive genes are actually alleles, the cross of an individual homozygous for the one gene with an individual homozygous for the other should result in children who are all abnormal. However, the distinction between linkage and independent assortment for human recessive genes presents serious practical difficulties which are insuperable if the genes are rare.

10.5. *The application of linkage data to other problems of genetical concern.*— The search for genetic linkages in man has been motivated primarily by the desire (1) to indicate the universality of the basic concept of linkage and (2) to use the data on genetic linkages to advantage in the detection of genet ic carriers and the estimation of genetic risk with respect to a given disease.

The purely theoretical implications of the first objective need no comment. The utility of the second objective, however, merits closer inspection.

The application of linkage methods to genetic prognosis envisages the demonstration of the linkage of a common marker gene, such as those responsible for the blood antigens, with one responsible for some pathological condition. The following example will serve to illustrate how such information could be put to use: Consider a mating where the mother was of type MN and was afflicted with Huntington's chorea, while the father was of type NN and normal. Assume that this mother, in turn, was the offspring of a mating where the mother was MM and a choreic and the father NN and normal. If now Huntington's chorea were to be linked with the MN antigens

THE GENETIC CONSEQUENCES OF ALLELISM, LINKAGE, AND INDEPENDENT ASSORTMENT

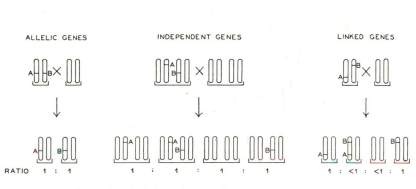

Fig. 10-2.—The results of matings of persons heterozygous for two dominant genes with normal individuals under the hypotheses of (*a*) allelism, (*b*) independent assortment, and (*c*) linkage.

with a recombination frequency of 5 per cent, blood-typing the offspring of the mating first considered would reveal the likelihood of any given offspring's developing Huntington's chorea. In this example the genes responsible for Huntington's chorea and the M antigen must be coupled in the mother. Consequently, if an offspring were of blood type NN, the only way in which he or she would develop chorea, barring mutation, would be if recombination occurred. This would obtain with a frequency of 5 times in 100; and hence the individual of blood type NN would run a risk of 1 chance in 20 of developing Huntington's chorea. Similarly, the individual of type MN would run a risk of 95 chances in 100 of developing the disease. If this type of prediction were possible, it could be of real service by permitting the anticipation of the carrier or of the manifest disease with a higher degree of accuracy than is presently possible.

Closer inspection of this technique reveals a number of deficiencies. The markers as well as the inherited pathologic traits must satisfy a number of criteria:

1. A large number of physiologically neutral markers must be available. These markers must be well distributed among the twenty-three autosomes, and it has been estimated that a minimum of twelve markers per chromosome is necessary to achieve the degree of prediction necessary for clinical medicine. Upward of 250 loci must be marked; at present, only a dozen or so satisfactory markers are known.

2. The markers should have a high frequency in the population. Obviously, the probability of the fortuitous combination in a given family of a rare marker and an even rarer pathology would be small.

3. Ideally, to minimize the sources of error and to be of maximum usefulness, the marker should be demonstrable in either the heterozygous or the homozygous condition.

4. The phase of linkage in the specific family in question must be known.

5. Furthermore, prediction is possible only in families where at least one member is doubly heterozygous.

6. An even more serious objection is the fact that many inherited diseases appear to have more than one genetic mode of transmission, e.g., choroidoretinal degeneration. Moreover, consistency in mode of transmission does not necessarily exclude the possibility that a given inherited disease may be due to entirely different genes in different families. For example, there exists an inherited abnormality of the red blood cells known as "ovalocytosis" (elliptocytosis), because of the characteristic shape of the red blood cell. This disease appears to be due to a dominant gene. Recently the linkage relationship between the blood-group markers and the gene or genes responsible for this disease have been investigated by Chalmers and Lawler (1953) and by Goodall et al. (1953). In two families these authors find no evidence of linkage between the gene responsible for ovalocytosis and any of the blood-group antigens. However, in a third family there appeared to be quite close linkage between the Rh locus or loci and ovalocytosis, with affected individuals always having an R_2 chromosome. The number of individuals tested in this family was sufficiently large to militate against chance as an explanation of the observed distribution. The authors conclude that "it is not possible to argue that linkages which are close enough to be of prognostic value in one family will necessarily be so in other families showing the same abnormality." It is obvious that under these circumstances prediction becomes extremely speculative.

In view of these considerations and the difficulty of initially demonstrating a linkage relationship in man, it is apparent that the application of linkage data offers little of practical value in the immediate future.

Bibliography

SPECIFIC REFERENCES

BAILEY, N. T. J. 1951. On simplifying the use of Fisher's u-statistics in the detection of linkage in man, Ann. Eugenics, 16:26–32.

———. 1951. The detection of linkage for partially manifesting rare "dominant" and recessive abnormalities in man, ibid., pp. 33–44.

BERNSTEIN, F. 1931. Zur Grundlegung der chromosomentheorie der Vererbung beim Menschen, mit besonderer Berücksichtigung der Blutgruppen, Ztschr. f. induct. Abstamm. u. Vererbungslehre, 57:113–38.

CHALMERS, J. N. M., and LAWLER, S. D., 1953. Elliptocytosis and blood groups. I. Families 1 and 2, Ann. Eugenics, 17:267–71.

FINNEY, D. J. 1940. The detection of linkage. I, Ann. Eugenics, 10:171–214.

———. 1941a. The detection of linkage. II, ibid., 11:10–30.

———. 1941b. The detection of linkage. III, ibid., pp. 115–35.

———. 1942a. The detection of linkage. IV, J. Hered., 33:157–60.

———. 1942b. The detection of linkage. V, Ann. Eugenics, 11:224–32.

———. 1942c. The detection of linkage. VI, ibid., pp. 233–45.

———. 1943. The detection of linkage. VII, ibid., 12:31–43.

FISHER, R. A. 1935a. The detection of linkage with dominant abnormalities, Ann. Eugenics, 6:187–201.

———. 1935b. The detection of linkage with recessive abnormalities, ibid., pp. 339–51.

———. 1936. Tests of significance applied to Haldane's data on partial sex linkage, ibid., 7:87–104.

GOODALL, H. B.; HENDRY, D. W. W.; LAWLER, S. D.; and STEPHEN, S. A. 1953. Elliptocytosis and blood groups. II. Family 3, Ann. Eugenics, 17:272–78.

HALDANE, J. B. S. 1934. Methods for detecting autosomal linkage in man, Ann. Eugenics, 6:26–65.

———. 1936. Search for incomplete sex-linkage in man, ibid., 7:28–57.

———. 1941. The partial sex-linkage of recessive spastic paraplegia, J. Genetics, 41:141–47.

HALDANE, J. B. S., and SMITH, C. A. B. 1947. A new estimate of the linkage between the genes for color-blindness and hemophilia in man, Ann. Eugenics, 14:10–31.

HOGBEN, L. 1934. The detection of linkage in human families, Proc. Roy. Soc. London, B, 114:340–63.

HOLT, H. A., et al. 1952. Linkage relations of the blood group genes in man, Heredity, 6:213–17.

KALLISS, N., and SCHWEITZER, M. D. 1943. Hereditary hemorrhagic diathesis—a case of partial sex-linkage in man, Genetics, 28:78.

KLOEPFER, H. W. 1946. An investigation of 171 possible linkage relationships in man, Ann. Eugenics, 13:35–71.

MACKLIN, M. T. 1952. Sex ratios in partial sex-linkage. I. Excess of affected females from consanguineous matings, Am. J. Human Genetics, 4:14–30.

MOHR, J. 1951. Estimation of linkage between the Lutheran and the Lewis blood groups, Acta path. et microbiol. Scandinav., 29:339–44.

NORTON, H. W. 1949. Estimation of linkage in Rucker's pedigree of nystagmus and color blindness, Am. J. Human Genetics, 1:55–66.

PENROSE, L. S. 1935. The detection of autosomal linkage in data which consists of pairs of brothers and sisters of unspecified parentage, Ann. Eugenics, 6:133–38.

———. 1938. Genetic linkage in human graded characters, *ibid.*, 8:233–38.

———. 1946. A further note on the sib-pair linkage method, *ibid.*, 13:25–29.

SNYDER, L. H. 1941. Studies in human inheritance. XX. Four sets of alleles tested for incomplete sex linkage, Ohio J. Sc., 41:89–92.

———. 1947. *In:* MULLER, H. J.; LITTLE, C. C.; and SNYDER, L. H., Genetics, medicine, and man. Ithaca: Cornell University Press.

SNYDER, L. H., and PALMER, D. M. 1943. An idiopathic convulsive disorder with deterioration, J. Hered., 34:207–12.

TAILLARD, W. 1951. Le linkage autosomique chez l'homme, Acta genet. et stat. med., 2:193–219.

WIENER, A. S. 1932. Method of measuring linkage in human genetics with special reference to the blood groups, Genetics, 17:335–50.

GENERAL REFERENCES

FISHER, R. A. 1950. Statistical methods for research workers. 11th ed. New York: Hafner Publishing Co.

MATHER, K. 1951. The measurement of linkages in heredity. 2d ed. London: Methuen & Co., Ltd.; New York: John Wiley & Sons, Inc.

MOHR, J. 1953. A study of linkage in man. "Opera ex domo biologiae hereditariae humanae Universitatis Hafniensis," Vol. 33. Copenhagen: E. Munksgaard.

Problems

1. Given the mating $TtWw \times TtWw$, where the genes T and W are dominant to their alleles, what are the expected proportions of the phenotypes TW,Tw,tW, and tw if in each parent the genes T and W are coupled with recombination frequency χ? If repulsed?

2. Mercaptobenzoselenazol (M.B.S.) is a compound like phenylthiocarbamide, in that it may or may not evoke a taste response in an individual. The ability to taste M.B.S. is assumed to be due to the presence of a dominant gene. Snyder (1941), in a study of 118 sib-pairs, finds the following distribution of tasters by sex:

TASTE M.B.S.	SEX		
	Like	Unlike	Total
Like........	49	44	93
Unlike.....	10	15	25
Total..	59	59	118

Is the capacity to taste M.B.S. linked with sex either completely or incompletely?

3. The following data on the M-N antigens and the sickling phenomenon were collected to determine the linkage relationship between these loci (Neel et al., 1952): "sksk" = normal; "Sksk" = sickle-cell trait; "SkSk" = sickle-cell disease.

Mating No.	Genotypes of Children								
	MM sksk	*MM* Sksk	*MM* SkSk	*MN* sksk	*MN* Sksk	*MN* SkSk	*NN* sksk	*NN* Sksk	*NN* SkSk
2.....							1	1	
6.....	1	1				1			
19....				1				2	
22....				1	1	1	2		
39....				1	2				1
48....				1	1	1			
53....						2			
57....					2	2			1
64....				1	2	1		2	
79....		3		1			1	1	
81....	1		1		1	1			
90.				2	1				
91....				1	2	2			
93..						2	1	1	1
98...				1	1		1	1	
99....				3	1		3		
110....						2			
111....	1	2		3	3		1	3	
113....			1		1				1
118....	3	1		1					
20-S....	1	1	1						
26-S....				1				1	
31-S....					1			2	
32-S....		1			1				

Test these data for linkage by the 2 × 2 method of sib-pairs.

4. Test the data of problem 3 for linkage by the 3 × 3 method of sib-pairs.

5. If two traits are correlated in a population, does this imply linkage? Explain.

CHAPTER 11

Mutation

MUTATION may be defined as a change in the genetic material which results in a new inherited variation. Geneticists have expended a great deal of time and effort in an attempt to determine exactly what takes place in a chromosome when a mutation occurs. Basic to an understanding of the nature of mutation is, of course, an understanding of the nature of the gene itself. Although we are still very far from this latter, there does seem to be general agreement in the hypothesis that genes are complex molecules or collections of molecules which, for the most part, act through initiating, and controlling the rate of, the intricate biochemical reactions which underlie the development and continued functioning of an organism. Mutation, then, results in the sudden origin of a gene with biochemical properties different from those possessed by the original gene. This may be due to an actual change in the chemical composition or organization of the gene itself or perhaps only to a change in the position of the gene with reference to other genes.

In this chapter we shall consider the nature of mutation and the techniques for determining the rate with which it occurs spontaneously in man. The present status of our knowledge concerning human mutation rates will be summarized. Finally, we shall consider the various factors which may modify human mutation rates.

11.1. *The nature of mutation.*—Mutation is certainly one of the most basic biological phenomena, on a level with cell respiration and cell division. Without mutation as a continuous source of new inherited variation, evolution would in time come to a standstill. Even in plants and animals as easily manipulated as corn and Drosophila, we are still far from an understanding of the causes of mutation. Nevertheless, even if we do not understand precisely what is involved when a mutation occurs, sufficient is known about the results and frequency of mutation in man to justify the present chapter.

Mutation may be either somatic or germinal. Germinal mutation is that which involves the developing eggs and sperm of the ovary and testis. Somatic mutation, on the other hand, involves the body (somatic) cells at any stage in development and leads to sectors of tissue differing in their genetic composition from the rest of the body. If a somatic mutation occurs at an early stage of development, as much as a half or a quarter or an eighth of the body may be involved. If the somatic mutation takes place before the germ cells have been differentiated and involves the region of the body from which they originate, then all or a sector of the gonad with its contained germ cells may be involved. Estimates of mutation rates, unless specifically stated otherwise, deal with germinal rather than somatic mutation.

The genes arising through mutation are, for the most part, less valuable to the organism than the already established genes. That is to say, an individual heterozygous for a dominant mutant gene, or homozygous for a recessive mutant, is usually handicapped in some way as contrasted to the normal individual who does not possess this gene. The reason for this is very simple. Mutation appears, for the most part, to involve a *random* change in the character of the gene. Sometimes such a random change may result in an improvement on the original; but, since most organisms represent a very

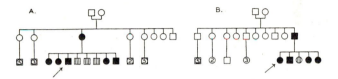

F<small>IG</small>. 11-1.—Two examples of the occurrence of dominant mutations in man: (*a*) neurofibromatosis and (*b*) multiple cartilaginous exostoses. Further explanation in text.

fine and efficient adaptation to their environments, the results of random change can only rarely be for the better. If a mutation does result in an improvement over the original type, this improved type, because of natural selection, will tend to replace the original. Natural selection works in terms of hundreds of generations and thousands of years. If only one mutation in a thousand or ten thousand results in an improvement over the original, this one will usually be sorted out—but the price of this slow improvement and evolution is the ultimate elimination through selection of individuals who receive the mutant genes with deleterious effects.

By this time a large number of examples of mutation in man has come under the scrutiny of various investigators. Figure 11-1 illustrates two of the instances which have come to the authors' attention. The first pedigree concerns neurofibromatosis, a disease characterized by numerous tumors involving the nerves and by the occurrence of areas of light-brown pigmentation over the body. The condition depends upon a dominant gene. In the family in question, a single child in a family of nine, the parents of whom were normal, exhibited the classical features of the disease and, as a proof of its genetic etiology, transmitted the disease to five of her eight children. It is difficult to explain these findings other than as due to the occurrence of a dominant mutation. The second pedigree concerns a man with numerous bony protuberances on his skeleton, a condition usually termed "multiple cartilaginous exostoses" and inherited as if due to a dominant gene. Both of this man's parents, as well as nine brothers and sisters, were normal, but four of the man's five children exhibited the same condition. Again, mutation seems the most likely cause of the findings.

11.2. *Sources of error in mutation-rate estimates.*—Mutation rates are usually expressed as the frequency of detectable mutation per genetic locus per generation. Considerable effort has been devoted to estimating the rate with which mutation occurs in man and other animals. Such estimates are open to several sources of error. Mutation can be recognized and its frequency measured with reasonable certainty only if it results in clear-cut departure from an antecedent phenotype, a departure whose inheritance can be analyzed in the simple terms of one-gene heredity. But there is ample reason to suspect that the results of mutation at some genetic loci are much more elusive in their phenotypic expression than at others and that, even at a given locus, by no means all the possible mutations are detected with equal ease. Furthermore, we are not likely to undertake a study of the rate of mutation of a particular gene unless the phenomenon occurs with sufficient frequency to render such a study feasible. *In other words, our present concepts of spontaneous mutation rates are based upon a biased sampling of some of the mutations of some of the many thousands of genes which each of the higher plants or animals possesses.* We shall return later to the nature of this bias.

The above-decribed source of bias is common to all studies of mutation, no matter what species is involved. The second type of bias which we should recognize at the outset is more specific for studies involving man. Identical —or, shall we say, clinically indistinguishable—phenotypes may arise as a result of mutation at any one of several loci. In such organisms as Drosophila, it is possible to allocate to each locus its share in the total picture by the appropriate genetic analyses. This is not the case for man. We can never be certain in man that the sudden appearance of a given trait is always due to mutation at the same locus or, in sporadic cases of certain diseases, that it is due to germinal, rather than somatic, mutation. However, on the basis of comparative genetics and developmental physiology, we can at least select for study specific phenotypes which, with a high degree of probability, are due to change involving a particular locus. Thus most, if not all, pedigrees of achondroplasia or hemophilia or retinoblastoma are probably due to mutation at the same locus, whereas this is not so for less specific entities, such as hypertension or diabetes.

It is convenient in mutation-rate studies to distinguish between mutation giving rise to recessive, incompletely recessive (or incompletely dominant), and dominant genes, each of these types of mutation presenting certain problems of rate estimation not raised by the other two. As we have already seen, the division of genes into these three categories is actually often quite artificial, being based on the present state of our knowledge rather than on any sound theoretical reason. However, the distinction into these three categories is a useful working arrangement as long as we bear in mind that it is often based on incomplete information and is arbitrary in the sense of delineating intergrading regions in a spectrum of dominance relationships.

Basic to all mutation-rate estimates is accurate information concerning the frequency of the phenotype associated with the gene in question and the selection pressures which bear upon this phenotype. We still have relatively little data concerning the frequency of various inherited diseases in man. Our knowledge of selection pressures is also quite fragmentary. Where a trait is lethal or near-lethal, it is relatively easy to estimate the amount of negative selection at work. But when we deal with characteristics less distinctly deleterious, we encounter a very involved issue. The frequencies of various inherited traits today reflect the selective pressures of one or two or even more hundreds of years ago, the outstanding exception to this generality involving dominant genes whose corresponding phenotypes are at a great reproductive disadvantage. If it is difficult to estimate present-day selection pressures, then it is doubly difficult to guess what the selective value of a given trait was several hundreds of years ago. And yet just such information is needed in estimating the rate of mutation of certain recessive genes.

11.3. *The study of mutation resulting in dominant autosomal genes.*—The simplest possible problem in the estimation of mutation rates involves mutation to a dominant gene which occurs so rarely, or whose corresponding phenotype is so strongly selected against, that the trait is seldom encountered and the possibility of the occurrence of homozygous individuals may be disregarded. One has first to select a suitable area for study and then to locate all the families in this area in which one or more individuals with the trait determined by the mutation in question have been born during a given period. By a comparison of the total number of affected persons with the total number of births in that area during that same period, an accurate prevalence figure is derived. At this point two alternative procedures are possible. The first (direct) procedure is to determine which of the affected persons are the only such occurrences in their families and presumably owe their origin to mutation. The actual mutation rate for nonsex-linked genes, since man is diploid, is half the prevalence (or potential prevalence) at birth of isolated

cases of the trait in question. Or, alternatively, as a second (indirect) procedure, one can calculate the fertility of affected persons in terms of the fraction of normal and then, assuming equilibrium, calculate how frequently new
cases must arise through mutation in each generation in order to maintain
in the population a constant proportion of persons with the trait. The
appropriate formula is

$$m = (1 - f) \, q \, ,$$

where

m = the mutation rate per gene per generation;
f = the relative fertility of individuals heterozygous for the particular gene
    under consideration, so that $(1 - f)$ is the selective disadvantage; and
q = the frequency of the gene, in this case, half the fraction given by the
    ratio of all cases born during a given period to all births during the
    same period.

The first procedure has the advantage that it involves no assumptions
concerning effective fertility, but the disadvantage that the estimate so derived may be too high, because included in the material are some individuals
who owe their trait to developmental accidents and poorly understood environmental factors rather than to mutation, and other individuals in whom
the trait is the result of somatic mutation. These disadvantages are of less
relative importance in the second procedure.

A study on achondroplasia in Denmark will serve as an example of both
the first and the second procedures (Mørch, 1941). Achondroplasia is a type
of dwarfism due to abnormally short arms and legs. The trait appears to be
due to a dominant autosomal gene of high penetrance. The death rate
among infants with achondroplasia is very high, but the adults are usually
quite vigorous people. Among a total of 94,075 births occurring in a large
hospital in Copenhagen over a period of some thirty years, there were 10
babies with achondroplasia. Eight of these babies were born to normal parents, while 2 of the infants had an achondroplastic parent. Accordingly, as
calculated by the direct method,

$$m = \frac{8}{94,073} \times \tfrac{1}{2} = 0.000042, \quad \text{or} \quad 4.2 \times 10^{-5}.$$

From Mørch's material it can be calculated that the expectation that an
achondroplastic child will have an achondroplastic offspring, as compared to
the expectation that the normal siblings of achondroplastics will have normal
offspring, is approximately $0.2(=f)$ (cf. Popham, 1953). The frequency of
achondroplasia at birth is 10/94,075, or 0.000106, and $q = 0.000053$. Accordingly, as calculated by the indirect method, $m = (1 - 0.2)0.000053 =$

0.000042, or 4.2×10^{-5}. It must be emphasized again that the latter calculation assumes equilibrium, i.e., that the number of achondroplastics in the population is remaining relatively constant.

Because of the high infant death rate and low marriage rate for achondroplasia, most of the cases of achondroplasia occurring in a population during a given period are isolated instances in the family concerned and hence are the direct result of mutation. Part of the agreement noted above between the direct and the indirect mutation-rate estimates for achondroplasia is due to the fact that the isolated cases, which form so large a proportion of the total cases, enter into both calculations. Among conditions where the at-birth expectation of reproduction is less impaired, the agreement between the two approaches will tend to be less because of the errors that enter into the estimate of the various factors used in the calculations. Quite aside from these errors of estimation, there is, as suggested above, an important biological source of error, namely, the possibility that mutation at two or even three different genetic loci may be responsible for the tendency to develop achondroplasia.

As we have already seen, completely dominant genes in man may be in the minority. Individuals homozygous for completely dominant genes will usually have an appreciable frequency in any given population only if the coefficient of selection against the corresponding phenotype is relatively small or zero. The formula given above may also be used to calculate the rate of mutation to a completely dominant gene which is sufficiently common that homozygous individuals are no longer a negligible factor, if the symbols are redefined so that $1 - f$ is now the coefficient of selection against either homozygote or heterozygote and q, the gene frequency, is now

$$1 - \sqrt{\frac{\text{the frequency of individuals neither homozygous}}{\text{nor heterozygous for the gene in question}}}$$

Penrose (1936) was the first to point out that, with regard to a trait produced by a dominant gene with rather deleterious effects, persons with negative family histories (who presumably were due to mutation) were inclined to be more severely affected than persons with positive family histories. This is explained on the basis that the difference between mildly and severely affected persons is often a matter of genetic modifiers. Severely affected persons are not likely to reproduce. It is therefore the mildly affected who propagate, transmitting to their offspring both the principal gene and (sometimes) the favorable modifiers.

11.4. *The study of mutation resulting in completely recessive, autosomal genes.*
—In the case of recessive genes we cannot, as for dominants, simply enu-

merate the isolated cases of a particular trait in a population and arrive thereby at an estimate of the rate of mutation. A completely recessive gene arising in consequence of a mutation cannot be detected with certainty at the time of its origin. Instead, a newly arisen recessive is usually transmitted through several or many generations of heterozygotes who appear phenotypically normal until, finally, from the marriage of two such heterozygotes, there results a homozygous person who is evidence for the occurrence of mutation at some time in the past. In a population in which the frequency of a particular recessive gene is not undergoing change, the rate of mutation to that gene is approximated by the formula

$$m = (1 - f) f(aa)$$

where

m has the same significance as previously;

$f =$ the relative fertility of individuals homozygous for the gene under consideration; and

$f(aa) =$ the frequency of the trait.

There are two reasons why this formula has to be applied with caution. The first reason is the assumption of complete recessiveness. If heterozygosity for a given gene confers upon its possessor even a very slight selective advantage, this will, because of the relative frequency of heterozygotes as compared to homozygotes, offset a considerable disadvantage which the same gene may confer upon the homozygote. Conversely, let us consider a gene which in the homozygous condition exerts a deleterious effect and which also has a very slightly harmful effect on viability or fertility when heterozygous. Now the rate of mutation necessary to maintain the gene frequency is greatly increased over that necessary to maintain a similarly defective homozygote whose heterozygote is truly neutral. There is accumulating evidence, mentioned in chapter 8, that many so-called "recessives" actually have some small effect when heterozygous. However, in human genetics it is seldom feasible at present to decide whether any particular gene affects average fertility by 5 per cent or less, because of the many variables which affect reproduction and combine to obscure the gene effect.

The second reason why the formula given above has to be applied with extreme caution is the error introduced by the assumption of what the geneticist terms "population equilibrium." A population is in equilibrium with respect to any particular gene if the rate with which the gene is arising through mutation is balanced by the rate with which the gene is being eliminated because of negative selection. The amount of inbreeding occurring in a population enters into the picture, because, as we have seen, the more the inbreeding, the more rapidly do individuals homozygous for recessive genes appear and become subject to natural selection. With rare possible exceptions, human populations are definitely not in genetic equilibrium at present. With increased population movement and changing social customs, there has almost certainly been, in the past several centuries, a decrease in the amount of inbreeding in most communities. One consequence of this will be an increase in the frequency of deleterious recessive genes, until a new equilibrium point is reached. Some hundreds of years are required to reach this new equilibrium. During this period of increase, the ratio of heterozygous to homozygous individuals which the Hardy-Weinberg law postulates will be disturbed, with a relative excess of heterozygotes. Consequently, a mutation rate estimated on the basis of the frequency of homozygous individuals will tend to be too low. These two reasons for being cautious in estimating the rate of mutation of recessive genes are so important that all the estimates available must be regarded as highly tentative.

11.5. *The study of sex-linked mutation.*—The estimation of the rate of mutation of sex-linked genes, whether completely dominant or completely recessive, raises certain special questions, since each female has two X-chromosomes, whereas males have only one. For a rare sex-linked dominant mutation,

$$m = (1 - f) \, q \, ,$$

where q, the frequency of the gene, assuming a rarity of the gene such that the occurrence of the homozygous females can be disregarded, is calculated from the formula

$$q = \frac{\text{number affected males} + \text{number affected females}}{\text{total number males} + 2 \, (\text{total number females})}$$

For recessive sex-linked genes, again of such rarity that the occurrence of homozygous females may be disregarded,

$$m = \tfrac{1}{3} \, (1 - f) \, x \, ,$$

where x = the frequency of affected males. The factor $\tfrac{1}{3}$ is introduced because, for every gene that manifests itself in a male, there are two in heterozygous females; selection acts only on the males.

11.6. *The study of mutation to incompletely recessive genes.*—As the third class of genes for which mutation rates can be calculated, we may consider incompletely recessive (or incompletely dominant) genes. Here the problem is more complicated because we must determine selection against both homozygote and heterozygote. If the selection to which the heterozygote is subject were large—either positive or negative—then the mutant gene would, in the course of a relatively few generations, either increase to the point where it would be relatively common or decrease to a point of rarity. It follows that where heterozygosity for an incompletely recessive gene does have some selective value, it will tend to be small, of a magnitude difficult to evaluate.

TABLE 11-1*

SUMMARY OF AVAILABLE ESTIMATES OF THE SPONTANEOUS
RATE OF MUTATION OF HUMAN GENES

Classification of Gene	Character Produced by Gene	Mutation Rate/ Gene/Generation	Author
Autosomal dominant	Epiloia	0.8–1.2×10^{-5}	Penrose (1936)
	Chondrodystrophy	4.2×10^{-5}	Mørch (1941)
		7.0×10^{-5}	Böök (1952)
	Pelger's nuclear anomaly	8.0×10^{-5}	Patau and Nachtsheim (1946)
	Aniridia	0.5×10^{-5}	Møllenbach (1947)
	Retinoblastoma	1.4×10^{-5}	Philip and Sorsby, unpublished
		2.3×10^{-5}	Neel and Falls (1951)
	Waardenburg's syndrome	3.7×10^{-6}	Waardenburg (1951)
Autosomal recessive	Microphthalmos and anophthalmos, with or without oligophrenia	1.0–2.0×10^{-5}	Sjögren and Larsson (1949)
	Albinism	2.8×10^{-5}	
	Congenital total color blindness	2.8×10^{-5}	Neel, Kodani, Brewer, and Anderson (1949)
	Infantile amaurotic idiocy	1.1×10^{-5}	
	Ichthyosis congenita	1.1×10^{-5}	
	Cystic fibrosis of pancreas	0.7–1.0×10^{-3}	Goodman and Reed (1952)
	Epidermolysis bullosa dystrophica lethalis	5.0×10^{-5}	Böök (1952)
	Amyotonia congenita	2.0×10^{-5}	
	Microcephaly	3.0×10^{-5}	Komai (1953)
Sex-linked recessive	Hemophilia	3.2×10^{-5}	Haldane (1947)
	Pseudohypertrophic muscular dystrophy	1.0×10^{-4}	Stephens and Tyler (1951)

* Modified from Neel (1952). We have omitted from this table estimates available for the incompletely recessive genes responsible for sickle-cell anemia and thalassemia because of the particular problems in the study of mutation rate presented by incompletely recessive genes (sec. 11.6).

11.7. *Summary of the various available estimates of the rate of mutation of human genes.*—In spite of the difficulties inherent in estimating the rate of mutation of human genes, first estimates are now available for 17 loci. These are given in Table 11-1. Where mutation gives rise to recessive genes, the estimates are all based on the assumption, discussed above, that the mutant genes in question are neutral when heterozygous.

Except for the estimate of Goodman and Reed (1952) with respect to cystic fibrosis of the pancreas (0.7×10^{-3}), the figures range from 1×10^{-4} to 4×10^{-6}. Omitting this one high estimate, not because it is any less accurate, but because it may represent a special situation, the average frequency of appearance of mutant genes with known phenotypic effects at each of the remaining 16 loci is about 3×10^{-5}. Approximately 48 germ cells per 100,000 should contain a mutation involving at least one of these 16 loci in such a manner as to result in the particular phenotypic effects listed in Table 11-1.

The question now arises as to how representative these estimates are of human mutation rates in general. There is no way at present of formulating a precise answer to this question, but certain generalizations seem pertinent. Mutation at any particular locus may be thought of in terms of (a) the total frequency of mutation at that locus; (b) the number of alternative forms of the gene which may occur at any locus, i.e., the number of multiple alleles; and (c) the ease with which the results of change at any locus can be detected with our present techniques. There are insufficient grounds at present for forming opinions as to the intercorrelations of these three aspects of the mutation picture. We assume that some genetic loci are more mutable than others, because we detect the results of mutation more frequently at these particular loci; but to what extent these apparent differences are real or to what extent they are due to the fact that the results of mutation at these particular loci are relatively easily identified is not now clear. Assuming that the differences are at least in part real, we do not know whether the more mutable loci tend to be associated with the greatest number of allelic forms or whether the more mutable loci tend to give rise to mutations with clear-cut effects relatively more or less frequently than the less mutable. Our present knowledge of mutation is limited, to an extent not usually recognized, by the techniques at our disposal. We tend to think of the frequency of mutation in terms of mutation giving rise to lethal genes or genes with clear-cut effects (readily detected even in Drosophila), because this is the kind of mutation which can be measured and studied; but it is obvious that one must be cautious in generalizing from this experience. Mutation does not always result in a phenotype which is sharply differentiated from normal. Mutation may result in genes which are known only because they modify the expression of other genes or which affect viability and fertility in subtle and ill-defined ways.

Just as most organisms can adapt themselves to wide changes in the external environment, so possibly they can adjust to the effects of minor genetic change in such a fashion as to make the detection of that change very difficult. A given genetic locus may mutate both to genes with clear-cut effects and to genes with poorly defined effects (e.g., the bobbed and white eye loci in Drosophila). It is clear that thus far in studies of human mutation rates, investigators have tended, even more than is the case for Drosophila and corn, to select for study loci at which mutations with clear-cut effects are occurring in numbers sufficient for study. If the proportion of mutations with clear-cut effects is the same for all loci, then there has clearly been a selection for the more mutable loci. But if, as seems equally likely, mutation at some loci is much more readily detected than at others, because of the nature of the functions of the locus, then it by no means follows that the studies to date have been based on the more frequently mutating loci. Since, further, by analogy with Drosophila, it seems that, in the case of the human genetic loci so far investigated, only a portion of the total mutation at that locus is being detected, it follows that our present very tentative estimates for these 17 loci are probably underestimates. Further speculation on this point seems profitless—we shall use the figure of 3×10^{-5} as the average mutation rate per locus per generation in man, simply because that is the average of present estimates.

In passing, it might be pointed out that the study of the frequency with which there occur mutations with small effects—"little mutations," if you will—is one of the most pressing problems confronting modern genetics. This is a type of inquiry for which some of the lower plants and animals are far better suited than man; a start has already been made on this problem in Drosophila. It has been difficult for biologists to see how the types one usually thinks of as mutants in, for example, Drosophila can supply the building blocks for evolution. But we now recognize that, in addition to mutations known because they result in marked changes in the animal, there are many more with individually unapparent (except in specially sensitized genetic systems) but collectively important effects. Although we can recognize that here is the origin of a store of genetic variability of great potential importance to the organism, we still have relatively little insight into the total amount of this kind of mutation. Unfortunately, it will be a very difficult task to demonstrate the total range of mutational effects at any one locus, since modern biochemistry is providing more and more examples of the complex fashion in which organisms may surmount metabolic problems and still achieve biochemical goals. The study of adjustment to change in the external environment, to which so much of biology is devoted, is much simpler than the observation of adjustment to genetically induced change in the internal environment. The appreciation of the extent to which this homeostatic adjustment to mutation may occur, and so conceal (at least temporarily) the effects of mutation and give rise to spuriously low estimates of mutation rate, may be expected to proceed slowly.

11.8. *"Total" mutation rates.*—In the second chapter we saw that the total diploid number of genes in man might be in the neighborhood of 40,000–80,000. With an average mutation rate of 3×10^{-5}, the average number of "new" mutant genes that any individual should receive from one or the other of his parents lies between 1.2 and 2.4. That is to say, on the average, according to these figures, each of us may well have received at least one mutant gene from one or the other of our parents. Some investigators will consider this estimate too high. However, even if the estimate is unduly high by a factor of 2 or even 4, this frequency still literally brings mutation into the life of each of us. It is no longer a rare phenomenon but a relatively common biological event.

More is known about mutation rates in Drosophila than is the case for any other animal. Here the average rate of mutation per gene per generation is thought to be about 1×10^{-5}; many reasons for dissatisfaction with this figure should be noted in passing, with the probability that the figure is too low rather than too high. The total diploid number of genes in Drosophila has been estimated at 10,000–20,000. On the basis of these assumptions, each fly will therefore average 0.1–0.2 mutation not present in either parent. But while Drosophila is, in all probability, less mutable than man in terms of average number of mutations per generation, the situation is different when viewed on a time scale. A pair of humans might in 25 years possess six to eight offspring, whereas in that same interval of time a pair of flies, averaging ten generations to the year and fifty pairs of offspring each generation, have the potential of 50^{250} pairs, a truly staggering figure and, of course, far greater than the total number of representatives of the human species who have ever lived. At the end of the 250 fly generations, which correspond in time to one human generation, each descendant fly should, on the average, assuming neutrality of the genes arising through mutation, have accumulated at least 250×0.10, or 25.0, mutations not present in the original pair of flies.

As far as the mammals are concerned, as much is known about the mutation rate of man as of any other species. In recent years some careful work on mice has been in progress. Russell's (1951) very preliminary figures indicate that, with respect to seven different genetic loci chosen "at random" for study, the spontaneous mutation rate per locus per generation approximates 1×10^{-5}. This appears lower than the average given above for man, but, as in the case of Drosophila, it must be remembered that the mouse has a much shorter life-cycle, averaging no more than half a year from birth to reproductive peak. At least one "mutable locus" is known in the mouse, the rate of appearance of new mutations at the T locus falling in the range of

1 × 10^{-3} per generation (Dunn and Gluecksohn-Waelsch, 1953). The total mutation rate per unit time would appear to be greater than for man.

11.9. *Induced mutation.*—In 1927, after a number of previous investigators had tried but failed to produce conclusive results, H. J. Muller clearly showed that the treatment of Drosophila with X-rays resulted in an increase in the frequency of recessive lethal mutations occurring in the sex chromosome (cf. sec. 8.1, p. 76). Since that time a great deal of effort has gone into the study of induced mutation in various species of plants and animals. It has been shown that other types of radiation, such as gamma or ultraviolet radiation, can affect mutation rates. It has also been shown that temperature and various chemicals can influence mutation rates. Among the chemicals which have been shown to be mutagenic are such diverse substances as certain of the mustard and lachrymator gases, phenol, urethane, carcinogenic hydrocarbons such as 1,2,5,6-dibenzanthracene and 20-methylcholanthrene, bile salts, formalin, hydrogen peroxide, copper sulfate, and manganous chloride. Much of the work in this field has been carried out with bacteria and Drosophila. In the case of the bacteria, the chemical to be tested is simply added to the nutrient medium in which they are growing. In the case of Drosophila, the chemicals are particularly effective when incorporated into a very fine "mist" (aerosol), which presumably is then carried to all parts of the body by the tracheal system.

Inasmuch as man has a relatively constant body temperature, it seems unlikely that fluctuation in the environmental temperature exerts an influence on human mutation rates. Moreover, because of the elaborate mechanisms which maintain the chemical constancy of the human body during life, it seems unlikely that exposure to various chemical agents exerts a very large influence on human mutation rates. But the situation with respect to certain types of radiation which can penetrate the body with ease is potentially quite different.

Radiation is usually defined as the process whereby energy is transmitted through space. It is convenient to visualize this transmission as due to rhythmic electric and magnetic oscillations, termed "waves." There exist many kinds of radiation, the various types differing in wave length. Figure 11-2 classifies radiation according to wave length. The types of radiation of particular concern to the geneticist because of a mutagenic effect are those with wave lengths less than approximately 10^{-6} cm., i.e., roentgen rays, gamma rays, and cosmic rays. The intensity of radiation of this wave length is customarily measured in *r* units.

One of the first questions to engage the attention of those interested in radiation genetics had to do with the relation between the amount of radiation applied to the fly and the amount of mutation produced. Figure

11-3 demonstrates that the amount of mutation is directly proportional to the amount of radiation applied. This relationship apparently holds good even at relatively small doses of irradiation delivered over a period of some days, i.e., there is no threshold effect (Uphoff and Stern, 1949). A given amount of irradiation has the same total effect if given in divided doses as when given in one continuous dose. In general, the effectiveness of any particular radiation or other physical agent (such as neutrons) is proportional

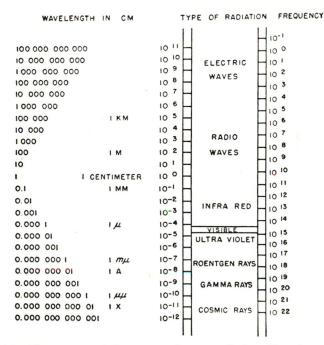

FIG. 11-2.—The spectrum of electromagnetic wave radiations. The column on the left gives wave length in centimeters, whereas the column on the right describes the oscillations in terms of frequency per second. The usual designations of the various wave lengths and their abbreviations are given in the central columns. (By permission from Physical foundations of radiology, by Drs. Glaser, Quimby, Taylor, and Weatherwax, copyright 1952, Paul B. Hoeber, Inc.)

to the density of ionization produced by the agent as it traverses the gonial tissues (cf. Lea, 1947, and Catcheside, 1948, for general summaries).

There are four chief sources of irradiation of wave length 10^{-6} or less to which modern man is exposed. These are cosmic radiation, radiation from the decomposition of naturally occurring radioactive isotopes, X-radiation, and radiation from the peacetime or military applications of atomic energy. A brief discussion of the possible mutagenic significance of each of these follows.

1. COSMIC RADIATION

Cosmic rays originate outside our planet. They possess extremely high penetrating power. However, the intensity of cosmic rays at sea-level is only about 0.0001 r per day, or 0.04 r per year. Although the dosage of cosmic radiation increases in proportion to the altitude, being double at elevations of a mile above sea-level at most of the altitudes in which man has established himself, cosmic rays deliver less than 0.08 r per year.

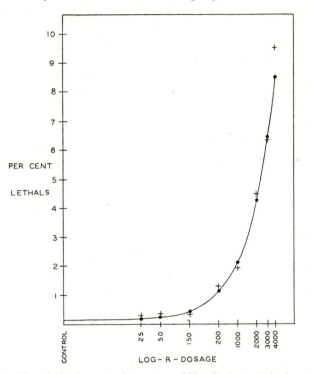

Fig. 11-3.—The relation between the amount of X-radiation and the frequency of sex-linked lethal mutations in Drosophila. Because of the large range in the experimental dosages employed, the data have been plotted logarithmically. The solid line represents the curve which best fits the various observations. (Modified by permission of Dr. W. P. Spencer and Genetics.)

2. RADIATION FROM DECOMPOSITION OF NATURALLY OCCURRING RADIOACTIVE ISOTOPES

Naturally occurring radioactive isotopes are found in soil, rocks, air, and water and also in the human body. Gamma radiation due to radioactive isotopes in the surrounding environment, of course, varies considerably from place to place but, in general, ranges between 0.0001 and 0.0003 r per day, or 0.04–0.11 *r* per year. Gamma radiation from the decomposition of natural isotopes incorporated into the various chemical compounds of the body amounts to about 0.0001 *r* per day, or 0.04 *r* per year; potassium is the chief source of this radioactivity. The total amount of annual radiation from sources 1 and 2 is thus only 0.12–0.23 *r*.

3. X-RADIATION

The preceding two sources of radiation have been with man since the dawn of human time. This is not the case for the two sources which remain to be discussed. X-rays, the first of these two sources, were introduced to medicine in 1895. Modern man is usually exposed to X-rays in consequence of their wide diagnostic and therapeutic uses. The approximate dosages in terms of *r* units commonly delivered in various diagnostic procedures are shown in the accompanying tabulation (Plough, 1952). Unless otherwise noted, the figures are for single exposures.

Type of X-ray	Dosage to Surface of Body (*r* Units)
Chest, posterior-anterior, 14 × 17-inch film	0.1
Chest, posterior-anterior, photofluoroscopic X-ray	1.0
Lumbar spine, anterior-posterior	1.5*
Lumbar spine, lateral	5.7*
Pelvis	1.1*
Pregnancy, anterior-posterior	3.6*
Pregnancy, lateral	9.0*
Abdomen	1.3*
Cardiac fluoroscopy and catheterization	10–140
Gastrointestinal series (6 films)	4.0*
Gastrointestinal fluoroscopy	10–20/minute*
Gall bladder	0.6*
Extremities (lower)	0.3*
Skull, posterior-anterior	1.3
Full-mouth dental films	5.0

Only the procedures indicated with an asterisk may result in direct irradiation of the gonads. From the standpoint of potential genetic effect, the procedures on that list to be most closely scrutinized are the X-ray examinations of the large and small intestine. Young individuals with gastrointestinal disease who are subject to repeated X-ray studies may, over a period of years, receive a total dose well in excess of 100 *r*, although only a fraction of this is delivered to the gonads.

The therapeutic uses of irradiation seldom present a genetic problem, since the bulk of the individuals receiving significant amounts of irradiation are either beyond the age of reproduction or, because of the nature of their disease, will not reproduce again. In passing, however, attention should be drawn to the occasional practice, now fortunately on the decline, of applying

200–300 *r* units of irradiation to the surface of the body overlying the ovaries in the treatment of certain types of menstrual disturbances and sterility.

4. THE PEACETIME OR MILITARY APPLICATIONS OF ATOMIC ENERGY

The safety measures which have been evolved in the various industrial installations in this country at which atomic weapons are being prepared or in which the peaceful applications of atomic energy are being explored are excellent. For example, at the famous Oak Ridge and Hanford plants in 1949, the average annual exposure was 0.2 *r*, while the average of the ten highest exposures in Oak Ridge workers was 4.2 *r* a year. Those "most exposed" workers received no more irradiation than would be received in the course of skull, chest, and spinal films. There would not appear to be any more important problem here than in the diagnostic use of X-rays in the practice of medicine.

This brings us to the subject of the genetic risks involved in the military use of atomic energy, as in the explosion of an atomic bomb. This is a question which has aroused considerable discussion ever since the atomic bombs burst over Hiroshima and Nagasaki in 1945. In every single plant and animal species adequately investigated, radiation has been found to induce mutation. There is thus no doubt that mutations were induced in the inhabitants of Hiroshima and Nagasaki in consequence of the atomic bombings. However, there are many "practical" difficulties interfering with the detection of the genetic effects on man of an atomic bombing. Also, there are problems in separating, in the first postbomb generation, the genetic from the nongenetic effects of maternal exposure to an atomic-bomb explosion. Finally, the amount of whole-body irradiation received by the survivors of such a bombing is, of course, variable but small by the standards of the radiation geneticist. The range, depending on distance from the explosion and amount of protective shielding, is from zero to as much as 600–700 *r*, the maximum amount of whole-body irradiation compatible with survival. Only a minority of survivors will receive the higher doses, the average being very much less. By analogy with what is known of the sensitivity of Drosophila and mouse genes to irradiation, geneticists could anticipate that at this dosage level there would probably be apparent in the first postbomb generation only very minor effects, so minor that they might well be obscured by various possible distorting factors as well as random fluctuations. Accordingly, while it is certain on a priori grounds that mutations have been produced by the ionizing radiation released by the explosion of an atomic bomb, there exist, under the circumstances which govern observations on human material, many obstacles to their demonstration.

For the last eight years, studies have been in progress in Hiroshima and Nagasaki on the characteristics of children born to the survivors of the atomic bombings, as well as to a suitable control group. Only a preliminary analysis of this work is available (Neel *et al.*, 1953). Attention has been concentrated on the sex ratio, the birth weight, the frequency of stillbirths, the frequency of gross malformations, the death rate during the first nine months of life, and body size at age nine months. Genetic theory suggests that irradiation of the mother should decrease the percentage of males among her offspring, owing to the fact that males will be affected by induced sex-linked mutations which will not affect the females. Radiation of the father, on the other hand, should result in a decrease in the percentage of females among his offspring, owing to the action of sex-linked dominant mutations. Any attempt to postulate the magnitude of the differences at a particular radiation level is rendered difficult by many factors, including the present unsatisfactory state of knowledge concerning the homologous segments of the X- and Y-chromosome. Inasmuch as mutations frequently interfere with metabolic sequences (cf. chap. 12), induced dominant mutations might be expected to decrease birth weight and growth rates. Finally, autosomal lethal and semi-lethal mutations, the frequency of which is well known to be increased by irradiation, might be expected to increase the frequency of stillbirths and gross malformations.

The actual amount of critical data which can be assembled in Hiroshima and Nagasaki is relatively small. The preliminary analysis of these data suggests a slight effect on the sex ratio and possibly an effect of maternal irradiation on the frequency of stillbirths. There is no clear effect on the frequency of malformation, birth weight, size at age nine months, and frequency of death during the first nine months. These effects, if real, are in a range compatible with extrapolation from the known genetic effects of irradiation on other animal species. In other words, the preliminary analysis suggests that, while there is no evidence that man is unusually sensitive to the genetic effects of irradiation, neither is there reason to feel that he has any particular "immunity."

11.10. *The geneticist views irradiation.*—In view of what is known of the results of mutation in general, the geneticist is forced to consider with concern any increase in the amount of irradiation to which mankind is subjected, justifiably only if a strong case can be made for its benefits, the more so since man is rapidly removing the normal counterbalance to mutation, namely, natural selection (cf. chap. 20). An evaluation of the precise genetic impact of an atomic bomb or any other form of irradiation requires much more extensive information than we now possess concerning spontaneous and radiation-induced mutation rates. The gaps in our present knowledge permit widely varying assumptions. For instance, whereas the present authors feel that the average spontaneous rate of mutation per locus per generation in man may, on the basis of our present information, be assumed to be greater than 1×10^{-5}, others have considered such an estimate as much too high, preferring figures of the order of 1×10^{-6} or 10^{-7} (Wright, 1950).

Preliminary figures on the sensitivity of mouse genes to irradiation indicate that the average rate of mutation per gene per r is in the neighborhood of $2–3 \times 10^{-7}$ (Russell, 1951); the Japanese data are compatible with this figure. If human genes have the same sensitivity to irradiation as mouse genes, then, inasmuch as the average adult of thirty has already received $3–6$ r from sources 1 and 2 described in section 11.9, the spontaneous mutation rate per locus per generation can scarcely be less than 1×10^{-6}, even on the assumption that all spontaneous mutation is due to ionizing radiation, and is, of course, even higher if, as seems very likely, ionizing radiation is only one of several factors responsible for the normal, "spontaneous" mutation rate. At a spontaneous mutation rate of 3×10^{-5} and an induced rate of 2×10^{-7} per gene per r, it will require 150 r to double the mutation rate, whereas at a spontaneous rate of 1×10^{-6} and a sensitivity of 3×10^{-7}, it will require only 3 r. Inasmuch as genetic considerations are one factor in setting tolerance limits for radiation exposure, there are both practical as well as theoretical reasons for wishing clarification of this question.

Bibliography

SPECIFIC REFERENCES

AUERBACH, C. 1949. Chemical mutagenesis, Biol. Rev. Cambridge Phil. Soc., **24**:355–91.

——. 1949. Chemical induction of mutations, Proc. Eighth Internat. Cong. Genetics, Stockholm, Hereditas suppl., pp. 128–41.

CATCHESIDE, D. G. 1948. Genetic effects of radiations, Adv. Genetics, **2**:271–358.

DEMEREC, M.; BERTANI, G.; and FLINT, J. 1951. A survey of chemicals for mutagenic action on *E. coli*, Am. Naturalist, **85**:119–36.

DUNN, L. C., and GLUECKSOHN-WAELSCH, S. 1953. Genetic analysis of seven newly discovered mutant alleles at locus T in the house mouse, Genetics, **38**:261–71.

HERSKOWITZ, I. H. 1951. A list of chemical substances studied for effects on Drosophila, with a bibliography, Am. Naturalist, **85**:181–99.

LEA, D. E. 1947. Actions of radiations on living cells. New York: Macmillan Co.

MØRCH, E. T. 1941. Chondrodystrophic dwarfs in Denmark. "Opera ex domo biologiae hereditariae humanae Universitatis Hafniensis," Vol. **3**. Copenhagen: E. Munksgaard.

MULLER, H. J. 1927. Artificial transmutation of the gene, Science, **66**:84–87.

PENROSE, L. S. 1936. Autosomal mutation and modification in man with special reference to mental defect, Ann. Eugenics, 7:1–16.

PLOUGH, H. H. 1952. Radiation tolerances and genetic effects, Nucleonics, 10: 17–20.

POPHAM, R. M. 1953. The calculation of reproductive fitness and the mutation rate of the gene for chondrodystrophy, Am. J. Human Genetics, 5:73–75.

RUSSELL, W. L. 1951. X-ray-induced mutations in mice, Cold Spring Harbor Symp. Quant. Biol., 16:327–36.

SPENCER, W. P., and STERN, C. 1948. Experiments to test the validity of the linear r-dose/mutation frequency relation in Drosophila at low dosage, Genetics, 33:43–74.

UPHOFF, D. E., and STERN, C. 1949. The genetic effects of low intensity irradiation, Science, 109:609–10.

GENERAL REFERENCES

Cold Spring Harbor Symposia on Quantitative Biology, Vol. 16: Genes and mutations. Lancaster, Pa.: Science Press, 1952.

EVANS, R. D. 1949. Quantitative inferences concerning the genetic effects of radiation on human beings, Science, 109:299–304.

HALDANE, J. B. S. 1947. The dysgenic effect of induced recessive mutations, Ann. Eugenics, 14:35–43.

———. 1948. The formal genetics of man, Proc. Roy. Soc. London, B, 135:147–70.

———. 1949. The rate of mutation of human genes, Proc. Eighth Internat. Cong. Genetics, Hereditas suppl., pp. 267–73.

McELROY, W. D., and SWANSON, C. P. 1951. The theory of rate processes and gene mutation, Quart. Rev. Biol., 26:348–63.

MULLER, H. J., 1947. The production of mutations, J. Hered., 38:259–70.

———. 1950. Our load of mutations, Am. J. Human Genetics, 2:111–76.

NEEL, J. V. 1952. The study of human mutation rates, Am. Naturalist, 86:129–44.

NEEL, J. V.; SCHULL, W. J.; McDONALD, D. J.; MORTON, N. E.; KODANI, M.; TAKESHIMA, K.; ANDERSON, R. C.; WOOD, J.; BREWER, R.; WRIGHT, S.; YAMAZAKI, J.; SUZUKI, M.; and KITAMURA, S. 1953. The effect of exposure to the atomic bombs on pregnancy termination in Hiroshima and Nagasaki: preliminary report, Science, 118:537–41.

WRIGHT, S. 1950. Discussion on population genetics and irradiation, J. Cell. & Comp. Physiol., 35:suppl. 1, 187–205.

Problems

1. Assume that in a homogeneous human population at equilibrium congenital total color blindness, a disease due to a recessive gene, has a frequency of 1:80,000. Individuals with this condition have very poor vision and in the past have probably been greatly handicapped from the reproductive standpoint. Assuming that their effective fertility is approximately half-normal, what is the

mutation rate necessary to maintain the frequency of the trait in the population?

2. Pseudo-hypertrophic muscular dystrophy is a sex-linked recessive trait; affected males fail to reproduce. Among a total of 63,000 males born in the state of Utah during a 10-year period, there were 18 affected with this condition. Calculate the mutation rate.

3. Assume that the average rate of mutation of 20 different loci in man is 1.2×10^{-5}. What is the probability that an individual will receive a mutant gene at some one of these loci from one or the other parent?

4. A young woman with a refractory peptic ulcer has had, over a 3-year period, five gastrointestinal X-ray series, receiving an average of 40 r to the skin of the abdomen each time, half of which may be assumed to penetrate to the ovaries. Assuming a haploid gene number of 40,000 and a mutation rate per gene per r of 2×10^{-7}, what is the expectation of induced mutation in this woman?

5. As a result of an atomic-bomb explosion, the 100,000 survivors of a city of 250,000 receive an average skin dose of 25 r. Utilizing the figures given in problem 4 for penetration, gene number, and mutation rate per gene per r, what is the total number of mutations to be expected in this population? Assuming a spontaneous mutation rate of 2×10^{-5} per gene per generation, how will the number of mutations due to irradiation compare with the number occurring spontaneously in the group?

CHAPTER 12

Physiological Genetics

PHYSIOLOGICAL genetics is that branch of genetics which is concerned with working out the intermediary steps whereby the presence of a given gene, as a member of a particular genotype and in a particular environment, is responsible for the appearance of a given trait. As stated earlier (p. 136), genes are thought to function through initiating and/or controlling the rate of progression of the various complex biochemical reactions which occur during the development and adult life of every living organism. It will be the purpose of this chapter to consider some of the evidence from man on which that belief is based. Between the primary effect of the gene and the end-result which the geneticist records, there may be many intermediate steps. Because the number and nature of these intermediate steps are usually very difficult to determine, conclusions as to the "primary effect" of a gene must be drawn with great care. Nevertheless, there are some inherited variations in man in which there is reason to believe that the measurable effect is rather close to the primary effect, and it is with these that we shall be especially concerned. We shall also stress some of the interpretive problems which arise in the physiological genetics of man.

12.1. *The biochemical genetics of Neurospora.*—In recent years the study of physiological genetics has progressed rapidly in consequence of recognition of the fact that some of the lower organisms are remarkably suited for such investigations. One of the most widely studied is *Neurospora crassa*, an ascomycete fungus which is capable of growing and reproducing on a simple medium containing only inorganic salts, carbohydrate, and a single vitamin, biotin. Since the cellular constituents of Neurospora include many complicated compounds, extending up to the level of complex proteins, Neurospora exhibits remarkable synthetic powers during the course of its life-cycle. By the use of irradiation, mutations have been produced which interfere with various links in the organism's chains of synthetic reactions. The precise results of these mutations may be studied with special techniques, and the point at which the mutation interferes with the normal activities of the cell can be identified. Through the careful analysis of the results of large numbers of these mutations, it is possible to reconstruct the genetically controlled

157

steps by which Neurospora normally synthesizes certain compounds. Conversely, the demonstration that there are sequences of reactions under rigid genetic control throws important light on the manner in which genes operate. Review papers on this general subject have been written by Beadle (1945*a*, *b;* 1946; 1949), Horowitz (1950), Tatum (1946, 1949), and Wright (1941).

The general techniques involved in the study of the genetic control of the synthesis of an essential cell constituent are as follows: Through irradiation, mutant strains are produced which are no longer capable of producing a given compound, which we may call *d*. Strains unable to synthesize *d* can grow when this substance is added to the medium. Among a dozen such strains, nine may be found which can grow when a substance chemically related to *d*, which we may designate as *c*, is added to the medium, while three cannot. For these three strains, it may be postulated that there is a reaction, $c \rightarrow d$, which has been interfered with in consequence of a mutation. Of the remaining nine strains, four are able to grow not only when *c* is added to the medium but also when a closely related compound, *b*, is added. The other five strains may be postulated to lack the ability to make the conversion $b \rightarrow c$. Of the four remaining strains, one can grow when, instead of *b* or *c*, the related compound *a* is added to the medium. The three strains which cannot grow in the presence of *a* but can when *b* or *c* is supplied must be postulated to lack the ability to effect the transformation $a \rightarrow b$. Finally, for the one remaining strain, which can grow if *a*, *b*, *c*, or *d* is added to the medium, tests of a dozen compounds related to *a* fail to bring to light a single one which can substitute for *a*. In other words, the precursor of *a* cannot be identified. The entire sequence may be written

$$\downarrow \quad \downarrow \quad \downarrow \quad \downarrow$$
$$\rightarrow a \rightarrow b \rightarrow c \rightarrow d \,,$$

the points known to be under genetic control being indicated by the vertical arrows.

Of the many metabolic sequences which have been studied in Neurospora, we will consider only one here. Nicotinic acid is a compound which is an essential vitamin for man. Neurospora, on the other hand, can normally synthesize this compound and so can thrive on a medium free of nicotinic acid. A series of steps has now been worked out whereby Neurospora can synthesize nicotinic acid from shikimic acid. These are shown in Figure 12-1. Each of these steps has been shown to be under genetic control, except for that from tryptophane to kynurenin. However, it should be pointed out that the compound (shikimic acid) indicated as the beginning of that particular chain is already a complicated amino acid, whose synthesis must likewise be assumed to have proceeded through a series of gene-controlled steps.

The transformation of a to a closely related compound b can in many instances be shown to be effected by a particular enzyme; in mutant strains incapable of bringing about this transformation, it can often be demonstrated that the enzyme in question is lacking. It has now been shown with respect to certain enzymes that mutation at any one of several different loci may result in the absence of the enzyme (Bonner, 1952). Accordingly, in the hypothetical sequence worked out above, each step may be under the control of several different genes.

In section 4.1 (p. 17) it was pointed out that certain mutations expressed themselves differently at high and at low temperatures. The study of biochemical genetics has suggested a possible explanation for this. There are

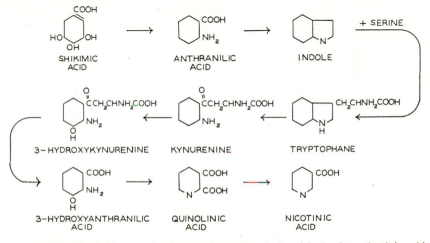

FIG. 12-1.—Probable steps in the transformation of phenylalanine into nicotinic acid by Neurospora. Genetic control has been established for every step indicated except that from tryptophane to kynurenin. (Modified, by permission, from Advances in genetics, chap. 3, by Dr. N. H. Horowitz, copyright 1950, Academic Press.)

now recognized a number of mutant strains in Neurospora which are able to synthesize a given compound, such as adenine or uridine, at a low temperature but not at a higher one. A similar situation could be the underlying basis for the variations in bristle number with temperature change in Drosophila which have been described earlier.

12.2. *The genetic control of certain aspects of food metabolism in man.*—
Turning now to man, the medical literature contains reports of an extensive array of inherited defects in the metabolism of carbohydrates, fats, proteins, and their precursors. The term "inborn errors of metabolism" was aptly applied to these defects many years ago by Garrod in a brilliant essay which

may still be read with profit (1908; see also 1923). More recently our knowledge of these errors has been reviewed by Taggart *et al.* (1950), Rapoport (1950), and, especially, Harris (1953). The following are representative of these inborn errors:

1. Carbohydrates: essential pentosuria (excretion of pentose sugars, such as arabinose or xyloketose, in the urine), and glycogenosis (accumulation of glycogen in the liver, heart, and spleen).

2. Fats: infantile amaurotic idiocy (accumulation of the lipid sphingomyelin in the brain cells), Gaucher's disease (deposition of abnormal amounts of the lipid kerasin in the cells of the reticuloendothelial system), and essential hypercholesterolemia (elevated blood cholesterol).

3. Proteins: cystinuria (marked excretion of amino acids, particularly cystine, in the urine), gout (an abnormality of nucleoprotein metabolism characterized by elevated blood uric acid and the deposition of sodium biurate in the tissues), and congenital porphyria (the presence of increased amounts of the porphyrins in the blood and urine).

In general, we know far less about genetically controlled metabolic sequences in man than in the more easily studied Neurospora. There is, however, one outstanding exception to this statement. This exception encompasses the errors associated with the metabolism of the amino acids, phenylalanine and tyrosine. The normal metabolism of these amino acids is envisaged as follows: The metabolic requirements for phenylalanine and tyrosine are obtained from dietary sources. Any excess phenylalanine is converted to tyrosine for further elaboration or degradation. Thus tyrosine is derived from two sources, namely, the diet and the conversion of phenylalanine. The latter source is known to provide the majority of the metabolic requirements of tyrosine. The latter, which may be viewed as the pivotal compound in this chain of reactions, in turn acts either as a precursor in the formation of proteins, melanin, and certain hormones such as thyroxine and noradrenalin or is degraded to carbon dioxide and water. The actual steps in this process are indicated in Figure 12-2. At the present time three inherited metabolic errors are recognized which interfere with this normal sequence of events; they are albinism, alcaptonuria, and phenylketonuria.

Albinism, due to a more or less complete absence of pigment from the body, has, because of the striking appearance of affected persons, been a recognized disease for many centuries. Normally melanin, the basic pigment of the human body, is derived from tyrosine by a series of complex steps which occur in specialized cells, the melanocytes. The elaboration of melanin in these cells depends upon (1) an enzyme, tyrosinase; (2) a suitable substrate, usually tyrosine or dopa (3,4-dihydroxyphenylalanine); and (3) molecular oxygen. Additional factors, such as temperature and pH, also influence the rate of melanogenesis. The enzyme tyrosinase, which is a copper protein complex, may be identified in pigmented individuals by appropriate histochemical techniques. In the complete albino, on the other hand, while the presence of melanocytes has been established, all efforts to demonstrate the enzyme have been unsuccessful. It seems reasonable to conclude that the typical, complete form of albinism is the result of the absence of the

PHENYLALANINE - TYROSINE METABOLISM IN MAN

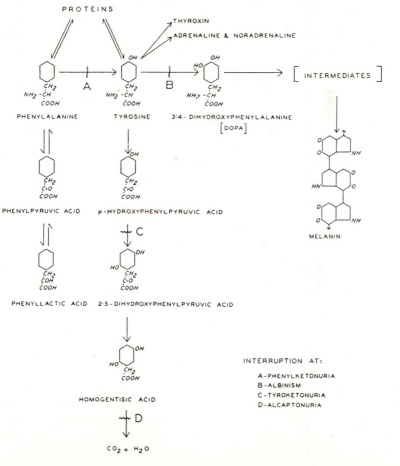

Fɪɢ. 12-2.—The probable phenylalanine-tyrosine metabolic sequence in man. The points at which this sequence appears to be interrupted by known inherited metabolic defects in man are indicated by {. The key to the specific defect involved is given in the lower right-hand corner of the figure. (With modifications after Crowe and Schull, Folia hered. et path., 1:259, 1952.)

enzyme tyrosinase. Furthermore, in visualizing the position of the enzymatic block, it seems appropriate to indicate the error as occurring at the first step which this enzyme is known to catalyze, that is, the conversion of tyrosine to dopa. In addition to the complete form of albinism, which is generally inherited as a simple recessive, other albinotic pigmentary variants are known, such as vitiligo and partial albinism. The biochemical relationship between these forms and the complete form is not known.

The metabolic defect associated with alcaptonuria was described by Boedeker in 1859. Garrod's demonstration in 1901 of an increased incidence of consanguinity among the parents of alcaptonurics led Bateson, in 1902, to suggest that alcaptonuria was inherited as a simple recessive. Alcaptonuria is thus not only the first recessive to be reported in man but the first biochemical mutant to be recognized in any species. In alcaptonuria, large amounts of homogentisic acid (2,5-dihydroxyphenylacetic acid) are present in the body tissues and fluids and in the urine. The oxidation of this acid in urine exposed to air results in a black pigment. This peculiarity generally leads to the recognition of the disease in early infancy. The normal individual when fed large amounts of homogentisic acid is able successfully to reduce this compound to carbon dioxide and water. The alcaptonuric individual is unable to do this, and it is believed that this inability results from the absence of an essential enzyme. Alcaptonurics may, in later life, develop ochronosis, a condition in which the cartilages acquire a black pigmentation, possibly due to the in vivo oxidation of the homogentisic acid. Alcaptonuria has also been reported in the rabbit and can be experimentally induced in mice fed a vitamin C–deficient diet. The congenital form, however, does not respond to vitamin C therapy.

Fölling in 1934 reported the third in the above-mentioned triad of metabolic errors. He pointed out that certain mental defectives excreted in their urine large amounts of another phenylalanine derivative known as phenylpyruvic acid. Studies of these defectives revealed that, in the families in which they occurred, the mental defect and the abnormal urinary constituent were always associated. From the ratio of affected children and the normalcy of the parents, it was soon obvious that both traits were due to one and the same recessive gene; but the relation that the metabolic error bears to the mental defect was and still is not clear. Initially, the error was assumed to be the inability to hydroxylate phenylpyruvic acid. This, however, left unaccounted for the facts that (1) although there are increased amounts of phenylalanine in the blood in this disease, phenylpyruvic acid is not found in the blood of phenylketonurics even on diets rich in this acid, and (2) phenylketonurics are always less pigmented than their normal sibs. Jervis (1947) suggested that the enzymatic block should more properly be represented as the inability to convert phenylalanine to tyrosine. This explanation would account for the high levels of phenylalanine and the absence of phenylpyruvic acid in the blood and would explain the presence of phenylpyruvic acid in the urine as a product of renal metabolism rather than as an unmetabolized normal intermediate. As has been subsequently pointed out, such an enzymatic block would also account for the reduced pigmentation (Lerner and Fitzpatrick, 1950). We have seen that the majority of the metabolic requirements of tyrosine are derived from the conversion of phenylalanine. If this conversion is blocked because of the absence of an enzyme, then the only tyrosine available for melanin and hormonal synthesis is that directly obtained from the diet. Since the latter is inadequate to meet the demands, competition for the available substrate could very easily account for the reduced pigmentation.

One further metabolic error associated with the metabolism of phenylalanine and tyrosine has been reported. In 1932 Medes described a condition called "tyrosinosis" in a middle-aged man with myasthenia gravis. The condition is characterized by large amounts of para-hydroxyphenylpyruvic acid in the urine. That this condition is not common is apparent from the fact that surveys of forty-odd thousand persons, as well as innumerable cases of myasthenia gravis, have failed to reveal another affected individual (Blatherwick, 1934). It has been suggested that this defect may also be an inherited abnormality. On the basis of the available evidence, it would seem premature, however, to accept wholeheartedly this hypothesis. In the single case reported, it is not clear whether or not the defect was congenital, as would be expected in the metabolic errors. Furthermore, it is now known that p-hydroxyphenylpyruvic acid can be found in the urine of individuals with pernicious anemia as well as in the urine of premature infants. In the premature infants, excretion of this acid will continue as long as vitamin C is withheld. The term "tyroketonuria" has recently been suggested for this more generalized presence of the ketone derivatives of tyrosine in the urine (Crowe and Schull, 1952).

Good progress is being made in the analysis of other inherited metabolic defects in man. The interested reader is especially recommended to the recent publications of the Coris (1952) on the nature of the defect in glycogen-storage disease. Most of the metabolic defects thus far studied involve relatively gross departures from normal, such as are attended by lack of pigmentation or the excretion of large quantities of abnormal urinary constituents. The development of rapid methods for analyzing large numbers of blood and urine samples for a variety of constituents through paper chromatography (to be mentioned below) will undoubtedly speed up the tempo of discovery.

12.3. *The genetic control of certain aspects of mineral metabolism.*—The genetically controlled biochemical differences between individuals are not limited to the metabolism of carbohydrate, fat, and protein. There are also differences in the metabolism of minerals and vitamins. One disease in which a derangement of mineral metabolism appears to play a prominent role, Wilson's disease, has already been discussed. A second interesting and inherited derangement of mineral metabolism is hemochromatosis (cf. Granick, 1949; Althausen *et al.*, 1951; Marble and Bailey, 1951; Peterson and Ettinger, 1953). The salient finding in this disease is the presence of large amounts of iron-containing pigment, hemosiderin, in various organs. The liver and pancreas are especially affected; diabetes is a frequent consequence of the deposits of hemosiderin in the liver. There is also a peculiar "bronzed" appearance to the skin. Studies with radioactive iron have suggested that abnormally large amounts of iron are absorbed from the gastrointestinal tract. Apparently, the body lacks a physiological mechanism for ridding itself of this excess iron, and it is deposited in the tissues.

It has been recognized for some years that there is a tendency for the siblings and occasionally a parent of affected individuals to exhibit the disease, the data suggesting that the condition was due to a dominant gene with incomplete penetrance. It has also been known that the disease was much more frequent in males than in females. The studies mentioned above on iron absorption have now supplied a rational explanation for these findings. One of the chief uses of iron in the body is as a component of the hemoglobin of the red blood cell. Blood loss results in iron loss. Because of her menstrual periods, woman has a source of iron loss not present in man and so has higher iron requirements. This is frequently a disadvantage, since if these higher requirements are not satisfied, anemia results. But in the case of women whose genotype predisposes them to the development of hemochromatosis, what is ordinarily a "disadvantage" may actually function as a protective mechanism, i.e., the loss of blood (and iron) through menstruation tends to counterbalance the increased absorption of iron.

Hemochromatosis thus not only illustrates how an inherited biochemical defect has far-reaching implications but also supplies insight into one mechanism which accounts for a different incidence of an inherited disease in males and females. A comparison of what is now known concerning hemochromatosis with our present understanding of Wilson's disease (sec. 6.10, p. 50) brings out another point of interest. In both these diseases there is an overabsorption of a mineral constituent of the diet. But in the case of hemochromatosis, this seems to be due to a primary defect in the mechanism which regulates iron absorption from the intestine, with a resulting high serum iron. In Wilson's disease, on the other hand, it seems that excess absorption of copper may well be a secondary defect, in response to an abnormally low serum concentration of this substance. These two diseases thus illustrate how what at first glance seem to be similar entities may be due to quite different biochemical mechanisms.

12.4. *Human metabolic differences as revealed by the comprehensive study of urine and blood.*—Human urine is the main pathway of excretion for the by-products of the body's chemical processes. A very considerable number of different compounds can be identified in the urine by the appropriate chemical tests. These include various inorganic salts, urea, amino acids, sugars, the degradation products of hormone metabolism, vitamins, etc. We have just considered briefly several metabolic disturbances which affect in striking fashion cer-

TABLE 12-1*

AVERAGE AMOUNTS OF SIX DIFFERENT COMPOUNDS PRESENT
IN URINE OF SIX INDIVIDUALS

COMPOUND	FEMALE		MALE		TWIN	
	A	B	C	D	E	F
Glucose	0.53	0.41	0.41	0.48	0.57	0.46
Alanine.	.084	.13	.065	.054	.064	.041
Glycine	.18	.21	.045	.027	.052	.041
Serine.	.071	.14	.045	.025	.056	.034
Glutamic acid. . .	.034	.013	.012	.026	.017	.012
Lysine.	0.132	0.007	0.035	0.0005	0.014	0.012

* The amounts of all compounds are given in terms of milligrams per milligram of urinary creatinine, the latter being chosen as a base because of the relative constancy of its excretion (data of Williams *et al.*, 1951).

tain urinary constituents. In recent years Dr. Roger Williams and his students (1951) have been exploring techniques suitable for the study of more subtle differences between individuals. The relative and absolute concentrations of a dozen or more substances excreted in the urine are analyzed simultaneously by the use of paper chromatography. It has now been shown that there are wide differences between individuals with respect to the amounts of certain of these compounds which are normally excreted in the urine. Table 12-1 summarizes the results of one recent study comparing six different individuals with respect to six normal urine constituents. Although there is variation with respect to all compounds, some, such as lysine, are much more variable than others. Similar differences have been observed with respect to the constituents of saliva. It is reasonable to assume that these variations reflect differences between individuals with respect to their chemical processes. To what extent these differences reflect variations between individuals in the way their kidneys function in the formation of urine and to what extent the differences are more basic is not now clear.

It is well known that diet may influence the nature of urinary excretion products. That dietary differences between individuals are not entirely responsible for the observed urinary variations is suggested by the fact that in the white rat, whose diet can be much better standardized than can man's, such variations are still observed. It follows that these variations probably are at least in part the result of innate metabolic differences between individuals. Thus far, genetic studies in this field are in their early stages. The last two columns in Table 12-1 summarize the findings on a pair of identical twins. It will be noted that their results are more similar than any other pair of results in the table, a finding which, of course, suggests the importance of genetic factors. Perhaps the most pertinent observations to date are those of Sutton and Vandenberg (1953), who studied the composition of the urine with respect to 19 organic compounds or groups of compounds in 16 sibling pairs, inmates of an institution for the feeble-minded. An analysis of these data revealed significant intra-pair resemblances with respect to 9 of the compounds. Since the urines were all collected within a brief period of time from an institutionalized population, the possible effects of different diets on these particular urinary constituents were minimized. It was the opinion of these investigators that when all 19 substances were considered simultaneously, at least three different patterns of excretion could be recognized. Further work along these lines promises to be exceedingly fruitful.

The possibility that these chemical differences may be correlated with measurable physical and psychological traits is being explored (Williams *et al.*, 1951). When a group of 7 overweight and 10 underweight individuals was studied with respect to 14 different substances in the urine, it was found that the urinary calcium and phosphate excretion of the obese tended to be higher than that of the underweight. Both magnesium and a variety of amino acids appear to be excreted in increased amounts in schizophrenics. Finally, a group of mentally deficient children appeared to excrete significantly more methionine sulfoxide, tyrosine, and proline than did normal children. It is not clear to what extent these differences are related to a causative mechanism and to what extent they are secondary, but they certainly indicate interesting paths for further exploration.

The blood stream, as the principal avenue of transport within the body, also reflects in many ways the metabolic activities of the body. Just as a study of urinary constituents may be expected to yield particular evidence concerning human differences in metabolic end-products, so the **study of blood** will yield special information concerning various intermediate metabolic compounds. Thus far no less than 150 compounds have been described and their exact concentration in the blood stream determined (Albritton, 1952). Many more compounds have been identified but not yet studied quantitatively. Little has been done in the way of studies on individual differences with respect to these various compounds and the genetic significance thereof.

12.5. *Complexities in the study of the biochemical mutants of man.*—Because of the complex nexus of metabolic interrelationships which exist in the human body, the study of the genetic nature of biochemical differences between individuals must be approached with caution. An increased amount of a given metabolite in either blood or urine may arise in several different ways, and genetic studies which fail to take this into account are of very limited value. The present and growing complexity of the situation is very well illustrated by what is now known concerning the excretion of the amino acid cystine in the urine (summary in Dent and Harris, 1951; Harris and Warren, 1953; Jackson and Linder, 1953). This substance is normally present in the blood plasma in small amounts and passes in the kidneys into the glomerular filtrate but is largely reabsorbed in the kidney tubules. Accordingly, in normal persons there is only a small amount of cystine in the urine. Increased amounts of cystine may be excreted under a variety of conditions.

1. We recognize, to begin with, individuals with the classical and most frequently encountered type of cystinuria, "essential cystinuria," in whom there is a tendency, because of the presence of excessive amounts of cystine in the urine, toward the formation of renal stones composed of this material. These individuals also excrete the amino acids lysine, ornithine, and arginine in excessive amounts. There also appears to be a slight increase in the isoleucine output, but a decrease in the amount of taurine excreted (Stein, 1951). The blood levels of these amino acids seem to be normal in such individuals, the defect therefore appearing to reside in the kidney.

2. Second, there are individuals who excrete cystine in relatively large amounts as part of a generalized tendency toward the excretion of increased quantities of many amino acids in the urine. Two subgroups are recognized. In individuals falling into the first of these subgroups there are evidences of severe renal dysfunction, with secondary rickets. The blood amino acid levels have not been found to be elevated, and the primary defect appears to be largely localized in the kidney (Fanconi's syndrome). The possibility that in this disease other biochemical mechanisms for dealing with cystine may also be deranged is suggested by the occurrence in some affected individuals of deposits of cystine in the tissues. The second subgroup is composed of indi-

viduals with Wilson's disease, where, as noted earlier, in addition to other findings, there is a generalized tendency toward the excretion of amino acids, including cystine, in the absence of any marked elevation of plasma amino acids. There is also an increased excretion of peptides with a terminal dicarboxylic amino acid residue; and it has been suggested that aminoaciduria is actually secondary to this, i.e., that in the renal tubules there is "competition" between peptides and amino acids for reabsorption, so that in the presence of abnormal amounts of peptides in the tubule, amino acid reabsorption is interfered with, in consequence of which increased amounts of amino acids appear in the urine (Uzman and Hood, 1952).

3. Third, there are persons in whom an excess of cystine and other amino acids in the urine is clearly a nonspecific consequence of severe hepatic, renal, or muscle disease. In persons with liver disease, for instance, there are increased amounts of amino acids in the plasma, and the increase in the urine is a secondary phenomenon, the kidney tubule being unable to reabsorb the excess amino acids that are present in the glomerular filtrate.

"Essential cystinuria," which has a frequency of about 1 in 600, definitely has a genetic basis. In some families the parents of affected persons excrete increased amounts of cystine. In other families the parents are normal, but there is an increased amount of consanguineous marriage among the parents in comparison with expectation for a recessive trait with a frequency of 1:600. This suggests that there are at least two forms, one inherited as an incomplete recessive, the other as a "true" recessive. Fanconi's syndrome, the second type of cystinuria discussed, appears to be due to a recessive gene, as does Wilson's disease. There are thus at least four different genetic entities of which cystinuria is one manifestation, in addition to the cystinuria of severe chronic liver or renal disease, which is, for the most part, the result of environmental factors. There seems little reason to doubt that the genetic basis for many other biochemical abnormalities will prove equally or even more complex.

12.6. *The significance to genetics of the concept of "alternate metabolic pathways."*—There is no estimate of the number of different organic compounds essential to the proper functioning of the average cell. The number is over 100, and very possibly in excess of 1,000. If the appearance of each compound is preceded by a series of reactions, such as those depicted in Figures 12-1 and 12-2, the complexity of the biochemical system which is under genetic control can be dimly visualized. In this connection it should be pointed out that modern biochemistry has developed the concept of "alternate metabolic pathways," wherein it is postulated that a given biochemical goal may be reached by any one of several different routes (cf. Potter and Heidelberger,

1950). By the same token, a particular compound may be the starting point for a number of different reactions. We must think of the genetically controlled reactions which take place in a cell not as a series of straight-line reactions, as depicted in Figure 12-1, but as a complex web of interrelated reactions which may be diagrammatically represented as in Figure 12-3. If, for instance, the transformation $a \rightarrow b$ pictured in the diagram represented a genetically controlled step in the synthesis of one of the nonessential amino acids, then a mutation affecting the quality or quantity of b could have an effect upon every protein in the body of which b was a constituent.

Furthermore, the possibility must be recognized that in the synthesis or degradation of a certain compound, apparently normal human beings may follow quite different pathways. Viewed in this light, the observation that what appears clinically to be one and the same disease has different modes of inheritance in different families becomes not only readily understandable

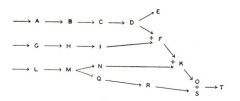

Fig. 12-3.—A diagram of the types of interrelated chemical reactions which may be postulated to occur in the cellular biochemistry of man.

but, indeed, the expected finding. It is a well-known fact that some individuals show rather marked reactions to small amounts of what is ordinarily a well-tolerated medication (drug idiosyncrasy). One wonders to what extent such reactions reflect inherited biochemical differences between individuals, which are brought to light only under these special circumstances. Conversely, systematic studies of the ability of individuals to metabolize a variety of chemical compounds not ordinarily (or only rarely) encountered in nature could shed light on unsuspected inherited biochemical attributes.

12.7. *The complex consequences of single-gene substitutions.*—Thus far we have focused attention on the genic control of biochemical reactions, without emphasizing the ramifications of these biochemical abnormalities, although some of the ramifications have been mentioned in passing. It is of interest now to consider just how a genetically determined biochemical abnormality may have far-reaching physiological consequences for an organism. The preceding section has pointed out the theoretical basis for this. Diabetes mellitus constitutes an actual case in point. It seems well established that this disease is often genetically determined, although the exact genetic mechanism is not clear. The fundamental abnormality consists of the failure of the cells of the islands of Langerhans in the pancreas to secrete the hormone insulin in sufficient amounts. The chain of consequences—or, as Grüneberg (1947) has termed it, the "pedigree of causes"—set in motion by this failure is depicted in Figure 12-4 (cf. Anderson, 1952). While this figure fails to give the complete story, sufficient is shown to enable the student to appreciate the complex consequences of failure of production of a single hormone. The uninitiated often think of the findings in diabetes mellitus as due to a simple failure of the body to burn sugar. It is important to point out

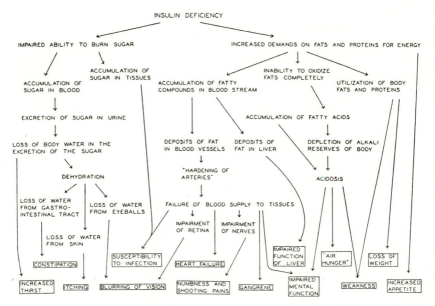

Fig. 12-4.—Some of the physiological consequences of the relative deficiency of insulin which occurs in diabetes mellitus.

in connection with Figure 12-4 that some of the most important findings in the disease are due to the efforts of the body to compensate for the loss of energy from sugar catabolism by increased catabolism of proteins and fats. This is, in a sense, the utilization of alternate metabolic pathways, as has been discussed above. Such an effort is successful up to a certain point, but beyond that the demands are excessive, and various serious consequences follow.

A second example of the far-reaching consequences of a simple defect can be drawn from the field of hematology. As mentioned in section 3.4 (p. 15), when the red blood cells of certain individuals are subjected to conditions of reduced oxygen concentration, they send out long projections, a phenomenon known as "sickling." Although most persons with this condition appear to suffer no ill consequences, in about every fortieth person who exhibits the sickling phenomenon there is a severe anemia, together with poor physical development, joint, muscular, and abdominal pains, and occasionally paralysis, as well as other findings. This condition is referred to as "sickle-cell disease," the mild, asymptomatic state being called the "sickle-cell trait." The sickling phenomenon is inherited, the sickle-cell trait being due to the heterozygous state for a gene which, when homozygous, is responsible for the serious sickle-cell disease (Neel, 1951). In persons with the sickle-cell trait it does not appear that the sickling phenomenon often, if ever, occurs under conditions encountered in the body. In sickle-cell disease, however, the cells actually sickle while in circulation, and the sickling is of a more extreme type. The two types of sickling are illustrated in Figure 12-5. Such sickle cells are de-

Fig. 12-5.—The sickling phenomenon as it is seen in individuals (A) heterozygous and (B) homozygous for the responsible gene.

stroyed much more rapidly than normal and, at the same time, tend to clump together and interfere with normal circulation. The results are depicted in Figure 12-6. Again, as for diabetes, the ultimate consequences are far removed from the original defect.

Recent studies, which have been reviewed by Itano (1953), have thrown important light on the biochemical basis for the sickling phenomenon. When an electric current is passed through a solution of protein, the protein molecules will slowly migrate toward either the anode or the cathode, the direction of the migration being determined by the pH of the solution. It has now been shown that in a charged field the hemoglobin molecules from patients with sickle-cell disease have a different rate of migration than do normal hemoglobin molecules, this being an indication of a biochemical difference between the two kinds of molecules. When the hemoglobin from a patient with the disease is broken down into heme and globin, the globin exhibits the same abnormal behavior as that observed in the intact molecule. The exact nature of the difference between the globin of normal hemoglobin and that of sickle-cell disease is not clear at present. A very exacting chemical analysis fails to reveal any clear difference in the amino acids which compose the two globins, although there is a suggestion that the number of sulfhydryl groups may differ in the two globins. The tendency at present is to regard the difference between the two globins as due to the way the constituent atoms are oriented in space rather than to qualitative chemical differences. Studies with the phase and electron microscopes on normal and sickle cells suggest that the sickling phenomenon is due to a tendency of the hemoglobin molecules, when in the "reduced" state due to a low oxygen tension, to arrange themselves in long chains which then align themselves side by side. Such is the

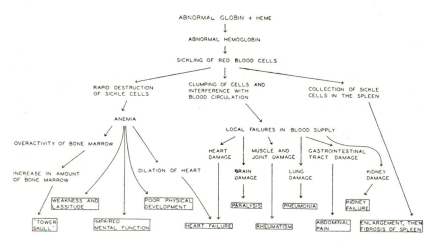

Fig. 12-6.—The chain of events which results from the presence of abnormal globin in sickle-cell disease.

"strength" of this tendency that the chains of abnormal molecules as they form actually distort the red-cell membrane, pushing it before them. In sickle-cell disease 80–100 per cent of the hemoglobin molecules are of this abnormal type, whereas in the sickle-cell trait only 25–45 per cent of the molecules are abnormal. Apparently, at these lower concentrations of abnormal hemoglobin, the tendency to form chains is not sufficiently strong to have serious consequences. Here, then, is an illustration of the genetic control of the structure of an important body protein. This is the first clear demonstration of the genetic control of protein structure, undoubtedly due in no small measure to the fact that this particular protein is so readily accessible and has such striking effects. It is a reasonable conjecture that, now that the path has been cleared, genetic variations in other proteins will soon be discovered.

12.8. *Unanswered questions.*—Although there is no doubt that in both plants and animals, including man, genes can and do act through the control of complex, intracellular biochemical reactions, there are still many unanswered questions concerning the details of this control. Among these are the following:

(1) What is the mechanism of the genetic control of biochemical reactions?

(2) What proportion of genetic effects is ultimately explicable in terms of biochemical deviations?

(3) Under what circumstances may biochemical genetic effects be reversed?

The first question, as to the mechanism of the genetic control of biochemical reactions, has excited a great deal of speculation. The rate of most biochemical reactions is under enzymatic control. For instance, the breakdown of foodstuffs in the digestive tract is due to a sequence of enzymatically controlled reactions. Genes have been demonstrated in some instances to act by controlling the amount of a particular enzyme present. Thus, as we have already seen, diabetes is due to a deficiency in a circulating enzyme, insulin. Some enzymes which cells produce are readily studied because they are secreted extracellularly in relatively large amounts by the cells in question. Other enzymes are much more difficult to study because they are intracellular in nature and because procedures which destroy the cell so as to render the enzyme accessible for study are liable to alter the amount and perhaps nature of the enzyme present. For this reason, the study of the relation between gene and intracellular enzyme proceeds slowly. But sufficient is already known to lead to the suggestion that some genes are actually single enzyme molecules or enzyme precursors. According to this hypothesis, gene reproduction results in the formation of two enzyme molecules where one existed previously. If this reproduction occurs during preparation for a cell division, the "new" molecule may be incorporated into the forming continuum, which is to become a chromosome. But if the molecular reproduction occurs during the interphase, the "new" molecule may diffuse out into the cytoplasm, there to catalyze a certain reaction; or the molecule may even escape from the cell entirely, to enter the circulation as a hormone or the digestive tract as a digestive enzyme.

The second question, as to the proportion of genetic effects ultimately explicable in terms of biochemical deviations, can only be a subject for speculation at present. The evidence that genes act through the control of biochemical reactions increases constantly, and it is tempting at this stage in our knowledge to infer that they *always* act that way. According to this viewpoint, such inherited morphological variations as polydactyly or brachydactyly are ultimately due to biochemical differences between individuals.

While this may prove to be the case, there is at present insufficient evidence for a sweeping generalization. On the other hand, it must be admitted that whenever it has been possible to subject to close scrutiny the ultimate effect of a gene, this effect has proved to be a biochemical one, and at the present time there is no other channel known through which genes can and do operate.

We turn now to the third question, the reversibility of genetically controlled biochemical defects. The crux of the matter here is the site of gene action. The pathological effects of some genes are manifest at a site distant from the original reaction. Thus in hereditary diabetes, when the islands of Langerhans fail to function properly, the consequences are felt by many tissues because of the endocrine function of the islands of Langerhans. When, in gout, there is a defect in uric acid metabolism, resulting in an excess of this substance in the blood stream, the consequences are urate deposits at various sites in the body. In pernicious anemia the primary hereditary defect is in the lining of the stomach, in consequence of which there is failure to absorb a substance necessary to the formation of red blood cells in the bone marrow. In all these cases, because the site of final action is distant from the site of the primary defect, the opportunity exists for the physician to intercede and in whole or part reverse the usual consequences of the genetic defect. Where, however, the consequences of a genetic defect are entirely intracellular, so that the cell in which a particular reaction fails is the one that suffers, then the opportunities for reversing the defect are more limited. Every living cell is bounded by a cell membrane which controls the entry and exit of various substances into and out of the cell. If the cell membrane will permit the passage into the cell of a compound normally formed within the cell, then the usual consequences of a biochemical defect may be reversed by appropriate substitution therapy. But if the cell membrane will not permit the passage into the cell of a substance normally formed there, but now deficient because of a genetic defect, then substitution therapy is impractical. However, the functioning of cell membranes is itself subject to various influences. It is accordingly too early to adopt an attitude of therapeutic defeatism toward the strictly "intracellular genetic effects."

Bibliography

SPECIFIC REFERENCES

ALTHAUSEN, T. L.; DOIG, R. K.; WEIDEN, S.; MOTTERAM, R.; TURNER, C. N.; and MOORE, A. 1951. Hemochromatosis: investigation of 23 cases, with special reference to etiology, nutrition, iron metabolism, and studies of hepatic and pancreatic function, Arch. Int. Med., **88**:553–70.

ANDERSON, C. E. 1952. The relation of some recent biochemical investigations to diseases of intermediary metabolism, North Carolina M. J., 13:189–96.

BLATHERWICK, N. R. 1934. Tyrosinosis: a search for additional cases, J.A.M.A., 103:1933.

CORI, G. T., and CORI, C. F. 1952. Glucose-6-phosphatase of the liver in glycogen storage disease, J. Biol. Chem., 199 (Part 2): 661–67.

CROWE, F. W., and SCHULL, W. J. 1952. Phenylketonuria: studies in pigment formation, Folia hered. et path., 1:259–68.

DENT, C. E., and HARRIS, H. 1951. The genetics of "cystinuria," Ann. Eugenics, 16:60–87.

FÖLLING, A. 1934. Über Ausscheidung von Phenylbrenztraubensäure in den Harn als Stoffwechselanomalie in Verbindung mit Imbezillität, Ztschr. f. physiol. Chem., 227:169–76.

GARROD, A. E. 1908. The Croonian Lectures on inborn errors of metabolism, Lancet, 2:1–7, 73–79, 142–48, 214–20.

GRANICK, S. 1949. Iron metabolism and hemochromatosis, Bull. New York Acad. Med., 25:403–28.

HARRIS, H., and WARREN, F. L. 1953. Quantitative studies on the urinary cystine in patients with cystine stone formation and in their relatives, Ann. Eugenics, 18:125–71.

ITANO, H. A. 1953. Human hemoglobin, Science, 117:89–94.

JACKSON, W. P. U., and LINDER, G. C. 1953. Innate functional defects of renal tubules, with particular reference to Fanconi syndrome: cases with retinitis pigmentosa, Quart. J. Med., 22:133–56.

JERVIS, G. A. 1947. Phenylpyruvic oligophrenia, Arch. Neurol. & Psychiat., 38:944–63.

LERNER, A. B., and FITZPATRICK, T. 1950. Biochemistry of melanin formation, Phys. Rev., 30:91–126.

MARBLE, A., and BAILEY, C. C. 1951. Hemochromatosis, Am. J. Med., 11:590–99.

MEDES, G. 1932. A new error of tyrosine metabolism: tyrosinosis; the intermediary metabolism of tyrosine and phenylalanine, Biochem. J., 26:917–40.

NEEL, J. V. 1951. The inheritance of the sickling phenomenon, with particular reference to sickle cell disease, Blood, 6:389–412.

PETERSON, R. E., and ETTINGER, R. H. 1953. Radioactive iron absorption in siderosis (hemochromatosis) of liver, Am. J. Med., 15:518–24.

POTTER, V. R., and HEIDELBERGER, C. 1950. Alternative metabolic pathways, Phys. Rev., 30:487–512.

STEIN, W. H. 1951. Excretion of amino acids in cystinuria, Proc. Soc. Exper. Biol. & Med., 78:705–8.

SUTTON, H. E., and VANDENBERG, S. G. 1953. Studies on the variability of human urinary excretion patterns, Human Biol., 25:318–32.

UZMAN, L. L., and HOOD, B. 1952. The familial nature of the amino-aciduria of Wilson's disease (hepatolenticular degeneration), Am. J. M. Sc., 223:392–400.

WILLIAMS, R. J., *et al.* 1951. Individual metabolic patterns and human disease: an exploratory study utilizing predominantly paper chromatographic methods. ("University of Texas Publications," No. 5109.) Austin.

GENERAL REFERENCES

ALBRITTON, E. C. (ed.). 1952. Standard values in blood. Philadelphia: W. B. Saunders Co.

BEADLE, G. W. 1945a. Biochemical genetics, Chem. Rev., 37:15–96.

———. 1945b. The genetic control of biochemical reactions, Harvey Lect., ser. 40, pp. 179–94.

———. 1946. Genes and the chemistry of the organism, Am. Scientist, 34:31–53.

———. 1949. Genes and biological enigmas, Science in Progress, 6th ser., pp. 184–249.

BONNER, D. M. 1952. Gene-enzyme relationships in Neurospora, Cold Spring Harbor Symp. Quant. Biol., 16:143–57.

BRICK, I. B. 1952. The clinical significance of aminoaciduria, New England J. Med., 247:635–44.

GARROD, A. E. 1923. Inborn errors of metabolism. 2d ed. London: H. Frowde and Hodder & Stoughton.

GRÜNEBERG, H. 1947. Animal genetics and medicine. New York: Paul B. Hoeber, Inc.

HARRIS, H. 1953. An introduction to human biochemical genetics. ("Eugenics Laboratory Memoirs," Vol. XXXVII.) London: Cambridge University Press.

HOROWITZ, N. H. 1950. Biochemical genetics of Neurospora, Adv. Genetics, 3:33–71.

RAPOPORT, M. 1950. Inborn errors of metabolism. *In:* NELSON, W. E. (ed.), Textbook of pediatrics. Philadelphia: W. B. Saunders Co.

TAGGART, J. V.; BONNER, D. M.; GOLDSTON, M.; MASON, H. M.; PINES, K. L.; and HEATH, F. K. 1950. Inborn errors of metabolism, Am. J. Med., 8:90–105.

TATUM, E. L. 1946. Induced biochemical mutations in bacteria, Cold Spring Harbor Symp. Quant. Biol., 11:278–84.

———. 1949. Amino acid metabolism in mutant strains of microorganisms, Federation Proc., 8:511–17.

WRIGHT, S. 1941. The physiology of the gene, Phys. Rev., 21:487–527.

CHAPTER 13

The Estimation of Genetic Parameters and Tests of Genetic Hypotheses

THE STUDY of human heredity is intimately dependent upon a sound statistical approach. This stems from a number of sources, but primarily from the fact that populations of human beings are essentially nonexperimental populations, i.e., they are not subject to manipulation by the investigator. As a consequence, in human genetics one must deal with collections of data which, from the standpoint of the experimental geneticist, leave much to be desired. These collections of data are largely unintelligible unless they can be reduced to, at most, a few numbers or facts which summarize the pertinent information. In the majority of cases the reduction of these collections of data poses either one or both of two major types of problems. One class of problems arises from the need to estimate from a set of data one or more parameters upon which the observed set of events is assumed to be functionally dependent. Otherwise stated, it is frequently desirable to be able to calculate from a set of data a quantity, termed an "estimate" or a "statistic," which may be taken as the value of a numerically unknown quantity, called a "parameter," characteristic of the population from which these data are assumed to be drawn. This is the general situation considered in the "estimation of genetic parameters." The second class of problems consists of determining the conformity of a set of observations to some genetic hypothesis. That is to say, the second class involves determining whether some known or assumed value of a genetic parameter is acceptable in the light of a particular set of observations. This is the general situation considered in "tests of genetic hypotheses." This distinction between the estimation of genetic parameters and the testing of genetic hypotheses is largely one of convenience, since in practical problems one is often the forerunner of the other.

The estimation of genetic parameters and the testing of genetic hypotheses are, in reality, only the application to genetics of the much more general statistical theories of estimation and tests of significance. Much of the present development of these latter theories is due to R. A. Fisher, on whose contributions we shall draw liberally in the succeeding pages. It will be our purpose in this chapter to present briefly the general problems of estimation and tests of significance as they impinge upon human heredity. A complete treatment of the diverse problems of estimation which confront the human geneticist is impractical here. We shall therefore limit our attention to those parameters most frequently estimated. No attempt will be made to prove the general formulas used.

13.1. *The general problem of the estimation of genetic parameters.*—The most frequent problems of estimation which arise in human genetics are, in the order of their occurrence, (1) the estimation of the frequency of a gene or genes in a population, (2) the estimation of the proportion of affected individuals from a given type of mating, and (3) the estimation of the frequency with which recombination occurs between linked genes. While in each of these instances the genetic parameter to be estimated is different, a statement of the underlying problem remains essentially the same, namely: We are given or we assume that a population is distributed in some form and that we could completely specify the distribution of individuals within this population if the value of some parameter (or parameters) were known. Available is a set of observations, a sample, from which we are required to calculate some value which may be taken as the true value of this unknown parameter.

As an illustration, suppose we wished to estimate the frequency of a gene a of a pair of alleles, A and a, lacking dominance. We recognize three events— the phenotypes AA, Aa, and aa—and we assume that these phenotypes are distributed according to the Hardy-Weinberg law, that is, trinomially. By hypothesis, the frequencies of these three phenotypes are p^2 (AA), $2pq$ (Aa), and q^2 (aa); and this population could be completely described if the value of p, or q, were known, where p is the frequency of the gene A and $q = 1 - p$ is the frequency of the gene a. Imagine, now, that among a randomly selected sample of N individuals n_{AA} were observed to be of type AA, n_{Aa} of type Aa, and n_{aa} of type aa. The problem of estimation consists in determining the value of q (or p) from the observed frequencies of AA, Aa, and aa. In this particular instance, a number of different possible estimates of q can be constructed. To cite but two, q might be estimated by either of the following methods: (1) Since the expected number of aa individuals is Nq^2, we may take as an estimate of q^2 the observed proportion of aa individuals and hence as an estimate of q the square root of this proportion, or $\sqrt{n_{aa}/N}$. (2) Among the N individuals observed, there are $2N$ genes at the locus in question. Of these $2N$ genes, $(n_{Aa} + 2n_{aa})$ are known to be of type a because of the phenotypes of the individuals. Therefore, an estimate of q is provided by $(n_{Aa} + 2n_{aa})/2N$. These two methods of estimating q will yield numerically equal results only if the product of the square roots of the frequencies of the two homozygous classes is equal to half the frequency of the heterozygous class. While this could obtain if the population were in equilibrium, the inescapable errors of sampling will generally cause these two methods to give rise to different numerical values from the same set of data. It becomes important, therefore, that we know which of the methods is the preferable one. To determine this, let us examine further the basic problem of estimation.

We note, first, that estimates are derived from samples. But any single sample may be quite improbable, and the estimate derived from it is, of necessity, a very poor indication of the true value of the parameter. It follows that the only infallible method of determining the value of a parameter involves a study of the entire population. This is generally neither feasible nor possible. What is desired, then, is a method of estimation which "on the whole" or "in the long run" will provide good estimates. But what precisely is a good estimate?

The first and minimal requisite of a good estimate is that, when derived from the entire population, it should be equal to the true value of the parameter, that is to say, the estimate should, in fact, be an estimate of the parameter in question. This requirement has been termed the "criterion of consistency." All estimates which are logically deduced from the theoretical distribution of the classes in the population may be expected to possess this prerequisite. Thus both the estimates of q in the example are consistent estimates. But since these estimates are not necessarily numerically equal, it is apparent that some additional criterion is needed, one which will permit the selection of the most appropriate estimate when more than one consistent estimate exists.

The second requisite of a good estimate concerns its precision, that is, how exact it is in affording an estimate of the parameter. Obviously, the best possible approximation, or the estimate with the greatest precision, is that estimate with the smallest error or variance. This requirement of minimum variance has been called the "criterion of efficiency." It can be shown that, for the vast majority of estimation problems, the so-called "method of maximum likelihood" gives rise to estimates having variances as small as, or smaller than, estimates derived by any other method. As an illustration of an exceptional case, the reader is referred to Smith and Haldane (1947). In the example previously given, the standard errors of the two estimates are $\sqrt{(1 - q^2)/4N}$ and $\sqrt{q(1 - q)/2N}$, respectively. The latter is the smaller, as is graphically illustrated in Figure 13-1. As we shall subsequently see, the latter estimate is the maximum-likelihood estimate and the most efficient estimate of q. In the succeeding pages we shall adopt the convention of indicating an efficient estimate by an asterisk (*) and an inefficient estimate by a prime (′).

An alternative way of thinking of efficiency is as follows: When a sample is drawn with the aim of estimating a parameter, we may visualize each successive observation as offering some constant amount of information about the parameter. Moreover, we may imagine that a given method of estimation will utilize all or only a portion of this information. Intuitively, we know that the precision of an estimate should increase as the amount of information which it utilizes increases; and so it may be shown to do. Since the most precise estimate in terms of the variance is the maximum-likelihood estimate,

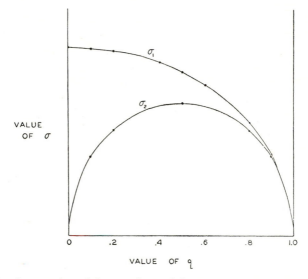

VALUE OF σ

VALUE OF q

Fig. 13-1.—A comparison of the magnitude of the standard errors associated with two different, consistent estimates for the case of a single pair of alleles lacking dominance.

it is convenient to define the total amount of information in a given sample with respect to a particular parameter as the inverse of the variance of the maximum-likelihood estimate. This has led to naming the amount of information the "invariance." The ratio of the amount of information utilized by a specific estimate to the amount of information yielded by the maximum-likelihood estimate affords a means of evaluating the comparative efficiencies of several different estimates.

13.2. *The method of maximum likelihood.*—The more common genetic situations for which estimates are desired are (1) for the estimation of gene frequency under the assumption of differing modes of inheritance, where the data consist of collections of unselected and unrelated individuals, or (2) for situations which necessitate introducing additional parameters, such as rates of inbreeding, selection, or mutation, into the equation of estimation. It is obvious that an exhaustive treatment of even one, much less both, of these classes of situations is impracticable. We shall therefore confine our attention to the following: (1) the simpler modes of inheritance and (2) the estimation of gene frequencies when rates of inbreeding are known. To derive estimates under these differing circumstances, we shall assume that the conditions for the Hardy-Weinberg law are fulfilled (cf. sec. 7.8, p. 68).

As has been stated, the most satisfactory method of estimation is the so-called "method of maximum likelihood." The foundations of and a description of the properties possessed by estimates derived by this method are due to R. A. Fisher (1922, 1925). This method has two main recommendations: (1) the intuitively appealing notion that the best estimate of a parameter is that value which gives the greatest probability (likelihood) to an observed set of events, and, more important, (2) the very desirable properties of consistency and efficiency which maximum-likelihood (M.L.) estimates almost always possess.

Briefly, the method of M.L. is as follows: Suppose that x is a random variable with a distribution dependent solely upon q. We may indicate this symbolically by $f(x; q)$. If we now assume that a sample of size n is to be randomly drawn from a population of x's, then the likelihood, say L, of obtaining a given sample will be

$$L = f(x_1; q) \, f(x_2; q) \ldots f(x_n; q) = \prod_{i=1}^{n} f(x_i; q) . \quad (13.2.1)$$

This probability, which is functionally dependent solely upon q, is called the "likelihood function." Now the principle of maximum likelihood states that the best estimate of q is afforded by the real positive root, if one exists, of the solution of

$$\frac{dL}{dq} = 0 ; \quad (13.2.2)$$

or, since L must be positive or at least not negative, the estimate of q may be obtained from the solution of

$$\frac{d(\log L)}{dq} = 0 . \quad (13.2.3)$$

The latter is generally the more convenient form to use.

The variance of this estimate is obtained by equating the expected value of the second derivative of the likelihood function to $-1/\sigma^2$, that is

$$\sigma^2 = -E \left(\frac{d^2 L}{dq^2} \right)^{-1} , \quad \text{or} \quad -E \left(\frac{d^2 \log L}{dq^2} \right)^{-1} , \quad (13.2.4)$$

where E is a general symbol for "expected value." As an alternative, one may calculate the variance from the following relationship:

$$\sigma^2 = \left\{ \sum_i \left[\frac{1}{e_i} \left(\frac{d\,e_i}{d\,q} \right)^2 \right] \right\}^{-1},$$ (13.2.5)

where e_i is the expected number in the ith class. In terms of information, the total amount of information with respect to q, I_{qq}, is defined as

$$I_{qq} = -E\left[\frac{d^2\,(\log L)}{d\,q^2} \right] = \sum_i \left[\frac{1}{e_i} \left(\frac{d\,e_i}{d\,q} \right)^2 \right].$$ (13.2.6)

13.3. *The estimation of gene frequencies.*—We shall examine, by way of illustration, two different problems of estimation and their solution by the method of M.L. These problems are the estimation of one and of two parameters from a sample. In each case the theory will be that immediately applicable to discrete distributions such as obtain for the distribution of phenotypes with respect to a discontinuous characteristic.

13.3.1. *The case of one parameter.*—In the one-parameter case we assert that a population could be completely specified if the numerical value of one parameter were known. Such is the case whenever genetic hypotheses involving but a single pair of alleles are considered. We shall examine four different types of genetic hypotheses involving one pair of genes, namely, an autosomal pair without dominance and one with dominance, and a sex-linked pair without dominance and one with dominance.

a) TWO AUTOSOMAL ALLELES WITH NO DOMINANCE

As an example, let us return to the problem posed in section 13.1, namely, estimating the frequency q of a gene a of a pair of alleles, A and a, lacking dominance. Given are N observations of which n_{AA} are of type AA, n_{Aa} are of type Aa, and n_{aa} of type aa. If we assume that the Hardy-Weinberg law obtains, then the probability of obtaining precisely n_{AA}, n_{Aa}, and n_{aa} individuals of the various phenotypes is given by that term in the expansion of the trinomial $(p^2 + 2pq + q^2)^N$, where the exponent of p^2 is n_{AA}, of $2pq$ is n_{Aa}, and of q^2 is n_{aa}, that is,

$$L = \frac{N!}{n_{AA}!\,n_{Aa}!\,n_{aa}!}\,(p^2)^{\,n_{AA}}\,(2\,p\,q)^{\,n_{Aa}}\,(q^2)^{\,n_{aa}}.$$ (13.3.1)

If the log of both sides of equation (13.3.1) is taken, then

$$\log L = \log k + n_{AA} \log p^2 + n_{Aa} \log 2pq + n_{aa} \log q^2,$$

where $k = N!/n_{AA}!\,n_{Aa}!\,n_{aa}!$. Now $\log L$ can be maximized by taking the derivative with respect to q and setting this derivative equal to zero. If we differentiate with respect to q, recalling that $p = 1 - q$, then

$$\frac{d\,(\log L)}{d\,q} = \frac{(2\,n_{aa} + n_{Aa})}{q} - \frac{(2\,n_{AA} + n_{Aa})}{1 - q} = 0. \qquad (13.3.2)$$

And for q we obtain the estimate

$$q^* = \frac{n_{Aa} + 2\,n_{aa}}{2\,N}; \qquad (13.3.3)$$

or, if we had differentiated with respect to p, we would have obtained

$$p^* = \frac{2\,n_{AA} + n_{Aa}}{2\,N}. \qquad (13.3.4)$$

The M.L. estimates of p and q are equal, therefore, to the frequency of the appropriate homozygous class plus half the frequency of the heterozygotes. In brief, the M.L. estimates are equivalent to simple gene counts in this case.

We determine the variance of q^* by differentiating equation (13.3.2), replacing the observed values in the various classes by the expected values, and equating the latter to $-1/\sigma^2$, thus:

$$\frac{d^2\,(\log L)}{d\,q^2} = - \left[\frac{2\,n_{AA} + n_{Aa}}{(1 - q)^2} + \frac{n_{Aa} + 2\,n_{aa}}{q^2} \right].$$

The expected values of AA, Aa, and aa are Np^2, $2Npq$, and Nq^2, respectively. When we replace n_{AA} by Np^2, n_{Aa} by $2Npq$, and n_{aa} by Nq^2, we find

$$\frac{d^2\,(\log L)}{d\,q^2} = - \frac{2\,N}{q\,(1 - q)};$$

hence

$$\sigma^2_{q*} = \frac{q\,(1 - q)}{2\,N}. \qquad (13.3.5)$$

In Table 13-1 are given the MN blood-type frequencies observed among 181 American Negroes as reported by Landsteiner and Levine (1929). If

TABLE 13-1*

The MN Blood Antigens of 181 American Negroes

	MM	MN	NN	Total
No. of individuals.	50	86	45	181
Proportion........	0.2762	0.4751	0.2486	0.9999

* After Landsteiner and Levine (1929).

we apply the M.L. method, we obtain $p^* = 0.5138$, and $q^* = 0.4862$; the standard error of q^* is 0.0263.

b) TWO AUTOSOMAL ALLELES WITH DOMINANCE

This situation differs from the preceding one, in that here only two phenotypes are recognized. As a result, the likelihood function is now defined as

$$L = \frac{N!}{n_{D-}!n_{dd}!}(p^2 + 2pq)^{n_{D-}}(q^2)^{n_{dd}}, \qquad (13.3.6)$$

where n_{D-} is the number of $D-$ individuals and n_{dd} the number of dd individuals observed among a sample of N, and p and q are the frequencies of
the dominant and recessive genes, respectively. We find that

$$q^* = \sqrt{\frac{n_{dd}}{n_{D-} + n_{dd}}} = \sqrt{\frac{n_{dd}}{N}} \qquad (13.3.7)$$

and

$$\sigma_{q^*}^2 = \frac{(1 - q^2)}{4N}. \qquad (13.3.8)$$

In the estimation of a recessive gene, then, it is necessary only to take the
square root of the sample proportion of dd individuals to obtain an efficient
estimate.

c) ALLELIC SEX-LINKED GENES

Either of the two preceding types of alleles, i.e., those with or those without dominance, may be sex-linked. When this occurs, since the human male
is an XY-organism, the problem of estimation is slightly altered. In general
data will be available from random samples of the two sexes, so we reason
as follows: Consider, first, the case of sex-linked alleles lacking dominance.
When this is the case, there will be three recognizable phenotypes in the female
but only two in the male. If p is the frequency of a gene A and q the frequency
of its allele a, then the frequencies of the phenotypes in the two sexes are
given by

$$p^2 \text{ (AA)} + 2pq \text{ (Aa)} + q^2 \text{ (aa)} = 1 \qquad \text{(female)},$$
$$p \text{ (AY)} + q \text{ (aY)} = 1 \qquad \text{(male)}.$$

Assume that a sample of N_f females and N_m males is drawn, where $N =$
$N_f + N_m$, and that we observe among the females n_{AA} of type AA, n_{Aa} of
type Aa, and n_{aa} of type aa, and among the males n_{AY} of type AY and n_{aY}
of type aY. The probability of obtaining this distribution of males and females
is the product of the probabilities of obtaining the observed distribution of
males and of females; hence the likelihood function is

$$L = \left[\frac{N_f!}{n_{AA}!n_{Aa}!n_{aa}!}(p^2)^{n_{AA}}(2pq)^{n_{Aa}}(q^2)^{n_{aa}}\right]$$
$$\times \left[\frac{N_m!}{n_{AY}!n_{aY}!}(p)^{n_{AY}}(q)^{n_{aY}}\right]. \qquad (13.3.9)$$

The estimates of p and q are obtained by maximizing this function; these estimates are

$$q^* = \frac{2 n_{aa} + n_{Aa} + n_{aY}}{2 N_f + N_m}, \quad p^* = \frac{2 n_{AA} + n_{Aa} + n_{AY}}{2 N_f + N_m}. \quad (13.3.10)$$

The variance may be found as indicated in Table 13-2; we note

$$\sigma^2_{q^*} = \frac{q(1-q)}{2 N_f + N_m}. \quad (13.3.11)$$

In the event that there is dominance among the sex-linked alleles, only two phenotypes will be recognizable in the female. Let p be the frequency of the dominant allele D, and q the frequency of its recessive counterpart. As-

TABLE 13-2

DETERMINATION OF AMOUNT OF INFORMATION IN CASE
OF SEX-LINKED GENES WITHOUT DOMINANCE

Genotype	Observed	Expected (e)	de/dq	$[(1/e)(de/dq)^2]$
AA........	n_{AA}	$N_f p^2$	$-2N_f(1-q)$	$4N_f$
Aa........	n_{Aa}	$N_f 2pq$	$2N_f(1-2q)$	$[2N_f(1-2q)^2]/pq$
aa........	n_{aa}	$N_f q^2$	$2N_f q$	$4N_f$
AY........	n_{AY}	$N_m p$	$-N_m$	N_m/p
aY........	n_{aY}	$N_m q$	N_m	N_m/q
Total...	N	N	0	$(2N_f/pq)+(N_m/pq)$

sume that n_{D-} females of type $D-$; n_{dd} females of type dd; n_{DY} males of type DY; and n_{dY} males of type dY are observed. The likelihood function will be

$$L = \left[\frac{N_f!}{n_{D-}! n_{dd}!} (p^2 + 2pq)^{n_{D-}} (q^2)^{n_{dd}} \right] \left[\frac{N_m!}{n_{DY}! n_{dY}!} (p)^{n_{DY}} (q)^{n_{dY}} \right]. \quad (13.3.12)$$

The equation of estimation, now, is

$$\frac{d(\log L)}{dq} = -\frac{2n_{D-}q}{(1-q^2)} - \frac{n_{DY}}{(1-q)} + \frac{2n_{dd}+n_{dY}}{q} = 0 ;$$

and, solving for q, we obtain

$$q^2 [2n_{D-} + 2n_{dd} + n_{DY} + n_{dY}] + n_{DY}q - [2n_{dd}+n_{dY}] = 0 ;$$

but $2(n_{D-} + n_{dd}) = 2N_f$ and $(n_{DY} + n_{dY}) = N_m$, whence, by solving the quadratic equation,

$$q^* = \frac{-n_{DY} + \sqrt{n_{DY}^2 + 4(2N_f + N_m)(2n_{dd}+n_{dY})}}{2(2N_f + N_m)}, \quad (13.3.13)$$

and $p^* = 1 - q^*$.

Here the amount of information with respect to q, that is, the invariance of q, is

$$I_{qq} = \frac{4 N_f}{1 - q^2} + \frac{N_m}{q (1 - q)} . \qquad (13.3.14)$$

To illustrate the estimation in this case, let us examine Waaler's observations on color blindness on 18,121 Oslo school children (1927). These data are presented in Table 13-3. We calculate $q^* = 0.0772$ and $\sigma_{q^*} = 0.00247$.

13.3.2. *The case of two parameters.*—In the preceding examples, all involving a single pair of alleles, it was necessary only to determine either p or q, since their sum must equal 1; hence only one independent parameter existed. This, however, is not always the case. For example, if we were interested in a set of alleles involving three genes, then, in order to specify the population completely, we must know the values of two parameters, that is, we must know

TABLE 13-3*

THE COLOR VISION OF 18,121 OSLO
SCHOOL CHILDREN

Children	Male	Female	Total
Color blind...	725	40	765
Normal......	8,324	9,032	17,356
Total....	9,049	9,072	18,121

* After Waaler (1927).

the frequency of two of the three genes. It matters not which two of the three gene frequencies are estimated, since the third must always be the difference between 1.0 and the sum of the other two. In this case, that of two independent parameters, the method of M.L. takes the following form: Suppose that x is a random variable with distribution function $f(x; q, r)$. Then the probability of obtaining a given sample of size n is dependent upon both q and r; whence

$$L = f (x_1; q, r) f (x_2; q, r) \ldots f (x_n; q, r) . \qquad (13.3.15)$$

Here the estimates of q and r are obtained by simultaneously solving the equations derived by the partial differentiation of L with respect to q and r; that is, by solving simultaneously

$$\frac{\partial L}{\partial q} = 0 , \qquad \frac{\partial L}{\partial r} = 0 , \qquad (13.3.16)$$

or

$$\frac{\partial (\log L)}{\partial q} = 0 , \qquad \frac{\partial (\log L)}{\partial r} = 0 . \qquad (13.3.17)$$

The determination of the variances of these estimates is slightly more complex because the estimates are correlated. The variances, however, may be shown to be

$$\sigma_q^2 = \frac{\text{cofactor } I_{qq}}{\Delta}, \qquad \sigma_r^2 = \frac{\text{cofactor } I_{rr}}{\Delta}, \qquad (13.3.18)$$

where Δ is the value of the determinant of the information matrix. The latter matrix is a matrix whose elements are the amounts of information with respect to q, r, and (q and r). Thus

$$\Delta = \begin{vmatrix} \sum_i \left(\frac{1}{e_i}\right)\left(\frac{\partial e_i}{\partial q}\right)^2 & \sum_i \left(\frac{1}{e_i}\right)\left(\frac{\partial e_i}{\partial q}\right)\left(\frac{\partial e_i}{\partial r}\right) \\ \sum_i \left(\frac{1}{e_i}\right)\left(\frac{\partial e_i}{\partial q}\right)\left(\frac{\partial e_i}{\partial r}\right) & \sum_i \left(\frac{1}{e_i}\right)\left(\frac{\partial e_i}{\partial r}\right)^2 \end{vmatrix} = \begin{vmatrix} I_{qq} & I_{rq} \\ I_{qr} & I_{rr} \end{vmatrix}$$

$$= I_{qq}I_{rr} - I_{qr}^2 . \qquad (13.3.19)$$

Now the cofactor is merely the element or determinant formed by deleting from the information matrix the row and column indicated by the subscripts of the element whose cofactor is to be determined. For example, in the two-parameter case, the cofactor of I_{qq} is I_{rr} (obtained by deleting the qth row and qth column from the matrix given in eq. [13.3.19]). Similarly we note that in this case the cofactor of I_{rr} is I_{qq}. From this and equation (13.3.18) it follows that the variances of q and r are

$$\sigma_q^2 = \frac{I_{rr}}{\Delta}, \qquad \sigma_r^2 = \frac{I_{qq}}{\Delta}.$$

a) THREE AUTOSOMAL ALLELES WITH NO DOMINANCE

Assume that a set of three alleles, say I^A, I^B, and I^O, exist, operative in a fashion such that no dominance relationships occur. Furthermore, assume that p, q, and r are the respective frequencies of these genes. Under these conditions, six recognizable phenotypes exist, and their frequencies are given by

$$p^2[AA] + 2pr[AO] + q^2[BB] + 2qr[BO] + r^2[OO]$$
$$+ 2pq[AB] = 1 .$$

However, we note that, since $q + r = 1 - p$,

$$q^2 + 2qr + r^2 = (1 - p)^2$$

and

$$2pq + 2pr = 2p(q + r) = 2p(1 - p) .$$

Therefore, we may write

$$p^2[AA] + 2p(1 - p)[AO, AB] + (1 - p)^2[BB, BO, OO] = 1 .$$

Now imagine that, among a sample of n individuals, n_{AA} are observed to be of type AA, n_{AO} of type AO, n_{BB} of type BB, etc. Accordingly, we may write the likelihood function as

$$L = \frac{N!}{n_{AA}! \, (n_{AO} + n_{AB})! \, (n_{BB} + n_{BO} + n_{OO})!}$$

$$\times (p^2)^{n_{AA}} [2p(1-p)]^{n_{AO}+n_{AB}} [(1-p)^2]^{n_{BB}+n_{BO}+n_{OO}} . \qquad (13.3.20)$$

Proceeding in the usual fashion, we find

$$\log L = \log k + 2n_{AA} \log p + (n_{AO} + n_{AB}) \log 2p(1-p)$$
$$+ 2(n_{BB} + n_{BO} + n_{OO}) \log (1-p) \, ;$$

whence

$$\frac{d(\log L)}{dp} = \frac{2n_{AA} + n_{AO} + n_{AB}}{p} - \frac{(n_{AO} + n_{AB}) + 2(n_{BB} + n_{BO} + n_{OO})}{1-p}$$

and

$$p^* = \frac{2n_{AA} + n_{AO} + n_{AB}}{2N} . \qquad (13.3.21)$$

It may be similarly shown that, had we proceeded from the relationships $[p + r = 1 - q]$ or $[p + q = 1 - r]$, we would have found that

$$q^* = \frac{2n_{BB} + n_{BO} + n_{AB}}{2N} , \qquad (13.3.22)$$

$$r^* = \frac{2n_{OO} + n_{AO} + n_{BO}}{2N} . \qquad (13.3.23)$$

The variances of p^*, q^*, and r^* can be readily derived as follows:

$$\frac{d^2(\log L)}{dp^2} = - \frac{(2n_{AA} + n_{AO} + n_{AB})}{p^2}$$
$$- \frac{(n_{AO} + n_{AB}) + 2(n_{BB} + n_{BO} + n_{OO})}{(1-p)^2} .$$

Substituting the expected values for the observed values, that is, Np^2 for n_{AA}, $2Npr$ for n_{AO}, etc., we obtain

$$\frac{d^2(\log L)}{dp^2} = - \frac{2N}{p(1-p)} , \qquad (13.3.24)$$

whence the variance of p^* is merely $p^*(1-p^*)/2N$. And the variances of q^* and r^* are $q^*(1-q^*)/2N$ and $r^*(1-r^*)/2N$, respectively.

Thus we observe that in a tri-allelic system, when all genotypes produce distinct, recognizable phenotypes, the maximum-likelihood estimates of the gene frequencies are simple gene counts. *This may be shown to be true, irrespective of the number of genes in the system, so long as no dominance relationships exist.* While there is at present no genetic system composed of three alleles without dominance which can be readily worked with, the trend of serological research suggests that such systems will be encountered in the future.

b) THREE AUTOSOMAL ALLELES WITH DOMINANCE

The estimation of the gene frequencies in a system of multiple alleles presents a number of difficulties. Two of these are the large number of different possible systems which may exist and the added difficulty in solving, in many cases, the maximum-likelihood equations. If we confine ourselves solely to tri-allelic systems, we find that the six genotypes may be grouped in 52 different ways, depending upon the dominance relationships which may be visualized (Cotterman, 1954). Potentially, each grouping represents a distinct genetic system; however, the majority of these potential systems is unrecognized in man at present.

As an illustration of one tri-allelic system involving a particular dominance relationship, let us consider the situation which obtains with respect to the ABO blood groups when only anti-A and anti-B sera are available, as is generally the case. Under these circumstances, four discrete serological reactions are possible, namely, $A+B-$, $A-B+$, $A+B+$, and $A-B-$. We say that these four serological reactions correspond to the phenotypes A, B, AB, and O, respectively, and postulate three genes. We assume (1) that these genes are operative in a fashion such that I^A and I^B are dominant to I^O, but no dominance exists between I^A and I^B; (2) that the three genes I^A, I^B, and I^O have frequencies p, q, and r; and (3) that the four phenotypes are distributed according to $[(p + q + r)^2]^N$. Imagine that a sample of N individuals is drawn from a population meeting these criteria and that n_A, n_B, n_O, and n_{AB} individuals of types A, B, O, and AB, respectively, are observed. The likelihood function, therefore, is

$$L = \frac{N!}{n_A! n_B! n_O! n_{AB}!} (p^2 + 2 p r)^{n_A} (q^2 + 2 q r)^{n_B} (r^2)^{n_O} (2 p q)^{n_{AB}}. \quad (13.3.25)$$

Proceeding in the standard fashion, we obtain the logarithm of this expression, which is

$$\log L = n_A \log (p^2 + 2pr) + n_B \log (q^2 + 2qr)$$
$$+ 2n_O \log r + n_{AB} \log 2pq + \log k .$$

Now, of these three gene frequencies, only two are independent because of the restriction that $p + q + r = 1$. Accordingly, if we make the substitution $r = 1 - p - q$ in the foregoing expression, and then partially differentiate the resulting equation with respect to p and q, we find

$$\frac{\partial \log L}{\partial p} = -\frac{2n_0}{1-p-q} + \frac{n_A + n_{AB}}{p} - \frac{n_A}{2-p-2q} - \frac{2n_B}{2-2p-q}, \quad (a)$$

$$\frac{\partial \log L}{\partial q} = -\frac{2n_0}{1-p-q} + \frac{n_B + n_{AB}}{q\mathrm{l}} - \frac{2n_A}{2-p-2q} - \frac{n_B}{2-2p-q}. \quad (b)$$

These equations may be solved by approximation techniques (see Mather, 1951). An alternative and ingenious solution, due to Bernstein (1930; see also Stevens, 1938), proceeds somewhat as follows: The maximum of the log expression on page 189 may also be obtained by solving the system of equations formed (1) by the restriction that $p + q + r = 1$, and (2) by equating the partial derivatives with respect to p, q, and r of the log expression plus some multiple, say λ, of the restriction, $p + q + r = 1$, to zero. If we adopt this approach, then we must obtain the partial derivatives with respect to p, q, and r of the expression

$$n_A \log p\,(p + 2r) + n_B \log q\,(q + 2r) + 2n_0 \log r + n_{AB} \log 2pq$$
$$+ \lambda\,(p + q + r).$$

After this differentiation, we are required to solve for p, q, r, and λ the following set of equations:

$$\frac{n_A + n_{AB}}{p} + \frac{n_A}{p + 2r} = -\lambda, \quad (a)$$

$$\frac{n_B + n_{AB}}{q} + \frac{n_B}{q + 2r} = -\lambda, \quad (b)$$

$$2\left[\frac{n_0}{r} + \frac{n_A}{p + 2r} + \frac{n_B}{q + 2r}\right] = -\lambda, \quad (c)$$

$$p + q + r = 1. \quad (d)$$

$$(13.3.26)$$

From $(13.3.26c)$ we note that, upon replacing n_0, n_A, and n_B by their expected values, namely, Nr^2, $N(p^2 + 2pr)$, $N(q^2 + 2qr)$, the expected value of λ is $-2N$. It follows, then, that

$$\frac{n_A + n_{AB}}{p} + \frac{n_A}{p + 2r} = \frac{n_B + n_{AB}}{q} + \frac{n_B}{q + 2r} = 2N;$$

or, if we divide each term in the numerators of this equation by N, which is equivalent to replacing n_A by the proportion of individuals of type A, say $\bar{A}$; n_B by the proportion of individuals of type B, say $\bar{B}$; etc., then

$$\frac{\bar{A} + \overline{AB}}{p} + \frac{\bar{A}}{p + 2r} = \frac{\bar{B} + \overline{AB}}{q} + \frac{\bar{B}}{q + 2r} = 2. \quad (13.3.27)$$

If we can find a solution to the relationship indicated in (13.3.27), then our problem is solved.

It had been suggested that consistent estimates of p, q, and r could be derived in the following manner: We might estimate the frequency of the gene, I , directly rom the proportion of individuals of phenotype O, thus

$$r^2 = \overline{O} \; ;$$

hence

$$r' = \sqrt{\overline{O}} \; . \tag{13.3.28}$$

The proportion of type A individuals is

$$\overline{A} = p^2 + 2 p r \; ;$$

and, if we add the proportion of type O individuals to both sides of this equation, then

$$\overline{A} + \overline{O} = p^2 + 2 p r + r$$

and

$$\sqrt{\overline{A} + \overline{O}} = p + r \; .$$

But, since $[p + q + r = 1]$, then $[p + r = 1 - q]$, and

$$q' = 1 - \sqrt{\overline{A} + \overline{O}} \; . \tag{13.3.29}$$

Similarly

$$p' = 1 - \sqrt{\overline{B} + \overline{O}} \; . \tag{13.3.30}$$

Now p', q', and r' are not the M.L. estimates, nor are they fully efficient. Moreover, these estimates will rarely add to 1. Bernstein solved this problem as follows: let D be the difference between 1 and their sum, that is

$$D = 1 - p' - q' - r'$$

$$= \sqrt{\overline{B} + \overline{O}} + \sqrt{\overline{A} + \overline{O}} - \sqrt{\overline{O}} - 1 \; . \tag{13.3.31}$$

In terms of p', q', and r', we find

$$\overline{A} + \overline{AB} = 1 - (\overline{B} + \overline{O})$$

$$= [1 - \sqrt{\overline{B} + \overline{O}}] [1 + \sqrt{\overline{B} + \overline{O}}]$$

$$= p' (2 - p') \; .$$

Similarly

$$\overline{B} + \overline{AB} = q' (2 - q') \; ;$$

$$\overline{A} = \cdot (\overline{A} + \overline{O}) - (\overline{O})$$

$$= [\sqrt{\overline{A} + \overline{O}} - \sqrt{\overline{O}}] [\sqrt{\overline{A} + \overline{O}} + \sqrt{\overline{O}}]$$

$$= (p' + D)(p' + D + 2r') \; ;$$

and

$$\overline{B} = (q' + D)(q' + D + 2r') \; .$$

Hence equation (13.3.27) may be rewritten as follows:

$$\frac{p'(2-p')}{p} + \frac{(p'+D)(p'+D+2r')}{p+2r} = \frac{q'(2-q')}{q}$$
$$+ \frac{(q'+D)(q'+D+2r')}{q+2r} = 2.$$

If we now set

$$p = p'\,[f(D)]; \qquad q = q'\,[f(D)]; \qquad r = \left(r'+\frac{D}{2}\right)[f(D)]$$

then

$$\frac{p'(2-p')}{p'\,[f(D)]} + \frac{(p'+D)(p'+D+2r')}{(p'+D+2r')\,[f(D)]} = 2$$

and

$$[f(D)] = \left(1+\frac{D}{2}\right).$$

Thus we observe that the M.L. equations are exactly solved by three estimates, namely,

$$p^* = p'\left(1+\frac{D}{2}\right), \tag{13.3.32}$$

$$q^* = q'\left(1+\frac{D}{2}\right), \tag{13.3.33}$$

and

$$r^* = \left(r'+\frac{D}{2}\right)\left(1+\frac{D}{2}\right), \tag{13.3.34}$$

which total not to 1 but to $(1 - D^2/4)$. While these estimates are not precisely M.L. estimates, they are fully efficient, and we may take as their variances the variances of the M.L. estimates. The latter may be derived with the aid of Table 13-4. We find that

$$\sigma_{r^*}^2 = \frac{I_{qq}}{\Delta}; \qquad \sigma_{q^*}^2 = \frac{I_{pp}}{\Delta}; \qquad \sigma_{r^*}^2 = \frac{I_{pp}+I_{qq}-2I_{pq}}{\Delta}, \tag{13.3.35}$$

where

$$I_{pp} = \frac{2N(4-4p-6q+6pq+2q^2-3pq^2)}{p(2-p-2q)(2-2p-q)},$$

$$I_{qq} = \frac{2N(4-6p-4q+6pq+2p^2-3p^2q)}{q(2-p-2q)(2-2p-q)},$$

$$I_{pq} = \frac{2N(4-4p-4q+3pq)}{(2-p-2q)(2-2p-q)},$$

and

$$\Delta = I_{pp}I_{qq} - I_{pq}^2.$$

TABLE 13-4

DETERMINATION OF VARIANCE OF ESTIMATES OF GENE FREQUENCIES UNDER SYSTEM OF THREE ALLELES WITH DOMINANCE RELATIONSHIPS AS POSTULATED FOR A, B, O BLOOD GROUPS WHEN TESTING WITH ANTI-A AND ANTI-B SERA

Class	Observed	Expected	$\partial e/\partial p$	$\partial e/\partial q$	$(1/e)[\partial e/\partial p]^2$	$(1/e)[\partial e/\partial p][\partial e/\partial q]$	$(1/e)[\partial e/\partial q]^2$
A.	n_A	$N(p^2+2pr)$	$2N(1-p-q)$	$-2Np$	$\dfrac{4N(1-p-q)^2}{p(2-p-2q)}$	$-\dfrac{4N(1-p-q)}{(2-p-2q)}$	$\dfrac{4Np}{(2-p-2q)}$
B.	n_B	$N(q^2+2qr)$	$-2Nq$	$2N(1-p-q)$	$\dfrac{4Nq}{(2-2p-q)}$	$-\dfrac{4N(1-p-q)}{(2-2p-q)}$	$\dfrac{4N(1-p-q)^2}{q(2-2p-q)}$
O.	n_O	Nr^2	$-2N(1-p-q)$	$-2N(1-p-q)$	$4N$	$4N$	$4N$
AB. . . .	n_{AB}	$N2pq$	$2Nq$	$2Np$	$\dfrac{2Nq}{p}$	$2N$	$\dfrac{2Np}{q}$
Total	N	N	0	0	$\dfrac{2N(4-4p-6q+6pq+2q^2-3pq^2)}{p(2-p-2q)(2-2p-q)}$	$\dfrac{2N(4-4p-4q+3pq)}{(2-p-2q)(2-2p-q)}$	$\dfrac{2N(4-6p-4q+6pq+2p^2-3p^2q)}{q(2-p-2q)(2-2p-q)}$

$$I_{pp} = \sum_i \frac{1}{e_i}\left[\frac{\partial e_i}{\partial p}\right]^2 \qquad I_{pq} = \sum_i \frac{1}{e_i}\left[\frac{\partial e_i}{\partial p}\right]\left[\frac{\partial e_i}{\partial q}\right]$$

$$I_{pq} = \sum_i \frac{1}{e_i}\left[\frac{\partial e_i}{\partial p}\right]\left[\frac{\partial e_i}{\partial q}\right] \qquad I_{qq} = \sum_i \frac{1}{e_i}\left[\frac{\partial e_i}{\partial q}\right]^2$$

$$\Delta = \begin{vmatrix} I_{pp} & I_{pq} \\ I_{pq} & I_{qq} \end{vmatrix} = I_{pp}I_{qq} - I_{pq}^2 .$$

In Table 13-5 are given Paddock's (1946) results of testing 870 Maori soldiers. If equations (13.3.28, 29, 30) are applied to these data, then $r' = 0.667832$, $p' = 0.318531$, and $q' = 0.016791$. If the adjustment is made, we find that $D = -0.003154$ and that $r^* = 0.665204$, $p^* = 0.318029$, and $q^* = 0.016765$. Note that, within the limits of rounding errors, $p^* + q^* + r^* = 1 - D^2/4$. In this case we calculate as the standard errors of p^*, q^*, and r^*, 0.012394, 0.00309, and 0.012531, respectively (see Table 13-6).

TABLE 13-5*

DISTRIBUTION OF THE A, B, O BLOOD GROUPS
AMONG 870 MAORIS

	O	A	B	AB	Total
No. of individuals.....	388	453	16	13	870
Proportion...........	0.4460	0.5207	0.0184	0.0149	1.0000

* After Paddock (1946).

TABLE 13-6*

CALCULATION OF STANDARD ERRORS OF ESTIMATES OF
GENE FREQUENCIES OF A, B, O ANTIGENS

Class	Observed	Expected	I_{pp}	I_{pq}	I_{qq}
O......	388	r^2N	3,480.000	3,480.000	3,480.000
A......	453	$(p^2+2pr)N$	2,937.321	−1,404.307	671.386
B......	16	$(q^2+2qr)N$	43.307	−1,718.347	68,182.271
AB.....	13	$2pqN$	91.725	1,740.000	33,007.483
Total	870	N	6,552.353	2,097.346	105,341.140

$$\Delta = (6,552.353)(105,341.140) - (2,097.346)^2 = 685,833,474.459$$

$$\sigma_{q*}^2 = \frac{I_{pp}}{\Delta} = 0.00000955 , \qquad \sigma_{q*} = 0.003090 ;$$

$$\sigma_{p*}^2 = \frac{I_{qq}}{\Delta} = 0.00015360 , \qquad \sigma_{p*} = 0.012394 \cdot$$

$$\sigma_{r*}^2 = \frac{I_{pp} + I_{qq} - 2I_{pq}}{\Delta} = 0.00015703 , \quad \sigma_{r*} = 0.012531 .$$

* Illustrated by data from Paddock (1946).

c) TWO PAIRS OF ALLELES

The problem of estimating two parameters may arise not only when one is dealing with a tri-allelic system but also when the genetic hypothesis asserts that a particular class of individuals arises as a consequence of possessing specified genes from two different pairs of alleles. For example, one might assert that a particular phenotype is the consequence of homozygosity for the genes a of a pair A and a, and b of a pair B and b. Estimation of the frequencies of a and b proceeds in a manner analogous to the general case in equation (13.3.2). As an illustration of estimation under a hypothesis of two sets of two factors, the reader is referred to Cotterman and Snyder (1937).

13.3.3. *The case of more than two parameters.*—With the exception of multi-allelic systems involving no dominance relationships, each genetic hypothesis postulating three or more parameters poses a new estimation problem. The solution of these problems is accomplished by an extension of the general case in equation (13.3.2). As an illustration of the estimation of three parameters, the reader is referred to Stevens (1938), who has treated the case of the A_1, A_2, B, O blood groups, assuming the availability of only anti-A_1, anti-A_2, and anti-B sera.

13.4. *The estimation of gene frequencies when rates of inbreeding are known.*—It is not always possible to obtain unselected sets of data, either of unrelated or of related individuals, and this is particularly true of the recessively inherited pathologic traits. Consequently, if gene-frequency estimates are desired, appropriate allowance must be made for the selected nature of the data. One method of estimation, in these circumstances, employs the relationship between consanguinity and the frequency of manifestation of a recessive trait (cf. sec. 7.9, p. 70). It is possible by this method to estimate the frequency of a gene in a population, provided that one knows the percentages of (1) affected individuals from consanguineous marriages and (2) cousin marriages as related to all marriages. The method (Dahlberg, 1929) is as follows: If c is the percentage of first-cousin marriages among all marriages, and $f(aa)$ is the frequency of affected individuals in the population, then, since affected individuals may arise from either first-cousin or nonfirst-cousin marriages, $f(aa)$ must be equal to the sum of these two components, or

$$f(aa) = (1 - c)\, q^2 + \frac{cq}{16}(1 + 15\, q).$$

Now the ratio of affected individuals derived from marriages between first cousins to that of the total number of affected individuals (k) will be

$$k = \frac{(cq/16)\,(1 + 15\, q)}{(1 - c)\, q^2 + (cq/16)\,(1 + 15\, q)}$$

or

$$k = \frac{c}{c + [16\,(1 - c)\, q]/[1 + 15\, q]};$$

rearranging and solving for q, we obtain

$$q = \frac{c(1-k)}{16k - 15c - ck}.$$ (13.4.1)

The use of formula (13.4.1) is subject to severe limitations, theoretical as well as biological. Such theoretical criteria as random mating are seldom satisfied. It is known that matings in man tend to occur within small and not entirely discrete population units which have been called "isolates." These isolates are subject to shifting migrational patterns. Moreover, different culturally determined preferences in the selection of a mate may exist in one isolate as opposed to another. Adequate compensation for these unmeasured, and largely unmeasurable, pressures on the isolates is presently impossible. In addition, the purely practical problem of reliably determining the incidence of consanguineous matings poses a difficulty because of errors on the part of the recorder as well as lack of knowledge of relationship on the part of the persons concerned.

The biological criteria which must be satisfied for the use of this formula are the existence of traits that are (1) clearly differentiated from the normal and (2) monogenic recessive in their inheritance. Only a limited number of inherited diseases in man satisfies these two prerequisites. An occasional finding in human inheritance is variation in the mode of transmission of a given disease from one family to the next (cf. sec. 7.6, p. 65). Furthermore, what appears to be the same recessive disease may be due to different non-allelic genes in different families. For example, Wibaut (1931) has pointed out that, in the case of retinitis pigmentosa, affected persons are far more often the result of consanguineous marriage than one would expect on the basis of formula (13.4.1). This finding is explicable if there are several different rare recessive genes which can produce this disease. Conversely, such a finding supplies some evidence that several different genes are involved and, in this particular case, permits us to subdivide still further the genetic etiology of retinitis pigmentosa. Despite these objections, formula (13.4.1) is of considerable value as a first approximation to the frequency of a given gene (see Neel *et al.*, 1949).

13.5. *The testing of genetic hypotheses.*—We turn now to the second class of problems mentioned in the introduction. For the major part, tests of genetic hypotheses are divisible into three groups, namely, tests of (1) the significance of a specified parameter value, (2) "goodness of fit," and (3) homogeneity.

1. Tests of the significance of a specified parameter value usually take the form of determining whether the evidence provided by a sample is consistent with some given value of the parameter. Such tests can generally be

carried out by the use of standard errors. The problem consists, essentially, of determining whether the specified value lies within an interval defined by the estimate of the parameter plus or minus some multiple of the standard error of this estimate. Examples of this particular type of testing will be given in chapter 14.

2. As has been previously mentioned, one of the principal uses of gene-frequency estimates is in testing the "goodness of fit" of a genetic hypothesis to an observed set of data. Here the problem assumes the following general form: We are given a sample of observations from a population whose parameters are either fully specified or are estimable from the data. We are required to determine the probability of obtaining such a sample of observations, by random sampling, from this population on the basis of the specified genetic hypothesis. If this probability is less than some acceptable value (generally 0.05 or 0.01), we assert that these data are not consistent with the genetic hypothesis.

3. Lastly, in tests of homogeneity our primary concern lies in ascertaining whether two or more populations may be assumed to be identical. This is to say, we are required to determine, from a sample to which two or more populations have contributed, whether the parameters of these populations are like or unlike values. In view of the numerous special tests which comprise these three groups, we shall consider only those tests most widely applicable to problems which arise in human genetics.

13.6. *Tests of "goodness of fit."*—The single test most useful in determining the "goodness of fit" of a hypothesis to a sample of observations is the chi-square test (cf. sec. 10.2, p. 120). More specifically, χ^2 is a satisfactory direct test in gene-frequency problems so long as the number of recognizable phenotypes exceeds the number of genes in the postulated system of alleles. Consider, for example, the case in which the genetic hypothesis is that of two autosomal alleles lacking dominance. Efficient estimates of these alleles have been shown to be

$$p^* = \frac{2\,n_{\mathrm{AA}} + n_{\mathrm{Aa}}}{2\,N} \quad \text{and} \quad q^* = \frac{n_{\mathrm{Aa}} + 2\,n_{\mathrm{aa}}}{2\,N}.$$

The agreement between an observed set of events and this hypothesis may be tested as follows: For each class or phenotype there will be some number, n_i, observed, and some number, e_i, expected. The latter may be calculated by the use of the estimates. The fit of the hypothesis to the data may then be tested, as shown in Table 13-7. To determine the significance of a χ^2, we must know the number of "degrees of freedom." In general, the number of degrees of freedom appropriate to a χ^2 test will be equal to the number of classes minus the number of restrictions placed on the data by the use of estimates, totals, subtotals, etc. In tests involving gene frequencies, the number of degrees of freedom is usually equal to the number of phenotypes minus the number of genes in the allelic system. In this example, since the number of phenotypes is three and the number of genes is two, χ^2 is based on one degree of freedom.

TABLE 13-7

TEST OF SIGNIFICANCE OF HYPOTHESIS OF AUTOSOMAL
ALLELES WITHOUT DOMINANCE

	PHENOTYPE			TOTAL
	aa	ab	bb	
Observed......	n_1	n_2	n_3	N
Expected......	$(p^*)^2N$	$2p^*q^*N$	$(q^*)^2N$	N
Difference.....	d_1	d_2	d_3	0

$$\chi^2 = \sum_i \frac{(n_i - e_i)^2}{e_i} = \sum_i \frac{d_i^2}{e_i}.$$

TABLE 13-8*

APPLICABILITY OF HYPOTHESIS OF AUTOSOMAL ALLELES WITH-
OUT DOMINANCE TO INHERITANCE OF MN ANTIGENS

	MM	MN	NN	Total
Observed.....	50	86	45	181
Expected.....	47.78	90.43	42.79	181
Difference....	2.22	− 4.43	2.21	0

$$\chi^2 = \frac{N(n_2^2 - 4n_1n_3)^2}{(2n_1 + n_2)^2(n_2 + 2n_3)^2} = \frac{181(-1604)^2}{(186)^2(176)^2} = 0.435.$$

$$df = 1.$$

* Data from Landsteiner and Levine (1929).

In Table 13-8 this method has been used to test the hypothesis of no dominance of the alleles of the MN blood groups on the data of Table 13-1. The number expected in each category is obtained by substituting the estimates, p^* and q^*, for the parameters, p and q. For example, the expected number of type MM individuals is $N(p^2) = N(p^*)^2 = 181(0.5138)^2 =$

47.78, where N is the total number of observations. In this example, χ^2 may be more simply calculated from

$$\chi^2 = \frac{N\,(n_{Aa}^2 - 4n_{AA}n_{aa})^2}{(2n_{AA} + n_{Aa})^2\,(n_{Aa} + 2n_{aa})^2}.\tag{13.6.1}$$

We calculate $\chi^2 = 0.435$ and note from Table 13-9 that for one degree of freedom the probability of obtaining discrepancies from the hypothesis of no dominance greater than the ones observed due to chance alone is greater than one time in two. We conclude that the data are consistent with the hypothesis. It may be pointed out here that, had the obverse obtained, that is, had the data not been readily compatible with the hypothesis, the statistical

TABLE 13-9*

TABLE OF χ^2 WHERE n IS NUMBER OF DEGREES OF FREEDOM AND P IS PROBABILITY THAT χ^2 WILL EXCEED SPECIFIED VALUE

n	$P=0.70$	$P=0.50$	$P=0.20$	$P=0.05$	$P=0.01$
1.......	0.148	0.455	1.642	3.841	6.635
2.......	0.713	1.386	3.219	5.991	9.210
3.......	1.424	2.366	4.642	7.815	11.345
4.......	2.195	3.357	5.989	9.488	13.277
5.......	3.000	4.351	7.289	11.070	15.086
6.......	3.828	5.348	8.558	12.592	16.812
7.......	4.671	6.346	9.803	14.067	18.475
8.......	5.527	7.344	11.030	15.507	20.090
9.......	6.393	8.343	12.242	16.919	21.666
10.......	7.267	9.342	13.442	18.307	23.209
20.......	16.266	19.337	25.038	31.410	37.566
30.......	25.508	29.336	36.250	43.773	50.892

* This table is abridged from Table III of Fisher, *Statistical Methods for Research Workers* (11th ed.; Edinburgh: Oliver & Boyd, Ltd.), by permission of the author and publishers.

model might be in error in either of two ways. First, the mode of inheritance might be one other than that postulated. Second, while the postulated mode of inheritance might be correct, conditions for a stable equilibrium, and hence for the legitimate application of the Hardy-Weinberg law, might not be fulfilled.

The χ^2 method may be applied to test such hypotheses as multiple alleles or sex-linked alleles, with or without dominance. For example, we may test the fit of the hypothesis of multiple alleles to the data on 870 Maoris given in Table 13-5. In this instance, since we recognize four phenotypes and since there are three genes in the system, the number of degrees of freedom is one. Here Stevens (1950) has shown that χ^2 may be reduced, for simplification in calculating, to

$$\chi^2 = 2N\left(1 + \frac{r^*}{p^*q^*}\right)D^2,\tag{13.6.2}$$

where D is as defined in formula (13.3.31). We find $\chi^2 = 2.166$, which is not significant; and the hypothesis is accepted as being a reasonable explanation of the distribution of the ABO antigens among these individuals. If hypotheses of sex-linked genes are tested, an additional degree of freedom is lost because of the use of the totals for each sex in calculating the expected number in each phenotype of each sex. In the case of sex-linked genes with dominance where we recognize four categories (affected males and females; normal males and females), the number of degrees of freedom is one, and in the absence of dominance the number is two.

If the general rule for the number of degrees of freedom is applied to the case of autosomal alleles with dominance, we find, since there are only two phenotypes and two genes, that there are zero degrees of freedom. The χ^2 test is thus inapplicable. This difficulty has been circumvented by a method suggested by Fisher (1939); unfortunately, this method requires information from two generations. Assume that such data are available; then a test may be constructed as follows: We may assume that mating is at random and arbitrarily divide all families into three groups, namely, those in which both parents are of the recessive phenotype dd; those in which one parent is dd and the other the dominant phenotype $D-$; and those in which both parents are $D-$. Furthermore, let $p =$ the frequency of the gene D, and $q = 1 - p$ the frequency of the gene d, and $D- = DD$ or Dd.

Consider, first, only families in which one parent is phenotypically $D-$ and the other dd. Two genotypically different matings, $DD \times dd$ and $Dd \times dd$, meet this requirement. We may, however, in some cases classify the type of mating by a knowledge of the offspring produced. For example, if the mating $D- \times dd$ produces a dd offspring, then we may conclude that the mating is, in fact, $Dd \times dd$. It does not necessarily follow, however, that if no dd children are produced, the mating is $DD \times dd$. The vagaries of sampling are such that we may expect some $Dd \times dd$ matings to produce only $D-$ offspring; the probability of such an event is a function of the number of children produced by the mating. Our problem consists of determining the expected number of matings of $DD \times dd$ and $Dd \times dd$ among a number of matings of $D- \times dd$.

The $D-$ parent in matings of $D- \times dd$ may be, as we have seen, either of genotype DD, with frequency p^2, or of genotype Dd, with frequency $2pq$. In the general population the frequency of $D-$ individuals is $(p^2 + 2pq)$, and consequently the probability that an individual is DD, given that he is $D-$, is

$$P(DD \mid D-) = \frac{p^2}{p^2 + 2pq} = \frac{p}{p + 2q};$$

similarly, the probability that an individual is Dd, given that he is $D-$, is

$$P(Dd \mid D-) = 1 - P(DD \mid D-) = 1 - \frac{p}{p+2q}.$$

If the $D-$ parent is DD, then all children will be $D-$; if, on the other hand, the $D-$ parent is Dd, we may expect some children to be of type dd. As has been indicated, the only way that we have of knowing that a given individual is heterozygous is by the occurrence of at least one child of type dd. We may, therefore, divide families from the mating of $D- \times dd$ into two groups: (1) all children $D-$ (the $D-$ parent may be either DD or Dd) and (2) some children of type dd (the $D-$ parent must be Dd). Let us examine these groups separately.

ALL CHILDREN $D-$

The probability that an offspring from the mating of $DD \times dd$ will be of type $D-$ is 1; and the probability that the same type of offspring will be born to the mating $Dd \times dd$ is $\frac{1}{2}$. The probability, therefore, that a $D-$ child is from the mating $DD \times dd$, given that the mating is $D- \times dd$, is

$$P \ (D-\text{ from mating } DD \times dd \mid D- \times dd) = (1) \ P \ (DD \mid D-) \ .$$

Similarly, the probability that a $D-$ child is from the mating $Dd \times dd$, given that the mating is $D- \times dd$, is

$$P \ (D-\text{ from mating } Dd \times dd \mid D- \times dd) = (\tfrac{1}{2}) \ P \ (Dd \mid D-) \ .$$

It follows that, given the mating $D- \times dd$, the probability that a child born to such a mating will be $D-$ is

$$P \ (D-\mid D- \times dd) = P \ (DD \mid D-) + (\tfrac{1}{2}) \ P \ (Dd \mid D-)$$

$$= \frac{p}{p+2q} + \left(\frac{1}{2}\right)\left(1 - \frac{p}{p+2q}\right).$$

Since the genotypes of successive offspring are independent, then, given the mating $D- \times dd$, the probability that s offspring from such a mating will all be $D-$ is

$$P \ (D-_s \mid D- \times dd) = \frac{p}{p+2q} + \left(\frac{1}{2}\right)^s \left(1 - \frac{p}{p+2q}\right). \quad (13.6.3)$$

Now, for any given size of family, the expected number of such families, all of whose offspring are $D-$, is

$$E \ (N_s) = N_s P(D-_s \mid D- \times dd) \ , \quad\quad (13.6.4)$$

where N_s is the observed number of families of size s. The total expected number of families, all of whose offspring are $D-$, is the sum of formula (13.6.4) for all sizes, s.

SOME CHILDREN dd

The expected number of such families is obtained by subtracting the total expected number of families, all of whose children are $D-$, from the total number of families observed, i.e.,

$$\sum_s N_s .$$

We may extend this argument to the case where both parents are $D-$. Briefly, we find the following: There are three genotypically different matings, $DD \times DD$, $DD \times Dd$, and $Dd \times Dd$. When at least one parent is DD, the probability of an offspring's being $D-$ is 1; and when both parents are Dd, the probability is $\frac{3}{4}$. Separating families into groups all children $D-$ and some dd, we note:

ALL CHILDREN $D-$

Given the mating $D- \times D-$, the probability that a child will be $D-$ is the sum of the probabilities that at least one parent is DD and $\frac{3}{4}$ the probability that both parents are Dd, or

$$P\,(D-|D- \times D-) = (1)\,P\,(\text{at least one } DD|\text{both } D-)$$
$$+ \tfrac{3}{4}P\,(\text{both } Dd|\text{both } D-)$$
$$= \Big[1 - \Big(\frac{2q}{p+2q}\Big)^2\Big] + \frac{3}{4}\Big(\frac{2q}{p+2q}\Big)^2,$$

and the probability that s children will all be $D-$ is

$$P\,(D_{-s}|\,D- \times D-) = \Big[1 - \Big(\frac{2q}{p+2q}\Big)^2\Big] + \Big(\frac{3}{4}\Big)^s \Big(\frac{2q}{p+2q}\Big)^2. \quad (13.6.5)$$

Thus the total expected number of families all of whose children are $D-$, when both parents are known to be $D-$, is

$$\sum_s N_s \Big\{ \Big[1 - \Big(\frac{2q}{p+2q}\Big)^2\Big] + \Big(\frac{3}{4}\Big)^s \Big(\frac{2q}{p+2q}\Big)^2 \Big\} .$$

SOME CHILDREN dd

The expected number of families is obtained by subtraction in a manner analogous to that outlined for the $D- \times dd$ mating.

Now the discrepancy between the observed and expected numbers of families with all $D-$ children or some dd children for each of the matings $D- \times D-$ and $D- \times dd$ may be evaluated by χ^2. There is one degree of freedom for each of the two types of matings. In Table 13-10 this method has been applied to data collected by Schiff and Sasaki (1932) on the secretor factor. They observed that, among 100 unrelated adults, 71 were secretors. We estimate from these data that the frequencies of the genes S and s are 0.4615 and 0.5385, respectively. With the aid of these estimates, the calculation of the expected number of families with no ss children proceeds as indicated in the table. We find that $\chi^2 = 1.4775$, which for two degrees of freedom is clearly not significant. Thus these data do not contradict the hypothesis of a single pair of alleles with S dominant to s.

TABLE 13-10*

APPLICATION OF THE FISHER METHOD OF TESTING
DOMINANT INHERITANCE

Mating	Size of Family	No. of Families	No. of Families with No ss Children	Probability that, Given Size of Family and Mating, There Will Be No ss Children	Expected No. of Families with No ss Children
$S \times S$	2	9	7	0.7856	7.0704
	3	3	3	.7167	2.1501
	4	3	3	.6650	1.9950
	5	3	2	.6263	1.8789
	6	2	1	.5972	1.1944
	7	1	1	.5754	0.5754
$S \times ss$	2	8	2	.4750	3.8000
	3	5	2	.3875	1.9375
	4	2	1	.3437	0.6874
	5	1	. . . .	.3219	0.3219
	6	1	. . . .	.3109	0.3109
	7	1	1	0.3054	0.3054

Mating	Class of Family	Observed No. of Families	Expected No. of Families	χ^2	D.F.
$S \times S$	All children S	17	14.864	1.0505	1
	At least one child ss	4	6.136		
$S \times ss$	All children S	6	7.363	0.4270	1
	At least one child ss	12	10.637		
Total	. . . .	. . . . . .	. . . .	1.4775	2

* Data from Schiff and Sasaki (1932).

13.7. *Tests of homogeneity.*—There remains but to examine the testing of homogeneity to complete this brief survey of some of the general procedures applicable to testing genetic hypotheses. As has been mentioned, the general problem consists of determining whether k samples may all be assumed to have been drawn from the same population. As an illustration of the simplest case in which a test of homogeneity is desired, consider the data in Table 13-11. Here are observations with respect to the frequencies of the MM,

TABLE 13-11*

DISTRIBUTION OF M-N ANTIGENS AMONG THREE
DIFFERENT TRIBES OF AMERICAN INDIANS

Tribe	MM	MN	NN	Total
Ute........	61	36	7	104
Navaho.....	305	52	4	361
Pueblo......	83	46	11	140
Total...	449	134	22	605

* Data from Boyd (1950).

TABLE 13-12

TEST OF HOMOGENEITY OF THREE AMERICAN
INDIAN TRIBES GIVEN IN TABLE 13-11

Tribe	Total Genes $(2N)$	M Genes (X)	$\stackrel{p}{(X/2N)}$	pX
Ute..........	208	158	0.760	120.080
Navaho.......	722	662	.917	607.054
Pueblo........	280	212	0.757	160.484
	1,210	$\Sigma X = 1,032$	....	$\Sigma pX = 887.618$

$$\bar{p} = 1,032/1,210 = 0.853 , \qquad \bar{q} = 1 - \bar{p} = 0.147 .$$

$$\chi^2 = \frac{\Sigma pX - \bar{p}\Sigma X}{\bar{p}\bar{q}} = \frac{7.322}{0.1254} = 58.389 ; \quad \text{D.F.} = 2 , \quad P < 0.001$$

MN, and NN blood types for three tribes of American Indians. Are the frequencies of the M and N genes comparable in these three tribes? That is to say, may these three tribes be regarded as but samples drawn from the same population? If so, then the best estimate of M, or N, is that derived from the total observations. A test which is fully satisfactory for these or comparable data is outlined in Table 13-12. We calculate $\chi^2 = 58.389$; and, for two degrees of freedom (one less than the number of samples), there is less than 1 chance in 1,000 of drawing from the same population three samples which differ more markedly than those observed. We conclude that these three tribes are dissimilar in the frequencies of the M and N genes.

A number of tests have been devised for testing the homogeneity of more than two samples where the sample distributions are dependent upon two or more parameters, such as is the case in the ABO blood groups (see Stevens, 1938, 1950). The use of one of these tests, the more conventional one, is illus-

TABLE 13-13a*

TEST OF HOMOGENEITY OF FREQUENCIES OF A, B, O BLOOD GROUPS
AMONG 2,694 INDIVIDUALS OF ICELANDIC, DANISH, OR SWEDISH ORIGIN

Population	No. Observed (n)	p^*	np^*	q^*	nq^*	r^*	nr^*
Iceland.....	800	0.19211	153.688	0.06321	50.568	0.74468	595.744
Denmark....	1,261	.28553	360.053	.07249	91.410	.64196	809.512
Sweden.....	633	0.30995	196.198	0.08283	52.431	0.60719	384.351
Total...	2,694	....	709.939	....	194.409	....	1,789.607

$$\bar{p} = 0.26353 \qquad \bar{q} = 0.07216 \qquad \bar{r} = 0.66429$$

SUMS OF SQUARES AND PRODUCTS

	p^*	q^*	r^*
p^*........	6.05228	0.83674	....
q^*........	0.83674	0.13702	....
r^*........	....	....	7.86902

* Data from Boyd (1950).

trated in Tables 13-13a and b. In this example, we are given the gene-frequency estimates based upon samples of individuals residing in Iceland, Denmark, and Sweden, and we are required to determine whether these groups are homogeneous with respect to the frequencies of the genes responsible for the ABO antigens. The procedure is as follows:

1. Weighted estimates of the frequency of the genes are obtained from the relationships

$$\bar{p} = \frac{\Sigma np^*}{\Sigma n}; \qquad \bar{q} = \frac{\Sigma nq^*}{\Sigma n}; \qquad \text{and} \qquad \bar{r} = \frac{\Sigma nr^*}{\Sigma n}.$$

2. The sums of squares and products of the estimates are then computed from

$$S_{pp} = \Sigma(p^*)(np^*) - \bar{p}\Sigma(np^*) ,$$

$$S_{qq} = \Sigma(q^*)(nq^*) - \bar{q}\Sigma(nq^*) ,$$

$$S_{rr} = \Sigma(r^*)(nr^*) - \bar{r}\Sigma(nr^*) ,$$

$$S_{pq} = S_{qp} = \Sigma(p^*)(nq^*) - \bar{p}\Sigma(nq^*) ,$$

where S_{pp}, S_{qq}, and S_{rr} are the sums of squares of p^*, q^*, and r^*, and S_{pq} is the sum of products of p^* and q^*. These terms may be logically set out in matrix form, as indicated in Table 13-13a. The sum of products involving r^* need not be calculated.

TABLE 13-13b*

TEST OF HOMOGENEITY OF FREQUENCIES OF A, B, O
BLOOD GROUPS AMONG 2,694 INDIVIDUALS OF
ICELANDIC, DANISH, OR SWEDISH ORIGIN

EVALUATION OF TOTAL INFORMATION

Population	I_{pp}	I_{pq}	I_{qq}
Iceland.....	7,169	1,947	23,544
Denmark....	11,300	3,069	37,111
Sweden.....	5,672	1,541	18,629
Total...	24,141	6,557	79,284

$$\sigma_{\bar{q}}^2 = \text{Variance of pooled estimate of } q = \frac{I_{pp}}{\Delta_t} = 1.29 \times 10^{-5},$$

$$\sigma_{\bar{p}}^2 = \text{Variance of pooled estimate of } p = \frac{I_{qq}}{\Delta_t} = 4.24 \times 10^{-5},$$

$$\sigma_{\bar{r}}^2 = \text{Variance of pooled estimate of } r = \frac{I_{pp} + I_{qq} - 2I_{pq}}{\Delta_t} = 4.83 \times 10^{-5}.$$

PARTITION OF χ^2

Source	D.F.	χ^2	P
Differences in r...	2	60.475	<0.001
Remainder.......	2	1.865	>0.30
Total........	4	62.340	<0.001

* Data from Boyd (1950).

3. The information with respect to p, q, and (p and q) in each sample is then obtained as indicated in formula (13.3.2), using $\bar{p}$ and $\bar{q}$ as the estimates of p and q. The total information in the three samples is obtained by summing the I_{pp}, I_{qq}, and I_{pq} obtained for each sample. The determinant, Δ_t, of the total information matrix and the variances of the pooled estimates of p, q, and r may be obtained as shown in Table 13-4.

4. A χ^2 testing the homogeneity of the samples may now be obtained from the following formula:

$$\chi^2 = \frac{S_{pp}I_{pp} + S_{qq}I_{qq} + 2S_{qp}I_{pq}}{\Sigma n}.$$

This total χ^2 will have degrees of freedom equal to the product of (the number of parameters) and (the number of sets of data minus 1). In this case there are two independent parameters and three sets of data; hence the number of degrees of freedom associated with the total χ^2 is 4. We calculate $\chi^2 = 62.340$ which for four degrees of freedom would be exceeded by chance less than one time in a thousand if, in fact, the Icelandic, Danish, and Swedish populations were homogeneous. Had the total χ^2 been nonsignificant, then the weighted estimates would be taken as the best estimates of the parameters, p, q, and r. And the variances of these estimates would have been those given in Table 13-13b.

Since the total χ^2 was significant, it is of interest to know whether the heterogeneity is due solely to differences in one ratio. On inspection we note that the estimates of r vary greatly; hence it would be of interest to know whether the heterogeneity is ascribable to differences in this gene. This can be accomplished by partitioning the total χ^2 into a component due to differences in r and a remainder. The χ^2 associated with differences in r may be obtained from

$$v^2 = \frac{S_{rr}}{(\sigma_r^2)(\Sigma n)}.$$

This χ^2 has degrees of freedom equal to one less than the number of samples, or, in this case, 2. We calculate, for this component, $\chi^2 = 60.475$ and note that the probability of exceeding this χ^2 due to chance is less than 1 in 1,000 if the populations are, in fact, homogeneous. The partitioning may be completed by determining, by subtraction, the component of the total χ^2 associated with the remainder. This component has degrees of freedom equal to the difference between the number of degrees of freedom of the total χ^2 and the number of degrees of freedom associated with the differences in r, or, in this example, 2.

The choice of which set of differences is tested is optional; had differences in p or differences in q been chosen, then

$$\chi^2 \text{ (differences in } p*) = \frac{S_{pp}}{(\sigma_{\frac{2}{p}}) \, (\Sigma n)} \, ,$$

$$\chi^2 \text{ (differences in } q*) = \frac{S_{qq}}{(\sigma_{\frac{2}{q}}) \, (\Sigma n)} \, .$$

The appropriate remainders in either of these cases would again be determined by subtracting from the total χ^2 the χ^2 associated with the differences either in p or in q. It should be noted, however, that it is not possible to partition the total χ^2 into independent components referable to differences in p, differences in q, and differences in r, at one time.

Bibliography

SPECIFIC REFERENCES

BERNSTEIN, F. 1930. Fortgesetzte Untersuchungen aus der Theorie der Blutgruppen, Ztschr. f. Abstamm.- u. Vererbungslehre, **56**:233–73.

BOYD, W. C. 1950. Genetics and the races of man. Boston: Little, Brown & Co.

COTTERMAN, C. W. 1937. Indication of unit factor inheritance in data comprising but a single generation, Ohio J. Sc., **37**:127–40.

———. 1953. Regular two-allele and three-allele phenotype systems. I, Am. J. Human Genetics, **5**:193–235.

———. 1954. Estimation of gene frequencies in non-experimental populations. *In:* KEMPTHORNE, O., *et al.*, Statistics and mathematics in biology, pp. 449–66. Ames: Iowa State College Press.

COTTERMAN, C. W., and SNYDER, L. H. 1937. Studies in human inheritance. XVII. The gene-frequency analysis of double recessive inheritance involving one autosomal and one sex-linked substitution, Genetica, **19**:537–52.

DAHLBERG, G. 1929. Inbreeding in man, Genetics, **14**:421–54.

FISHER, R. A. 1922. On the mathematical foundations of theoretical statistics, Phil. Trans. (A), **222**:309–68.

———. 1925. Theory of statistical estimation, Proc. Cambridge Phil. Soc., **22**:700–725.

———. 1939. *In:* TAYLOR, G. L., and PRIOR, A. M., Blood groups in England. III. Discussion of the family material, Ann. Eugenics, **9**:18–44.

———. 1946. The fitting of gene frequencies to data on rhesus reactions, Ann. Eugenics, **13**:150–55.

———. 1947. Note on the calculation of the frequencies of rhesus allelomorphs, *ibid.*, **13**:223–24.

———. 1951. Standard calculations for evaluating a blood-group system, Heredity, **5**:95–102.

HALDANE, J. B. S., and SMITH, C. A. B. 1947. A new estimate of the linkage between the genes for colour-blindness and hemophilia in man, Ann. Eugenics, 14:10–31.

LANDSTEINER, K., and LEVINE, P. 1929. On the racial distribution of some agglutinable structures of human blood, J. Immunol., 16:123–31.

NEEL, J. V.; KODANI, M.; BREWER, R.; and ANDERSON, R. C. 1949. The incidence of consanguineous matings in Japan, with remarks on the estimation of comparative gene frequencies and the expected rate of appearance of induced recessive mutations, Am. J. Human Genetics, 1:156–78.

PADDOCK, R. F. 1946. New Zealand blood-type distributions. Preliminary investigations, Ann. Eugenics, 13:4–6.

RACE, R. R., and SANGER, R. 1951. Inheritance of the human blood group antigen Jkᵃ, Nature, 168:207.

SCHIFF, F., and SASAKI, H. 1932. Der Ausscheidungstypus, ein auf serologischem Wege nachweisbares Mendelndes Merkmal, Klin. Wchnschr., 11:1426–29.

SNYDER, L. H., and YINGLING, H. C. 1935. The application of the gene frequency method of analysis to sex-influenced factors, with special reference to baldness, Human Biol., 7:608–15.

STEVENS, W. L. 1938. Estimation of blood-group frequencies, Ann. Eugenics, 8:362–75.

———. 1950. Statistical analysis of the A-B-O blood groups, Human Biol., 22:191–217.

———. 1952. A-B-O system in mixed populations, *ibid.*, 24:12–24.

WAALER, G. H. M. 1927. Über die Erblichkeitsverhältnisse der verschiedenen Arten von angeborener Rotgrünblindheit, Ztschr. f. Abstamm.- u. Vererbungslehre, 45:279–333.

WIBAUT, F. 1931. Studien über Retinitis pigmentosa. I. Statistische Gründe für die Annahme, dass die R. p. in genetischem Sinn keine Einheit darstellt., Klin. Monatsbl. f. Augenh., 87:298–307.

GENERAL REFERENCES

FISHER, R. A. 1949. The design of experiments. 5th ed. Edinburgh: Oliver & Boyd.

———. 1950. Statistical methods for research workers. 11th ed. New York: Hafner Publishing Co.

MATHER, K. 1951. The measurement of linkage in heredity. 2d ed. ("Methuen Monographs.") New York: John Wiley & Sons, Inc.

Problems

1. Verify the gene-frequency estimates based on Waaler's data (Table 13-3). Do Waaler's data differ significantly from the hypothesis of a single pair of sex-linked genes with dominance?

2. A study of 382 Filipinos revealed 99 to be MM, 192 MN, and 91 NN. Calculate the frequency of the gene N by the two methods given in sec. 13.1. How much more efficient is the maximum-likelihood estimate?

3. Snyder and Yingling (1935) have suggested that baldness may be explained by postulating a pair of genes, B and b, operative such that BB and Bb males are bald but only BB females are bald. Derive the maximum-likelihood estimate of the frequency of b, assuming random samples of N_f females and N_m males. Derive the variance of this estimate.

4. Cotterman (1937) has devised a method of testing unit factor inheritance, utilizing pairs of siblings. Apply this method to the data on the secretor factor of Schiff and Sasaki (1932) presented in Table 13-10.

5. In problem 3, chapter 8 (p. 94), attention was called to the fact that originally the A, B, O blood groups were thought to be due to two independent pairs of genes. Derive efficient estimates of the frequencies of the genes a and b under this assumption. Derive the variances of these estimates.

CHAPTER 14

The Analysis of Family Data

In the preceding chapters we have arbitrarily distinguished between the results to be expected in families segregating for rare and for common genes with simple modes of inheritance. In man, however, owing to the relatively small size of the average family, the theoretical distribution of genotypes and phenotypes indicated earlier is rarely realized within any one family. Consequently, to test the hypothesis of a dominant or a recessive mode of inheritance, we must be able to combine data derived from a number of different families. The greater the number of families combined, the less likely are random departures from expectation to obscure the fundamental ratios. These families may vary as to both mating type and number of children. In the actual analysis of family data we are usually able, either because of conscious selection or because of the existence, for all practical purposes, of only one type of mating which will yield pertinent information, to confine our attention to the results of one type of mating at a time. This necessity for pooling data from families of varying size demands statistical techniques which not only offset these difficulties but also take into account systematic biases which may be introduced by the method of collection of data. In this chapter we propose to examine the effect of differing systems of ascertainment in the analysis of family data.

14.1. *Ascertainment.*—The primary object in collecting family data is to compare the proportion of affected children actually observed with some theoretical proportion based upon the type of mating and the suspected mode of inheritance. To this end, we may collect family data, in general, in only two ways: (1) we may select families *at random without reference to the type of offspring produced;* (2) we may select families *which contain at least one affected child.*

It should be borne in mind that the selection "at random" referred to in the first method does not mean haphazard selection. Random selection of families implies sampling from a population of families in such a way that each family in the population has an equal chance of being chosen. If we are interested in a rare inherited trait, then a random selection of all families will necessarily lead to a preponderance of families in the data which could yield no information, since the majority of families would not possess the rare gene. Consequently, in random selection we are interested, in fact, in a random selection of families from a population of families of a specific mating type. It is feasible to delimit this population only when the mating type is recognizable because of the parental phenotypes. If the trait in question were due to a rare dominant gene, then the most common mating would involve a heterozygote and a normal, recessive individual; and it would be possible, as we shall see, to construct a sampling scheme which would give rise to a random selection of such matings. If, on the other hand, the gene were a rare recessive, we should be unable to distinguish between marriages which are incapable of producing affected children and those which, while involving two heterozygous parents, do not happen to produce at least one affected child. In families selected because of one or more affected offspring, when the parents are normal, the affected child serves to identify the mating type.

"Ascertainment," per se, refers merely to the method of data collection. For example, when a family is brought to the attention of and recorded by the investigator through random selection, the family is said to be ascertained at random; similarly, when a family is brought to the attention of the investigator through an affected individual, the family is said to be ascertained through the affected individual. In the latter case this affected individual is known as the "propositus" or the "proband." It follows that a family may be ascertained through affected individuals as many times as there are such individuals, but obviously it can be ascertained only once through random selection.

Ascertainment through affected individuals may be complete or incomplete. It is said to be complete if it is almost certain that all affected individuals will come under observation independently. That is to say, each affected individual is discovered because he or she is affected and not because of the discovery of the abnormality in a sibling. As an example, consider the case of a family of two affected children, both of school age. Assume that one child is seven and the other ten years of age. If we surveyed all seven- and ten-year-old children and discovered the seven-year-old affected child during the survey of the seven-year-olds while the ten-year-old affected child was discovered during the survey of the ten-year-olds, then these siblings would have been ascertained independently. On the other hand, if only a survey of seven-year-olds was made, then only the seven-year-old affected child would be selected. Now if this child provided information that he had an older affected sibling, we would have learned of the affected ten-year-old only because of a younger affected sib, and hence the ten-year-old would not have been ascertained independently. If the latter situation obtained for a number of families, then ascertainment would be incomplete.

As we shall see later in the chapter, the definition of the degree of ascertainment for a given trait is not always easy. Whenever this problem arises and medical records are to be the starting point for the study, a consideration of the nature of the disease itself may be helpful. Where medical procedures offer significant alleviation of the consequences of a genetic effect, ascertainment is more likely to be complete than where they do not. Thus splenectomy is of great value in hereditary hemolytic icterus; and, once a single member of a family in which this disease occurs has undergone surgery, he is likely to urge other members of the family to seek medical attention. On the other hand, little can be done for multiple neurofibromatosis; and, once this fact has been established by an individual, he is likely to dissuade other affected persons in his family from seeking medical attention.

14.2. Random selection of families of a specific mating type.—In Table 14-1 are listed the sex and type of offspring of 31 families on record with the

TABLE 14-1

DISTRIBUTION OF PHENOTYPE AND SEX BY SIZE OF
FAMILY FOR THIRTY-ONE FAMILIES WHERE
ONE PARENT HAS OPALESCENT DENTINE

SIZE OF FAMILY	NO. OF FAMILIES	MALE OFFSPRING		FEMALE OFFSPRING	
		Normal	Affected	Normal	Affected
2.........	13	9	4	7	6
3.........	5	5	4	4	2
4.........	4	5	3	3	5
5.........	6	7	7	9	7
7.........	1	3	1	1	2
8.........	1	2	2	2	2
10.........	1	1	4	2	3
Total...	31	32	25	28	27

Heredity Clinic of the University of Michigan. These families have been selected so as to have two common denominators: (1) each family contains at least two children, and (2) one parent in each family has opalescent dentine, a condition in which the teeth are extremely translucent, discolored, and friable. Opalescent dentine, while not extremely rare, is an uncommon finding. It does not appear to be the result of any dietary deficiency. The occurrence of a number of families with more than one affected member suggested that this condition may be inherited. If it is inherited and if we assume that opalescent dentine results from a single-gene substitution, what is the mode of transmission of this gene?

The distribution of affected individuals among these families suggests that the gene may be a simple dominant. If this is the case, and since the disease is uncommon, then the most frequent mating which would yield information on the mode of inheritance would involve a heterozygote with the normal, recessive homozygote. A 1:1 ratio between affected and normal offspring would be expected from such matings.

Now these families could, initially, be incorporated into a genetic study because of contact with an affected parent or an affected child. In practice, either or both of these possibilities may occur; for example, adults alone would come to the attention of a dentist whose practice consisted of the operation of an industrial clinic; similarly, only the offspring would be observed at a children's clinic; or both parents and/or offspring could be the general dental practitioner's introduction to the family.

Let us consider what might happen under the following type of sampling. Assume that a mass dental survey of adults is initiated; furthermore, assume that this survey is exhaustive, that is, that every affected adult comes under observation. We could then determine every marriage in which one parent had opalescent dentine. This would be tantamount to the random selection of families derived from the mating of a heterozygote with the normal homozygote. We will ignore those sibships of which the parents are members and focus attention only on the offspring of these parents. An examination of the children produced by each of these matings would reveal families in which as few as none or as many as all of the children had opalescent dentine. Table 14-1, with the single exception that we have discarded all one-child families for convenience, could serve as a model for what might be obtained.

Among the 112 offspring listed in Table 14-1, 52 had opalescent dentine. On the basis of our hypothesis that the gene is a simple dominant and that all the marriages involve a heterozygous parent and a normal parent, we would have expected 56 affected children. The significance of this departure from the theoretical ratio of 1:1 may be tested by χ^2 as indicated in Table 14-2. It will be noted that a $\chi^2 = 0.571$ was obtained for the total cases and that only one degree of freedom exists. Reference to Table 13-9 (p. 199) indicates that in more than three cases out of ten a larger χ^2 would be expected as due to chance alone, even though the hypothesis of a simple dominant were true. This is clearly not a significant departure from the hypothesis. It is of interest to know whether the data are homogeneous, that is, whether the ratio of 1:1 obtains in both sexes. A simple test of homogeneity can be made by determining the χ^2 for each sex separately. The difference between the sum of such χ^2's and the χ^2 obtained from the total is a χ^2 with degrees of freedom equal to the difference between the sum of the degrees of freedom associated with each subset and those associated with the total χ^2. We note that in this case this χ^2, which is termed the "heterogeneity χ^2," is equal to 0.307, with one degree of freedom. Clearly, the sexes are homogeneous, and we may assert that these data are in agreement with the behavior expected from a simple autosomal dominant.

14.3. *Ascertainment through affected individuals.*—Rarely does the opportunity arise to select families at random in the study of the inheritance of a rare trait, and, when it does arise, it is generally in connection with rare dominant

TABLE 14-2

TEST OF HYPOTHESIS OF A 1:1 RATIO OF AFFECTED AND NORMAL
OFFSPRING AMONG THIRTY-ONE SIBSHIPS WHERE
ONE PARENT IS AFFECTED

Offspring	Normal	Affected	Total	χ^2	D.F.	P
Males:						
Observed......	32.0	25.0	57 }	0.860	1	....
Expected......	28.5	28.5	57 }			
Females:						
Observed......	28.0	27.0	55 }	0.018	1	....
Expected......	27.5	27.5	55 }			
Total:						
Observed......	60.0	52.0	112 }	0.571	1	> .30
Expected......	56.0	56.0	112 }			

HOMOGENEITY WITHIN SEXES

Source	χ^2	D.F.	P
Sum.............	0.878	2	>0.50
Total...........	0.571	1	>0.30
Heterogeneity....	0.307	1	>0.50

genes. More frequently, our introduction to families exhibiting an inherited or possibly inherited pathologic entity occurs because of an affected member who happens to come to medical attention. When ascertainment occurs through affected individuals, allowance must be made for two sources of bias in the data: (1) the omission, in recessive heredity, of families with heterozygous parents, who, however, lack an affected child, and (2) such inequalities as may exist in the probabilities of recording families with differing numbers of affected individuals. We will consider first the case where ascertainment is complete and then the case where it is incomplete.

1. COMPLETE ASCERTAINMENT

Friedreich's ataxia is a degenerative neurological disorder of the dorsal and lateral tracts of the spinal cord characterized by ataxia (the loss of muscular co-ordination), dysarthria (inarticulateness of speech), pes cavus (an excessive curvature of the sole of the foot), nystagmus, and a lateral curvature of the spine. The age of onset of this disease and its distribution within families suggest the occurrence of both a dominant and a recessive form (cf.

TABLE 14-3*

DISTRIBUTION BY FAMILIES OF 100 INDIVIDUALS WITH FRIEDREICH'S ATAXIA

Family No.	No. of Children	No. Affected	Probands	Family No.	No. of Children	No. Affected	Probands	Family No.	No. of Children	No. Affected	Probands
1.....	4	1	1	19...	3	1	1	39...	2	1	1
1A...	6	1	1	20...	3	2	2	40...	7	1	1
2.....	5	2	1	20A..	5	2	2	41...	8	1	1
3.....	11	5	1	21...	5	2	2	42...	3	1	1
4.....	9	2	1	22...	4	1	1	43...	2	1	1
5.....	6	3	1	23...	3	1	1	44...	5	2	2
6.....	10	5	5	24...	4	1	1	45...	4	1	1
6A...	5	1	1	25...	5	1	1	46...	2	1	1
7.....	2	2	2	26...	3	3	1	47...	3	1	1
8.....	6	2	2	27...	6	2	1	48...	2	1	1
8A...	3	1	0	28...	3	1	1	49...	2	1	1
9.....	4	1	1	29...	9	3	2	50...	4	2	1
10.....	9	2	1	30...	5	1	1	51...	10	1	1
11.....	7	2	1	31...	3	1	1	51A..	2	1	0
12.....	5	2	2	32...	4	1	1	53...	8	1	1
13.....	1	1	1	33...	3	1	1	54...	9	1	1
14.....	12	6	4	34...	1	1	1	55...	6	1	1
15.´...	1	1	1	35...	5	1	1	56...	7	1	1
16.....	6	3	1	36...	2	1	1	57...	1	1	1
17.....	4	2	2	37...	4	1	1	58...	11	1	1
18.....	6	3	2	38...	3	1	1	58A..	7	1	0
Totals .	122	48	32	......	83	29	25	......	105	23	20
Grand total	.....	.....	.....	.....	.....	.....	.....	.....	310	100	77

* Used by permission of Dr. T. Sjögren and Acta Psychiatrica et Neurologica. The number of probands indicates the number of times the family was independently ascertained.

Table 7-1, p. 66). Sjögren (1943) studied all the cases of "hereditary ataxia" reported in the major hospitals of Sweden over approximately a 40-year period ending in 1941. He found 188 cases of the disease distributed among 118 families. Of these 188 cases, 100 were from 63 families in which neither parent was affected and where the disease was confined to one sibship, suggesting recessive inheritance. These apparently recessive cases are indicated in Table 14-3. To determine whether these data are compatible with the hypothesis of a single recessive gene, we must assume that, whatever the mode of inheritance is, it is the same in all families. This assumption is basic to all the methods which follow; and, as we have seen in the case of choroido-retinal degeneration, this assumption may not always be justified.

The simplest method of analyzing these data is one which is variously called the "method of Apert," the "direct method," or the "a priori method" (Ludwig and Boost, 1940). The basis for this method is as follows: If we are dealing with a rare recessive gene, then the majority of affected individuals will arise from matings involving two heterozygous parents. From such a mating the probability, p, of an affected child is $\frac{1}{4}$, and the probability, q, of a normal child is $(1 - p)$ or $\frac{3}{4}$. As previously stated, the binomial $(p + q)^s$ describes the probabilities of obtaining none, one, . . . , s affected offspring in s-child sibships; and the probability that a family of size s will contain at least one affected child is precisely $(1 - q^s)$, where q^s is the probability of no affected children among s. In the event, therefore, that one investigates only those sibships with at least one affected child, the relative frequency of affected children to be expected among these sibships is $p/(1 - q^s)$. Consequently, the expected number of affected children, $E(a_s)$ in s-child families is merely s times as much, or

$$E(a_s) = \frac{sp}{1 - q^s}.$$

The expected total number of affected children is

$$E(a) = \sum_s \frac{sp}{1 - q^s} f_s, \tag{14.3.1}$$

where f_s is the frequency of families of size s.

The test of the agreement of the data with the hypothesis consists of a comparison of the observed and expected total number of affected with the standard deviation of the latter. The standard deviation of the expected total number of affected is the square root of the total variance obtained by summing the separate variances for the various sizes of family as frequently as these sizes occur. The separate components of the total variance are obtained from

$$\sigma_s^2 = \frac{spq}{1 - q^s} - \frac{s^2 p^2 q^s}{(1 - q^s)^2}. \tag{14.3.2}$$

The expected number of affected children and the variances for the more common sizes of family and values of p are given in Table 14-4.

If we apply this method to Sjögren's data, excluding families of size one, we find that the total number of affected offspring to be expected on the basis of a simple recessive is 100.49 and the standard deviation is 6.24 (see Table 14-5). The difference between the total number of affected children

TABLE 14-4*

EXPECTED NUMBER OF AFFECTED OFFSPRING AND
VARIANCES FOR VARYING SIZES OF FAMILY AS
DETERMINED BY A PRIORI METHOD, WHERE p
IS PROBABILITY OF AFFECTED OFFSPRING

Family Size	$p=0.5$	σ_s^2	$p=0.25$	σ_s^2
2......	1.333	0.222	1.1428	0.122
3......	1.715	0.490	1.2973	0.263
4......	2.134	0.782	1.4628	0.420
5......	2.581	1.082	1.6389	0.592
6......	3.047	1.379	1.8248	0.776
7......	3.527	1.667	2.0196	0.970
8......	4.015	1.945	2.2225	1.172
9......	4.509	2.215	2.4328	1.380
10......	5.005	2.478	2.649	1.592
11......	5.503	2.737	2.871	1.805
12......	6.001	2.992	3.098	2.020
13......	6.5	3.245	3.329	2.234
14......	7.0	3.497	3.563	2.446
15......	7.5	3.748	3.801	2.658

* Used by permission from An introduction to mathematical genetics, by Dr. Lancelot Hogben, copyright 1946, W. W. Norton and Co., Inc.

TABLE 14-5

ANALYSIS OF SJÖGREN'S 59 SIBSHIPS WITH ONE
OR MORE CASES OF FRIEDREICH'S ATAXIA BY
A PRIORI METHOD (ONE-CHILD FAMILIES OMIT-
TED)

Family Size	No. of Families	No. Affected Observed	No. Affected Expected	Variance
2.....	8	9	9.142	0.976
3.....	11	14	14.270	2.893
4.....	9	11	13.165	3.780
5.....	9	14	14.750	5.328
6.....	7	15	12.774	5.432
7.....	4	5	8.078	3.880
8.....	2	2	4.445	2.344
9.....	4	8	9.731	5.520
10.....	2	6	5.298	3.184
11.....	2	6	5.742	3.610
12.....	1	6	3.098	2.020
Total	59	96	100.493	38.967*

* Standard deviation = $\sqrt{38.967}$ = 6.242.

observed, 96, and the total number of expected children is less than one standard deviation. This is clearly a nonsignificant deviation.

An objection to this method is that one specifies in advance the value of p, and then tests for agreement with the hypothesis. It is actually more correct to estimate the true proportion of affected children from the data and then test this proportion for agreement with the proportion to be expected according to one or more different genetic hypotheses (e.g., $\frac{1}{16}$, $\frac{1}{4}$, $\frac{9}{16}$, etc.). Haldane (1932) has developed a method of analysis based on the method of maximum likelihood which permits us to determine whether an interval of specified length around the estimate includes the parameter. To develop this test, which has been termed the "a posteriori method," we reason as follows (the notation used is that devised by Haldane): The data consist of a number of families of size s ($s = 2, \ldots, S$), containing r affected offspring ($r = 1, \ldots, s$). Let p be the proportion of affected children, $q = 1 - p$ the proportion of normal children, and a_{rs} the number of families of size s of whom r are affected. Now among all families of size s, the probability that such a family will contain r affected is that term in the binomial with r as the exponent of p, that is,

$$\binom{s}{r} p^r q^{s-r}.$$

The data, however, do not consist of a sample of all families but rather only of those families with at least one affected child. Since the probability that a family of size s will contain at least one affected child is $(1 - q^s)$, the probability that a family of size s containing at least one affected will actually contain r is

$$\frac{\binom{s}{r} p^r q^{s-r}}{1 - q^s}.$$

The data consist of a_{rs} such families for each pair of values of r and s, and we are required to estimate p so as to maximize the likelihood of the observed set of values. The function to be maximized is

$$L = \prod_{s=2}^{S} \prod_{r=1}^{s} \left[\frac{\binom{s}{r} p^r q^{s-r}}{1 - q^s} \right]^{a_{rs}}.$$

The standard method of maximizing such a function has been given in section 13.2 (p. 180). We note that

$$\log L = \sum_{s} \sum_{r} \left[r \log p + (s - r) \log q - \log(1 - q^s) \right] a_{rs} + \text{constant},$$

and, differentiating the log expression with respect to p, we obtain

$$\frac{d \, (\log L)}{dp} = \sum_s \sum_r \left[\frac{r}{pq} - \frac{s}{q} - \frac{s \, q^{s-1}}{1-q^s} \right] a_{rs}.$$

We solve as follows: equating to zero, we have

$$\sum_s \sum_r \left[\frac{r}{pq} - \frac{s}{q} - \frac{s \, q^{s-1}}{1-q^s} \right] a_{rs} = 0 \, ;$$

whence

$$\sum_s \sum_r \frac{r \, a_{rs}}{pq} = \sum_s \sum_r \left[\frac{s}{q} + \frac{s \, q^{s-1}}{1-q^s} \right] a_{rs} = \sum_s \sum_r \frac{s \, a_{rs}}{q \, (1-q^s)}.$$

Now $\sum_s \sum_r r \, a_{rs}$ is the total number of affected children, say R, and $\sum_r a_{rs}$ is the number of s-child sibships irrespective of the number of affected, say n_s; hence summing over r and s and eliminating the q common to all denominators, we obtain

$$\frac{R}{p} = \sum_s \frac{s \, n_s}{1-q^s}. \qquad (14.3.3)$$

The test of the agreement of the data with the hypothesis consists of a comparison of the theoretical proportion of affected with the estimate of p based on the sample. The significance of any discrepancy between these two proportions is judged by the standard deviation of the estimated proportion. The latter may be shown to be

$$\frac{1}{\sigma} = \sqrt{ \frac{R}{p^2 q} - \sum_s \frac{s^2 n_s q^{s-2}}{(1-q^s)^2} }. \qquad (14.3.4)$$

To obtain the maximum-likelihood estimate of p from Sjögren's data, we note that, by expanding equation (14.3.3), we have

$$\frac{R}{1-q} = \frac{2 n_2}{1-q^2} + \frac{3 n_3}{1-q^3} + \frac{4 n_4}{1-q^4} + \ldots + \frac{12 n_{12}}{1-q^{12}};$$

whence

$$R = \frac{2 n_2}{1+q} + \frac{3 n_3}{1+q+q^2} + \frac{4 n_4}{1+q+q^2+q^3} + \frac{5 n_5}{1+q+q^2+q^3+q^4}$$
$$+ \ldots + \frac{12 n_{12}}{1+q+q^2+q^3+q^4+\ldots+q^{10}+q^{11}}.$$

If the values obtained by Sjögren are substituted in the above formula, then

$$96 = \frac{16}{1+q} + \frac{33}{1+q+q^2} + \frac{36}{1+\ldots+q^3} + \frac{45}{1+\ldots+q^4} + \frac{42}{1+\ldots+q^5}$$
$$+ \frac{28}{1+\ldots+q^6} + \frac{16}{1+\ldots+q^7} + \frac{36}{1+\ldots+q^8} + \frac{20}{1+\ldots+q^9}$$
$$+ \frac{22}{1+\ldots+q^{10}} + \frac{12}{1+\ldots+q^{11}}.$$

This may be solved for q by iteration or trial and error. The latter method, while slightly less accurate, is generally quicker. If, as the first trial, we substitute the value $q = 0.76$, then the right side of the above equation sums to 98.433, which is clearly in excess of the observed number of affected, and hence our trial estimate of q is too small. If, as the second trial, we substitute the value $q = 0.77$, then the right side sums to 96.401, and again the trial value is too small. When the value $q = 0.78$ is used, a sum equal to 94.401 is obtained. Thus it is apparent that q lies between 0.77 and 0.78. Linear interpolation yields $q = 0.772$, and hence $p = 0.228$. The standard error of this estimate is 0.0295, and it is readily apparent that the theoretical proportion lies within the usually accepted interval of the estimate plus or minus two standard errors. We may assert that the data do not depart from expectation on the basis of a single autosomal recessive gene.

The reader should note that formulas (14.3.1) and (14.3.3) and (14.3.2) and (14.3.4) are algebraic equivalents. Accordingly, the distinction between the a priori and the a posteriori method is essentially a logical one. In the a priori method the calculation of the proportion of affected involves the expected value; that is to say, one assumes some value of the parameter in the calculation of the proportion of affected. No such assumption is necessary in the a posteriori method.

2. INCOMPLETE ASCERTAINMENT

Two general methods of collecting data may be visualized which would give rise to incomplete ascertainment. One might, for example, have elected to study only those families with Friedreich's ataxia, where the propositi were of the same age and were reported to hospitals in the same year. This would lead to ascertaining only one affected individual in each family, if we ignore multiple births, and only in those families in which the affected individual was within the prescribed age. This method has been called "single ascertainment" or "single selection," for obvious reasons. On the other hand, one might have elected to study all the cases reported over a longer period of time but only for a selected number of hospitals. We may assume that all individuals with Friedreich's ataxia come to the attention of some hospital during the course of life and that related individuals are most likely to come to the attention of the same hospital. Thus, if this scheme of collecting data were employed, one might ascertain a number of affected individuals of the same family independently, probably not all, however, and it is quite unlikely that all the affected in all families would be ascertained. In the event of multiple selection, we may know exactly how many affected members of the same family were propositi if we were the collecting agents; or, if the data were collected by others, we may have only a vague idea as to the probability of any affected individual's being ascertained.

If ascertainment is thought of in the probability sense, then it follows that this probability may be any value between zero and 1. If it is 1, that is, if it is certain that an affected individual will be ascertained, then ascertainment is complete. The other limiting value, zero, is obviously never realized, but single selection may be visualized as being an instance when the probability tends toward zero. Any value intermediate between these limiting situations would correspond to multiple selection. When ascertainment is incomplete but when the probability of ascertaining an affected individual is constant, then the greater the number of affected individuals in a family, the greater the probability that that family will be recorded. That is to say, we are more likely to discover and record a family with, say, three affected individuals than a family with only one such person. This presents a new source of bias in addition to the omission of those matings which fail to produce an affected child. The methods appropriate for obviating these biases will be developed separately for single and multiple selection.

2a. *Single selection:* Consider an s-child family with r affected. This family could be ascertained by single selection of an affected individual in r different ways. Now if k is the probability of ascertaining an affected individual, then $(1 - k)$ is the probability that a single individual will not be ascertained. It follows that the probability that a family containing r affected persons will not be recorded is $(1 - k)^r$ and that the probability that a family with r affected individuals will be recorded at least once is $[1 - (1 - k)^r]$. In the whole population of s-child families, the probability of a family of size s with r affected being ascertained at least once is the product of the probabilities of a family with r affected persons being recorded at least once and of an s-child family having r affected, or $[1 - (1 - k)^r] \binom{s}{r} p^r q^{s-r}$. If k tends to zero, as in single selection, then the probability of a family of size s with r affected being ascertained at least once tends to $kr\binom{s}{r} p^r q^{s-r}$, and we see that the probability of recording a family is proportional to the number of affected individuals. If we let a_{rs}, R, and n_s be the number of s-child sibships with r affected, the total number of affected recorded $\left(\text{i.e., } \sum_s \sum_r r\, a_{rs}\right)$, and the total number of s-child sibships $\left(\text{i.e., } \sum_r a_{rs}\right)$, respectively, then the expect-

ed number of s-child sibships with r affected is the probability that such a family will be recorded at least once *times* some constant, c, to be determined, that is,

$$E(a_{rs}) = ck\,r\binom{s}{r}p^r q^{s-r},$$

which may be rewritten as

$$E(a_{rs}) = ck\,sp\binom{s-1}{r-1}p^{r-1}q^{s-r}.$$

Now the expected number of s-child sibships irrespective of the number of affected members is

$$E\left(\sum_r a_{rs}\right) = E(n_s) = ck\,sp\sum_r\binom{s-1}{r-1}p^{r-1}q^{s-r};$$

but

$$\sum_r\binom{s-1}{r-1}p^{r-1}q^{s-r} = (p+q)^{s-1} = 1,$$

and hence

$$E(n_s) = cksp.$$

Therefore, we may write

$$E(a_{rs}) = n_s\binom{s-1}{r-1}p^{r-1}q^{s-r}.$$

If the method of maximum likelihood is now applied, we obtain, as the likelihood function,

$$L = \prod_s\prod_r\left[\binom{s-1}{r-1}p^{r-1}q^{s-r}\right]^{a_{rs}};$$

whence

$$\log L = \sum_s\sum_r\left[(r-1)\log p + (s-r)\log q\right]a_{rs} + \text{constant},$$

where, again, the constant may be ignored, since it does not include p. Now

$$\frac{d(\log L)}{dp} = \sum_s\sum_r\left[\frac{r}{pq} - \frac{1}{p} - \frac{s}{q}\right]a_{rs} = 0.$$

Solving, we obtain

$$\sum_s\sum_r r\,a_{rs} = \sum_s\sum_r a_{rs} + p\left[\sum_s\sum_r s\,a_{rs} - \sum_s\sum_r a_{rs}\right]$$

and

$$p = \frac{\sum_s \sum_r r\, a_{rs} - \sum_s \sum_r a_{rs}}{\sum_s \sum_r s\, a_{rs} - \sum_s \sum_r a_{rs}}.$$

However, $\sum_s \sum_r r\, a_{rs} = R$; $\sum_s \sum_r a_{rs}$ is the total number of sibships re-

corded, say N; and $\sum_s \sum_r s\, a_{rs}$ is the total number of children recorded, say

T. Thus

$$p^* = \frac{R - N}{T - N}. \qquad (14.3.5)$$

The variance of this estimate may be obtained as follows: Since

$$\frac{d\,(\log L)}{dp} = \sum_s \sum_r \left[\frac{r}{pq} - \frac{1}{p} - \frac{s}{q} \right] a_{rs} ,$$

$$\frac{d^2\,(\log L)}{dp^2} = \sum_s \sum_r \left[\frac{r}{p\,q^2} - \frac{r}{p^2 q} + \frac{1}{p^2} - \frac{s}{q^2} \right] a_{rs} ;$$

or

$$\frac{d^2\,(\log L)}{dp^2} = \frac{R}{p\,q^2} - \frac{R}{p^2 q} + \frac{N}{p^2} - \frac{T}{q^2} = \frac{(N - T)}{p\,q}.$$

Now

$$\sigma^2 = \left\{ -E\left[\frac{d^2\,(\log L)}{dp^2} \right] \right\}^{-1} ,$$

and, by replacing p and q by their expected values, we obtain

$$\sigma^2 = \frac{(T - R)\,(R - N)}{(T - N)^3}. \qquad (14.3.6)$$

The test of the significance of a deviation of the estimated from the theoretical proportion is the same as in the case of complete ascertainment, that is, the deviation is judged by the standard error of the estimated proportion. The conventional level for asserting a significance is a deviation of two or more standard errors from the estimate.

2b. *Multiple selection:* As was previously indicated, we may visualize two distinctly different sets of data involving multiple selection. In one, each propositus or proband is clearly identified, and hence we know the number of independent ascertainments in each family. In the alternative situation, such as is generally the case if one utilizes data available in the medical literature, the number of independent ascertainments is not known, and we have, at most, only a vague indication of the probability of ascertaining an affected individual. Methods appropriate for handling data obtained in the first fashion have been developed by Fisher (1934) and Bailey (1951a). These methods, however, are neither simply developed nor simply applied, and, since the alternative situation is presently more common, we shall limit our attention to the one case.

The method appropriate to the analysis of data when ascertainment is known to be incomplete but the number of independent ascertainments is unknown was developed by Haldane (1938). We have seen that the probability that a family of size s will be ascertained at least once is

$$[1 - (1 - k)^r](\tbinom{s}{r})p^r q^{s-r},$$

where k is the probability that a single individual will be recorded. The expected number of s-child sibships with r affected is, now, some constant *times* the preceding probability, or (using the same notation)

$$E(a_{rs}) = c[1 - (1 - k)^r]\binom{s}{r}p^r q^{s-r},$$

where c is the constant to be determined. Now the total number of s-child sibships will be the sum of such sibships with $r = 1, \ldots, s$ affected, or

$$n_s = c\sum_{r=1}^{s}\binom{s}{r}[1 - (1 - k)^r]p^r q^{s-r}.$$

We solve for c as follows:

$$n_s = c\left[\sum_{r=1}^{s}\binom{s}{r}p^r q^{s-r} - \sum_{r=1}^{s}(1 - k)^r\binom{s}{r}p^r q^{s-r}\right];$$

but

$$\sum_{r=1}^{s}\binom{s}{r}p^r q^{s-r} = \sum_{r=0}^{s}\binom{s}{r}p^r q^{s-r} - q^s = 1 - q^s,$$

and

$$\sum_{r=1}^{s}\binom{s}{r}(1 - k)^r p^r q^{s-r}$$

is equal to

$$(1 - kp)^s - q^s ;$$

whence

$$n_s = c[1 - (1 - kp)^s]$$

or

$$c = \frac{n_s}{1 - (1 - kp)^s}.$$

Thus

$$E(a_{rs}) = \frac{\binom{s}{r}[1-(1-k)^r]\,p^r q^{s-r}n_s}{1-(1-kp)^s}.$$

Our data consist of a set of values of a_{rs} for different values of r and s; hence the likelihood function is

$$L = \prod_s \prod_r \left\{ \frac{\binom{s}{r}[1-(1-k)^r]\,p^r q^{s-r}}{1-(1-kp)^s} \right\}^{a_{rs}},$$

or

$$\log L = \sum_s \sum_r \{\, r\log p + (s-r)\log q - \log[1-(1-kp)^s]\,\}\, a_{rs} + \text{constant}.$$

Now

$$\frac{d(\log L)}{dp} = \sum_s \sum_r \left[\frac{r}{pq} - \frac{s}{q} - \frac{ks(1-kp)^{s-1}}{1-(1-kp)^s}\right] a_{rs} = 0,$$

$$\sum_s \sum_r \frac{r\,a_{rs}}{pq} = \sum_s \sum_r \left\{\frac{s\,[1-(1-kp)^{s-1}(1-k)\,]}{q\,[1-(1-kp)^s]}\right\} a_{rs},$$

$$\frac{R}{p} = \sum_s \sum_r \left[\frac{1-(1-kp)^{s-1}(1-k)}{1-(1-kp)^s}\right] s\,a_{rs},$$

$$\frac{R}{p} = \sum_s \left[\frac{1-(1-kp)^{s-1}(1-k)}{1-(1-kp)^s}\right] s\,n_s, \qquad (14.3.7)$$

where $n_s = \sum_r a_{rs}$.

The test of the agreement of the data with the hypothesis again consists of a comparison of the theoretical proportion of affected with the estimate of p based on the data. The significance is judged by the standard error of the estimate, where the standard error of the estimate is the rather cumbersome formula

$$\frac{1}{\sigma} = \sqrt{\frac{\sum\limits_s s\,n_s}{pq} + \frac{k(q-p+kp^2)}{pq}\sum_{s=2}\frac{(1-kp)^{s-2}s\,n_s}{1-(1-kp)^s} - k^2\sum_{s=2}\frac{(1-kp)^{s-2}s^2 n_s}{[1-(1-kp)^s]^2}}.$$

$$(14.3.8)$$

The reader may verify for himself that the two cases, $k = 1$ and $k = 0$, are, in fact, limiting cases of this more general case.

If we re-examine Sjögren's data (Table 14-3), it is patent that ascertainment was not complete, since only 73 of the 96 cases were independently recorded. These data would suggest, however, a fairly high probability of ascertaining an affected individual. Suppose that we assume that k, the probability of ascertaining an affected individual, was 0.80. Under this assumption, the maximum-likelihood estimate obtained by solving equation (14.3.7) by trial and error and interpolating between the two closest values is found to be $p^* = 0.211$, with standard error equal to 0.0287. The discrepancy between the theoretical and estimated proportions is $0.250 - 0.211$, or 0.039, and is clearly not significant.

14.4. *The nature of genetic data.*—It is all too common in the literature of medical genetics to come upon a collection of laboriously acquired data presented with a fine disregard for the principles incorporated in the last two

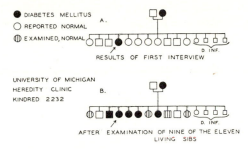

Fig. 14-1.—A pedigree of diabetes mellitus before and after glucose tolerance studies

chapters. On the other hand, one may also encounter elaborate statistical presentations of data whose accuracy is very much open to question. It cannot be too strongly emphasized that the most refined statistical techniques cannot overcome the biases and inaccuracies introduced by poor data. Hearsay evidence is always suspect. All too frequently, the propositus for the kindred is either unaware of, or unwilling to talk about, the presence of a similar or other defect in the other members of the kindred. Figure 14-1 illustrates how far astray hearsay evidence may lead one. Figure 14-1, *a*, is a portion of a pedigree of diabetes mellitus supplied by a sincere, interested, and intelligent propositus. Figure 14-1, *b*, depicts the true occurrence of diabetes mellitus in the sibship at that particular time, as discovered by the administration of a glucose tolerance test to most of the members of the propositus' sibship. Four new cases of diabetes mellitus were brought to light by the test. Some of these individuals had mild symptoms but had not yet consulted a physician. This underreporting of the number of affected persons in a kindred where a disease of insidious onset is concerned can be avoided only by firsthand contacts with the family.

14.5. *The testing of complex data for agreement with a simple genetic hypothesis.*
—In chapter 6 we pointed out that not infrequently individuals may fail to
show the effects of heterozygosity for a gene which usually behaves as a
dominant, or the effects of homozygosity for a gene which is usually regarded
as a recessive ("incomplete penetrance"). We have also pointed out that a
given hereditary disease may have a variable age of onset. For instance, individuals with the dominantly inherited type of peroneal atrophy may develop the disease any time between the first and sixth decades (mean =
18.95 years, S.D. = 13.59 years) (Bell, 1935). In the present chapter we have
discussed the various types of ascertainment of hereditary disease and the
statistical procedures appropriate to each type. There are diseases with a
hereditary element in whose study one may encounter all three of these statistical problems, i.e., incomplete penetrance, a variable age of onset, and an
uncertain degree of ascertainment. In addition, the possibility that the disease is genetically heterogeneous hangs constantly over the investigator's
head. Diabetes mellitus appears to be an example of such a disease. It is a
relatively simple matter to demonstrate an increased incidence of diabetes
mellitus among the relatives of diabetics. But the further step, of testing the
observed results for agreement with expectation on some simple Mendelian
hypothesis, presents serious problems. It is in part because different authors
make different assumptions concerning these variables that we have so many
theories concerning the inheritance of diabetes mellitus (for a review of the
literature see Harris, 1950; Steinberg and Wilder, 1952). The genetic approach to a collection of medical data can, on occasion, contribute impor-
tantly to clarifying clinical concepts. On the other hand, there are times
when the proper genetic analysis of a given disease must await certain medical advances. It is just as important that the student of human genetics become familiar with the limitations of his approach and the data with which
he works as that he learn all the "tricks of the trade."

Bibliography

SPECIFIC REFERENCES

BAILEY, N. T. J. 1951*a*. The estimation of the frequencies of recessives with incomplete multiple selection, Ann. Eugenics, **16**:215–22.

BELL, J. 1935. On the peroneal type of progressive muscular atrophy, Treas. Human Inherit., **4**:69–140.

FISHER, R. A. 1934. The effects of methods of ascertainment upon the estimation of
frequencies, Ann. Eugenics, **6**:13–25.

HALDANE, J. B. S. 1932. A method for investigating recessive characters in man,
J. Genetics, **25**:251–55.

———. 1938. The estimation of the frequencies of recessive conditions in man, Ann.
Eugenics, **8**:255–62.

HARRIS, H. 1950. The familial distribution of diabetes mellitus: a study of the relatives of 1,241 diabetic propositi, Ann. Eugenics, 15:95–119.
HOGBEN, L. 1946. An introduction to mathematical genetics. New York: W. W. Norton & Co., Inc.
SJÖGREN, T. 1931. Die juvenile amaurotische Idiotie—klinische und erblichkeitsmedizinische Untersuchungen, Hereditas, 14:197–426.
———. 1943. Klinische und erbbiologische Untersuchungen über die Heredoataxien, Acta psychiat. et neurol., Suppl. XXVII. Copenhagen: E. Munksgaard.
STEINBERG, A. G., and WILDER, R. M. 1952. A study of the genetics of diabetes mellitus, Am. J. Human Genetics, 4:113–35.

GENERAL REFERENCES

BAILEY, N. T. J. 1951b. A classification of methods of ascertainment and analysis in estimating the frequencies of recessives in man, Ann. Eugenics, 16:223–25.
LUDWIG, W., and BOOST, C. 1940. Vergleichende Wertung der Methoden zur Analyse recessiver Erbgänge beim Menschen, Ztschr. f. menschl. Vererb.- u. Konstitutionslehre, 24:577–619.

Problems

1. Among 98 families where one parent had neurofibromatosis, it was found that, of the 221 offspring, 65 males were normal and 55 were neurofibromatous, 49 females were normal and 52 were neurofibromatous. Are these data compatible with the hypothesis that neurofibromatosis is due to a single, rare, dominant gene?

2. Calculate the estimate of the proportion of affected in the data of Table 14-3, assuming single selection. Under this circumstance, is the hypothesis of a single, rare, recessive gene tenable?

3. Sjögren (1931), in a study of juvenile amaurotic idiocy, ascertained 108 cases of the disease distributed among 52 families as follows:

Size of Family	No. of Families	No. of Affected	Size of Family	No. of Families	No. of Affected
2	3	5	8	7	25
3	7	10	9	2	6
4	10	19	10	0	0
5	7	11	11	1	3
6	6	9	12	1	3
7	7	13	13	1	4

Assuming complete ascertainment, are these data consistent with the hypothesis of a single, rare, recessive gene?

4. If ascertainment is assumed to be only 50 per cent complete, are the data in problem 3 consistent with the hypothesis of a single, rare, recessive gene?

5. If the probability of ascertaining a "sporadic" case of inherited disease is higher than the probability of ascertaining a "familial" case, i.e., one of several affected in a family, what would be the effect on mutation estimates?

CHAPTER 15

Population Genetics

15.1. Methods available for evaluating the biological relationship of different human groups.—From earliest recorded history man has been fascinated by the question of his origin and the manner in which the society to which he belongs is related to other societies. The classical methods of approach to the problem are discussed below.

1. PHYSICAL ANTHROPOLOGY

In essence, the physical anthropologist has been concerned with developing quantitative techniques for describing similarities and differences between groups of people. His techniques may be divided into two types, according to whether they are concerned with the measurement of the perishable soft tissues—skin and hair color, eye color, etc.—or the less perishable hard tissues—the skeleton. Both types of techniques may be applied to living populations, but only the second type to populations known entirely from skeletal remains. Of the various measures used in describing individuals and groups of people, perhaps the most widely employed is the cephalic index, which may be defined as the breadth of the skull expressed as a percentage of the length. A skull with a cephalic index below 75 is termed "dolichocephalic." A skull with an index over 80 is termed "brachycephalic." Skulls with intermediate values are "mesocephalic."

2. CULTURAL ANTHROPOLOGY

The approach of the cultural anthropologist consists in an attempt to measure origin and relationship on the basis of similarity in cultural pattern, the use of certain materials for certain purposes, the manner in which these materials are used, religious tradition, etc.

3. LINGUISTICS

Those employing the linguistics approach are concerned with the more plastic spoken and the less changeable written language—similarities in roots, syntax, etc.

230

4. DOCUMENTATION

Documentation, the way of the historian, depends upon an evaluation of the written records concerning human populations. Where such records exist and are reliable, they are one of the best approaches to the problem—but, unfortunately, documentary records carry us back no more than several thousand years.

Recently there has been added to these classical techniques a powerful new approach, only now in the midst of a rapid evolution, its full impact yet to be felt. In a previous chapter the steps whereby we can estimate the frequency with which various genes occur in a population have been discussed. These estimates enable us, given the occurrence of suitable inherited traits, to characterize populations in terms of the percentage representation of specific genes. The recent spectacular developments in the field of serological genetics, referred to in chapter 8, provide us with exactly the material most amenable to these techniques, i.e., a series of inherited traits which can be precisely and rapidly studied in large groups of people and which, in contrast to the characters often employed by the physical anthropologist, are little, if at all, influenced by environmental variables. This new approach—the study of the precise genetic composition of groups of people and the factors which bring about change in this composition—is usually termed "population genetics."

The fact that human populations might differ as regards the frequency of various blood-group genes was first brought out by the Hirszfelds in 1919. Since that time there has been a constantly accelerating stream of information on this subject. At the present time a truly staggering amount of work of this type is under way. Much of this has been summarized in the excellent book by Boyd (1950), on which we shall draw heavily in the following discussion. The danger for the student is that he may become so immersed in a welter of statistics concerning the frequency of various traits in various groups that he will lose sight of the real contributions of the data. In what follows, we shall make no attempt to summarize the work to date in any detail but shall concentrate on examples of the results which have emerged thus far. We shall consider, first, the kinds of gene-frequency differences known to exist between groups of people, then the problems which arise in the interpretation of these differences, and, finally, the present status of attempts at synthesis.

For didactic purposes we shall distinguish between genes which are of "universal" distribution, being observed in all or almost all populations studied thus far (sec. 15.2), and genes of more restricted distribution, apparently largely confined to specific groups (sec. 15.3). In a later section we shall consider the question of whether this distinction recognizes basic differences between the genes involved or is more one of convenience in the light of our present knowledge. We will confine our attention to genes which have a frequency of at least 5 per cent in some population; no attempt will be made to discuss the poorly studied distribution of the rare genes responsible for pathologic traits.

15.2. *Genes of universal distribution.—*

1. THE A, B, O BLOOD GROUPS

The A and B serological reactions are, in all probability, of great antiquity. Similar serological reactions have been observed in a variety of primates, and it seems a reasonable postulate that very primitive man could be classified as A, B, O, or AB. A simple tabulation of the results of studies of A, B, O frequencies in various populations carried out before 1938 requires seventy-five pages (Boyd, 1939). Table 15-1 contains the results of some of the many studies to date. It will be observed that different groups of people vary widely in the frequency with which the three genes responsible for these serological reactions are represented. At one extreme we find certain American Indian tribes in which O genes occur to the virtual exclusion of A and B, whereas at the other extreme there are groups in which the ratio of $A:B:O$ genes is approximately $1:1:2$. Despite considerable local fluctuation, the significance of which we will discuss later, there is a certain regularity in the distribution of the genes the world over. This is brought out in Figures 15-1 to 15-3. In these maps, of course, it is the blood-group frequencies of the "native" populations which have been plotted rather than those of the more recent immigrants. For instance, the findings for Australia and North and South America concern the aborigines and Indians rather than the present-day inhabitants.

The most regular pattern of distribution is shown by the gene B. The highest incidence is observed in Central Asia and in northern and central India, with the incidence decreasing in all directions as we move out from this center. There is a secondary "high" in the Nile Valley region of Africa. In Australia and North America the gene is practically nonexistent and, where found, may be a relatively recent introduction. There also appears to be a low frequency of the B gene in South America. Gene A has a somewhat more universal distribution, with, according to Lundman (1948), four more or less distinct highs: (1) the whole of Europe with a great part of the Near East excepting Arabia, but including Egypt and the northern half of the Somali Peninsula; (2) the Australian continent, but not New Guinea; (3) the outer (eastern) part of Polynesia; and (4) Japan and also a great part of China. There are rather localized highs among the Tibetans, the Bushmen in

TABLE 15-1*

Frequencies of Blood Groups O, A, B, and AB in Typical Populations

Population	Place	No. Tested	O	A	B	AB	p	q	r
Low A, Virtually No B									
Amer. Indians.... (Utes)	Montana	138	97.4	2.6	0	0	0.013	0.0	0.987
Amer. Indians.... (Toba)	Argentina	194	98.5	1.5	0	0	.007	0	.993
Amer. Indians.... (Kwakiutl)	B. Columbia	123	85.4	12.2	2.4	0	0.063	0.013	0.926
Moderate A, Virtually No B									
Amer. Indians.... (Navaho)	New Mexico	359	77.7	22.5	0	0	0.125	0	0.875
High A, Little B									
Amer. Indians.... (Blackfeet)	Montana	115	23.5	76.5	0	0	0.515	0	0.485
Australian aborigines............	West Australia	243	48.1	51.9	0	0	.306	0	.694
Basques..........	San Sebastián	91	57.2	41.7	1.1	0	.239	0.008	.756
Polynesians.......	Hawaii	413	36.5	60.8	2.2	0.5	.382	.018	.604
Eskimo..........	CapeFarewell	484	41.1	53.8	3.5	1.4	.333	.027	.642
Australian aborigines............	Queensland	447	58.6	37.8	3.6	0	0.216	0.023	0.766
Fairly High A, Some B									
West Georgians...	Tiflis	707	59.1	34.4	6.1	0.4	0.198	0.038	0.769
English..........	London	422	47.9	42.4	8.3	1.4	.250	.050	.692
Belgians..........	Liége	3,500	46.7	41.9	8.3	3.1	.257	.058	.684
Swedes..........	Stockholm	633	37.9	46.1	9.5	6.5	.301	.073	.616
Icelanders........	Iceland	800	55.7	32.1	9.6	2.6	.190	.062	.747
Danes...........	Copenhagen	1,261	40.7	45.3	10.5	3.5	.290	.078	.638
Armenians.......	From Turkey	330	27.3	53.9	12.7	6.1	.379	.110	.523
Lapps...........	Finland	94	33.0	52.1	12.8	2.1	.323	.078	.574
Melanesians......	New Guinea	500	37.6	44.4	13.2	4.8	.293	.099	.613
Germans.........	Berlin	39,174	36.5	42.5	14.5	6.5	0.285	0.110	0.604

* Condensed by permission from Genetics and the races of man, by Dr. William C. Boyd, copyright 1950, Little, Brown and Co. In this table the frequencies of the blood groups O, A, B, and AB are given in percentages, but the gene frequencies are given as straight frequencies, following common practice. The symbol p represents the frequency of the gene A; q, the frequency of the gene B; and r, the frequency of the gene O.

TABLE 15-1—*Continued*

POPULATION	PLACE	No. TESTED	O	A	B	AB	p	q	r
				High A, High B					
East Georgians....	Tiflis	1,274	36.8	42.3	15.0	5.9	0.283	0.113	0.607
Bulgarians........	Sofia	6,060	32.1	44.4	15.4	8.1	.308	.123	.567
Italians..........	Sicily	540	45.9	33.4	17.3	3.4	.213	.118	.678
Siamese..........	Bangkok	213	37.1	17.8	35.2	9.9	.148	.257	.595
Finns............	Háme	972	34.0	42.4	17.1	6.5	.285	.126	.583
"Berbers"........	Algiers	300	39.0	37.6	18.6	4.6	.251	.134	.625
Japanese.........	Tokyo	29,799	30.1	38.4	21.9	9.7	.279	.172	.549
Madagascans.....	Madagascar	266	45.5	27.5	22.5	4.5	.180	.151	.674
Russians.........	Near Moscow	489	31.9	34.4	24.9	8.8	.250	.189	.565
Abyssinians......	Abyssinia	400	42.8	26.5	25.3	5.0	.178	.172	.654
Egyptians........	Cairo	502	27.3	38.5	25.5	8.8	.288	.203	.523
Bogobos.........	Philippines	302	53.6	16.9	26.5	3.0	.107	.163	.732
Chinese..........	Huang-Ho River	2,127	34.2	30.8	27.7	7.3	.220	.201	.587
Iraqis............	Baghdad	386	33.7	31.4	28.2	6.7	.226	.208	.581
Pygmies..........	Belgian Congo	1,032	30.6	30.3	29.1	10.0	.227	.219	.554
Asiatic Indians....	Goa	400	29.2	26.8	34.0	10.0	.208	.254	.540
Javanese.........	Ampelgading	450	30.4	24.7	37.3	7.6	.190	.271	.552
Buriats..........	North Irkutsk	1,320	32.4	20.2	39.2	8.2	0.156	0.277	0.570

South Africa, Congo Pygmies, Negritos in the Philippine Islands, and some "hill tribes" of southern India.

As noted in chapter 8, there appear to be two different allelic genes which may produce the A-reaction, designated as A_1 and A_2. Although the majority of studies on the blood groups of populations have not distinguished between the presence of A_1 and A_2, a few have, with interesting results. On the basis of the available evidence, it appears that gene A_2 does not occur in the Australian aborigines, in China or Japan, in the American Indians, or in the natives of the islands of the Pacific. On the other hand, the frequency of the A_2 gene among western Europeans is in the neighborhood of 6–7 per cent.

As will be apparent from Table 15-1 and Figures 15-1 to 15-3, the A, B, O frequencies are in some instances actually quite similar for groups which, by other anthropological standards, are not closely related. Thus Boyd (1950) has pointed out that the A, B, O frequencies of Cantonese Chinese, Russians from Kazan, and Katanga Negroes from the Belgian Congo are all essentially the same. Since the three groups in question are quite different from one another, what does this imply with respect to the utility of the blood groups in this context? The answer is very simple. Just as no one anthropometric measurement can serve as a basis for classifying the relationships of all the groups of people inhabiting the world, so no one gene frequency is adequate. Rather, as we shall see, a battery of gene frequencies is necessary.

2. THE MNS-TYPES

The results of selected studies on the frequencies of the two genes, M and N, responsible for the MN blood types are shown in Table 15-2. The frequency of the M gene varies from as high as 91.3 per cent in Greenland Eskimos to as low as 16.0 per cent in Australian aborigines. The frequency of the gene N, of course, has a reciprocal relationship to the frequency of M. The S-s reaction, described earlier as serving to subdivide the MN-types, has been so recently discovered that very few studies of its geographic rela-

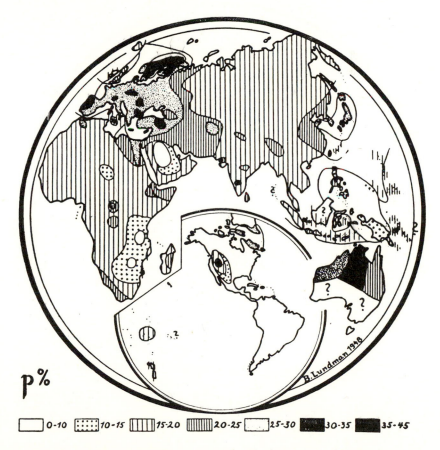

FIG. 15-1.—The world-wide frequency of the blood-group gene $A(p)$. (By permission from Umriss der Rassenkunde des Menschen in geschichtliches Zeit, by Dr. B. Lundman, copyright 1952, E. Munksgaard; also Evolution, Vol. **2**, 1948.)

tionships are yet available. That it, too, will vary significantly is suggested by the fact that, whereas the frequency of the four genes, or linked complexes, MS, Ms, NS, and Ns in Englishmen was found to be 24.72, 28.31, 8.02, and 38.95 per cent, respectively; in Australian aborigines the frequencies are 0.0, 25.56, 0.0, and 74.44 per cent, respectively (Race and Sanger, 1950).

3. THE RH-TYPES

The discovery of the Rh-types is also so relatively recent that there still exist very large gaps in our knowledge of the distribution of the various Rh

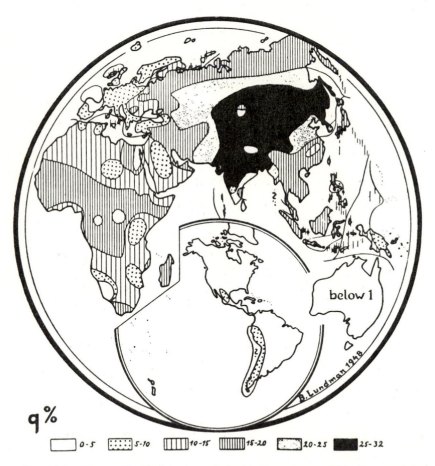

FIG. 15-2.—The world-wide frequency of the blood-group gene $B(q)$. (By permission from Umriss der Rassenkunde des Menschen in geschichtliches Zeit, by Dr. B. Lundman, copyright 1952, E. Munksgaard; also Evolution, Vol. 2, 1948.)

genes. However, it is already clear that very significant differences do exist from one population to the next. For instance, the r (cde) gene varies in frequency from zero or near zero in Mexican Indians, Australian aborigines, and Chinese to 55 per cent among the Basques. The R^o (cDe) gene, on the other hand, has a frequency of 5 per cent or less in Englishmen, Basques, Australian aborigines, Hindus, Chinese, and Mexican Indians, but 60 or more per cent in African Negroes and Pygmies.

4. OTHER BLOOD TYPES

The most recently discovered blood types have so far been little utilized in anthropological studies. That very considerable variation will also be en-

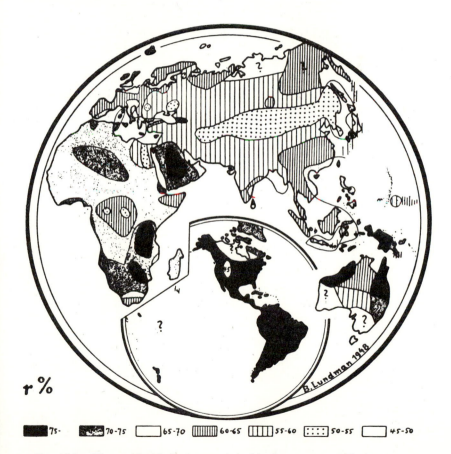

$r\%$

■ 75-   70-75   65-70   60-65   55-60   50-55   45-50

FIG. 15-3.—The world-wide frequency of the blood-group gene $O(r)$. (By permission, from Umriss der Rassenkunde des Menschen in geschichtliches Zeit, by Dr. B. Lundman, copyright 1952, E. Munksgaard; also Evolution, Vol. 2, 1948.)

TABLE 15-2*

Studies on Frequency of MN Blood Types

Population	Place	No. Tested	M	MN	N	m	n
Populations with Low N and therefore High M							
Eskimo............	East Greenland	569	83.5	15.6	0.9	0.913	0.087
Amer. Indians......	New Mexico	361	84.5	14.4	1.1	.917	.083
(Navaho)							
Aleuts............	Aleutian Islands	132	67.5	29.4	3.2	.822	.179
Arabs, Bedouin.....	Near Damascus	208	57.5	36.7	5.8	.758	.241
(Rwala)							
Amer. Indians......	New Mexico	140	59.3	32.8	7.9	0.757	0.243
(Pueblo)							
Populations with Low M (High N)							
Australian aborigines	Queensland	372	2.4	30.4	67.2	0.176	0.824
Papuans...........	Papua	200	7.0	24.0	69.0	.190	.810
Fijians............	Fiji	200	11.0	44.5	44.5	.332	.667
Ainu..............	Shizunai	504	17.9	50.2	31.9	0.430	0.570
Populations with "Normal" M and N Frequencies							
Negroes...........	New York City	730	....	....	28.1	....	0.530
Basques...........	Spain	91	23.1	51.6	25.3	0.489	.511
Filipinos..........	Leyte, Samar, etc.	382	25.9	50.3	23.8	.510	.490
Egyptians.........	Cairo	502	27.8	48.9	23.3	.522	.477
English...........	London	422	28.7	47.4	23.9	.524	.476
Lapps............	Inari, Finland	56	28.6	48.2	23.2	.527	.473
Indonesians.......	Java, etc.	296	30.4	45.6	24.0	.532	.468
Danes............	Copenhagen	2,023	29.1	49.5	21.4	.538	.461
Germans..........	Berlin	8,144	29.7	50.7	19.6	.550	.449
Russians..........	Leningrad	701	32.0	46.7	21.3	.553	.446
Japanese..........	Tokyo	1,100	32.4	47.2	20.4	.560	.440
Syrians...........	Meshghara	306	30.7	52.0	17.3	.567	.433
Armenians........	From Turkey	339	34.2	45.4	20.4	.569	.431
Chinese...........	Hong Kong	1,029	33.2	48.6	18.2	.575	.425
Ukrainians........	Kharkov	310	36.1	44.3	19.6	.583	.417
Welsh............	Northern towns	192	30.7	55.3	14.0	.583	.416
Estonians.........	Estonia	310	34.8	49.7	15.5	.596	.403
Iraqis............	Baghdad	387	37.0	47.0	16.0	.605	.395
Arabs, Bedouin.....	Near Mosul	206	36.9	49.5	13.6	.616	.383
(Jabour)							
Finns.............	Karjala	398	45.7	43.2	11.1	0.673	0.327

* Condensed by permission from Genetics and the races of man, by Dr. William C. Boyd, copyright 1950, Little, Brown and Co. M, MN, and N stand for the percentages of the M, MN, and N blood types; *m* stands for the frequency of the gene resulting in the M reaction; and *n* stands for the frequency of the gene resulting in the N reaction.

countered with reference to these traits is indicated by the fact that, whereas the gene responsible for the Kidd blood-group antigen has a frequency of 47.7 per cent in white Americans (Allen, Diamond, and Niedziela, 1951), in a small sample of African Negroes the frequency was 78.2 per cent (Ikin and Mourant, 1952). Even more striking, the frequency of the Duffy reaction, due to the Fy^a gene, has been found to vary from zero in pure-bred Brazilian Indians to 99 per cent in Chinese (Pantin and Junqueira, 1951).

5. OTHER GENETIC TRAITS OF VALUE TO THE ANTHROPOLOGIST

Although in the wide sense all the various measurements traditionally used by the physical anthropologist in his attempts to describe groups of people are at least in part genetically determined, most of these are genetically complex and subject in their expression to environmental influences. There are, however, a few other traits of wide distribution which have a relatively simple genetic basis, and have been incorporated into population surveys. These include those discussed below.

a) *Ability to taste phenylthiocarbamide:* In 1932 it was discovered that, whereas some people taste the chemical known as phenylthiocarbamide (PTC) as an intensely bitter substance, others find it more or less tasteless (Fox, 1932). There is considerable difference between "tasters" in the concentration at which they can first identify the presence of the chemical. Also, some "nontasters" actually can taste the chemical in very concentrated form. Furthermore, there are significant differences between the sexes in their ability to taste the compound. When, however, for either sex one plots the concentration at which the chemical is tasted against the number of persons tasting at each concentration, there is observed a bimodal curve with few intermediates. The ability to taste appears to depend on a dominant gene. Occasional exceptions are not enough to invalidate the use of the gene for gene-frequency studies. The percentage of tasters varies from 98 in Ramah Navaho to 67 in western Europeans.

b) *Secretor:* Some individuals have the A, B, O blood-group substances in their various bodily secretions—saliva, gastric juice, urine, etc.—whereas others do not. This difference is apparently due to a single gene, with the ability to secrete the substances due to a dominant gene, S. Among the American Utah Indians the S gene has a frequency of 100 per cent, whereas among Negroes living in New York the frequency of the S gene is 38 per cent.

c) *Color blindness:* There are several types of color blindness, the most common of which, red-green color blindness, appears to depend on a single, recessive, sex-linked gene. The frequency of this gene appears to vary from approximately 1 per cent in Eskimos, Melanesians, and Navahos, to 8 per cent in Europeans and white Americans.

15.3. *Genes of restricted distribution.*—The genes which have been discussed thus far in the chapter have a very wide distribution. By contrast, there are several genes whose corresponding phenotype is readily identified which exhibit a more marked restriction to certain peoples and regions. Two of the better known of these are the genes responsible for thalassemia and for the sickling phenomenon.

1. THALASSEMIA

The condition known as "thalassemia" occurs in two forms, a severe anemia, known as "thalassemia major" (Cooley's anemia), and a much milder abnormality, known as "thalassemia minor" (Cooley's trait). Children with thalassemia major, with very few exceptions, die before the age of reproduction. Individuals with thalassemia minor, on the other hand, although their blood cells are somewhat abnormal in appearance, have not yet been demonstrated to have an impaired state of health. Genetic studies have shown that thalassemia major is due to homozygosity for the same gene which when heterozygous produces thalassemia minor. The distinction between the major and minor forms of thalassemia is usually readily made, although occasionally, owing to genetic and environmental modifiers, intermediates are seen (reviews in Silvestroni and Bianco, 1949; Neel, 1951; Bianco, Montalenti, Silvestroni, and Siniscalco, 1952).

In the years since thalassemia major was first recognized as a distinct clinical entity, most of the case reports appearing in the medical literature have been based upon people inhabiting a restricted geographical area and such descendants of theirs as had in relatively recent times spread throughout the world. This restricted geographical area comprises the countries occupying the northeast sector of the Mediterranean coast—Italy, Greece, Turkey, and Syria—and the islands of this same sector of the Mediterranean Sea. Although there have been no actual surveys of the number of children affected with thalassemia major in such countries as Italy and Greece, a small survey carried out in Rochester, New York, on persons of southern Italian and particularly Sicilian derivation suggested that about one in every 2,400 babies born to this group had the disease. But while there have been no surveys on the frequency of thalassemia major in Italy, extensive work has been carried out on the frequency of thalassemia minor. The findings are summarized in Figure 15-4. It is apparent that in Italy there are two peaks as regards the frequency of the heterozygote with thalassemia minor, one in the Po River Valley (Ferrara, Rovigo), and the other in southern Italy, Sicily, and Sardinia. It is interesting in this connection that Mourant (1950) has reported that the inhabitants of Ferrara and of Sardinia (no figures on Sicily) are also characterized by a relatively high frequency of the CDe (R^1) chromosome and a relatively low frequency of cde (r).

That the frequency of thalassemia minor may be even higher in some other parts of the Mediterranean Basin than in Italy is suggested by the studies of Banton (1950), who observed that on the island of Cyprus approximately 20 per cent of the inhabitants had hematological findings consistent with the diagnosis of thalassemia minor. Very recently (Minnich *et al.*, 1954) it has been reported that thalassemia major has a very considerable frequency in

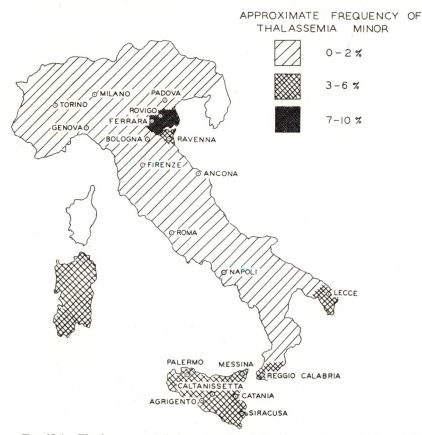

APPROXIMATE FREQUENCY OF
THALASSEMIA MINOR

0 - 2 %

3 - 6 %

7 - 10 %

FIG. 15-4.—The frequency of thalassemia minor in Italy, according to the data of Silvestroni and collaborators (1952).

Thailand. This unexpected finding, which raises some very interesting questions, awaits confirmation.

2. THE SICKLING PHENOMENON

The genetics of the sickling phenomenon were discussed in chapter 12. Studies on a peculiar anemia first observed in the American Negro led to the recognition of the occurrence in American Negroes of a gene which, when heterozygous, resulted in the asymptomatic sickle-cell trait, but which, when homozygous, resulted in a severe and often fatal anemia (Neel, 1951). The distribution of the sickling phenomenon in Africa today is shown in Figure 15-5. These figures refer to the total of heterozygotes and homozygotes. Although there are wide differences between tribes in relatively close geographical proximity, in general we may speak of a broad belt occupying the central half of Africa in which the frequency of the phenomenon is in the

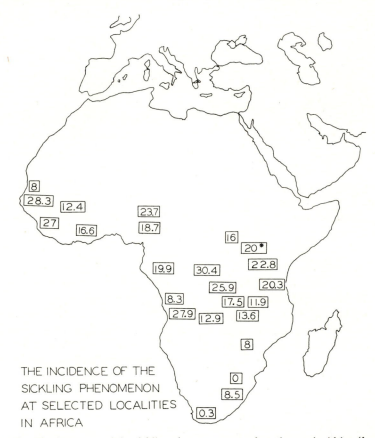

THE INCIDENCE OF THE
SICKLING PHENOMENON
AT SELECTED LOCALITIES
IN AFRICA

Fig. 15-5.—The frequency of the sickling phenomenon at selected areas in Africa (from Neel, 1951*b*, with additions). The entry of 20 per cent indicated by the asterisk is of especial interest because this particular study, carried out by Lehmann and Raper (1949), indicates the heterogeneity of the African "Negro" with respect to this trait. In this study the average frequency of the sickling phenomenon in four tribes speaking a Hamitic-type language varied between 0.8 and 3.9 per cent, whereas in seven different tribes speaking a Nilotic-type language the frequency varied between 21.0 and 28.0 per cent, and in eleven Bantu-language tribes the average frequency of the sickling phenomenon varied all the way from 2.0 to 45.0 per cent.

neighborhood of 15–20 per cent, with very much lower values in North and South Africa. There are rather striking fluctuations from tribe to tribe. Thus Lehmann and Raper (1949), in their comprehensive studies in the Upper Nile region, found that in four different tribes speaking a Hamitic type of language, the frequency of individuals whose blood cells sickled varied between 0.8 and 3.9 per cent; in seven different Nilotic-language-group tribes the frequency varied between 21.0 and 28.0 per cent; whereas in tribes speaking a Bantu-type language, the frequency of sicklers varied all the way from 2.0 to 45.0 per cent.

Outside Africa and populations largely of African derivation, the gene has, with a few scattered exceptions, been observed with any frequency only in (1) Greeks in several different villages and cities (Choremis, Zervos, Constantinides, and Zannos, 1951; Caminopetros, 1952) and (2) certain Dravidian-speaking inhabitants of southern India, in the terminology of Von Eickstedt the "Veddids" and to a much lesser extent their "Indid" and "North Indid" neighbors. Among 201 Veddids from three different tribes, the frequency of persons exhibiting the sickling phenomenon was 31.3 per cent (Lehmann and Cutbush, 1952). The question of the ethnological relationship of these Indians to the African Negro at once arises. However the R^o gene, which, as noted above, is found in some 60 per cent of African Negroes, was not detected in any of the 156 Veddids examined in this respect, although 9 per cent of this 156 had the usually very rare R^z gene. In addition, the Veddids differed from other southern Indians (and Negroes) in a relatively high incidence of the blood-type gene A and a low incidence of gene B. Their MN-type frequencies did not differ markedly from other Indians, but the N gene was less frequent than in Negroes. Figures regarding the frequency of the R^o gene are not available for Greeks, so that no conclusions can be reached as to whether the findings in Greece are due to a Negro admixture. It is of great interest, however, that the R^o frequency figures for the Veddids provide no evidence of affinity with the Negro and suggest either an independent origin of the trait or the introduction of the trait from this group or some common ancestor *into* the Negro.

15.4. *Problems of interpretation.*—It is one thing simply to catalogue the gene frequencies of the various groups of people inhabiting the world, it is quite another to attempt to interpret the findings. This is the old problem of passing from the descriptive to the interpretive phase of data. In the first flush of anthropological enthusiasm for the methods of population genetics, there was a tendency to treat the blood-group proportions as something fixed and immutable and hence capable of supplying the key to any number of anthropological riddles. The possibility arose of a system of classifying all mankind into groups solely on the basis of the present-day gene frequencies, with each group capable of being assigned its exact position on the basis of relatively small gene-frequency differences from other groups. It is now becoming increasingly apparent that over the course of generations there are many factors operative to alter gene frequencies. If a given population were to be subdivided into two groups, these factors, operating over a period of several hundred generations, are capable of bringing about marked differences in the relative frequencies of various specific genes. Recognition of this is leading to a more cautious interpretation of gene-frequency findings. These factors will now be discussed briefly, in order to provide the background against which the significance of the data regarding genetic differences between populations may be evaluated.

1. SELECTION

The evidence concerning the selective forces to which the blood groups are subjected has been summarized by Race and Sanger (1950), Mollison (1951), and Levine, Vogel, and Rosenfield (1953). Hemolytic disease in newborn infants due to maternal-fetal blood antigen incompatibilities, best known for the Rh antibodies but occurring, although more rarely, for the Kell, S, and A, B, O groups, is spectacular evidence of selection involving the blood groups. There is evidence that selection may also be at work in a more subtle fashion. Thus Waterhouse and Hogben (1947) have summarized all the findings in the literature concerning the blood types of the children of O mothers and A fathers and of A mothers and O fathers. In the former type of marriage there is thought to be a very significant deficiency of A children, amounting to almost 25 per cent. This would represent a fetal death rate of 8 per cent of A children, or 3 per cent of all conceptions. The most apparent explanation is that the O mother pregnant with an A fetus elaborates antibodies responsible for the death of the fetus at a relatively early stage of pregnancy. That the situation may be somewhat more complicated is suggested by the finding of Sanghvi (1951), to the effect that the percentage of male births was significantly less among A babies born to A mothers than among O infants born to O mothers or B infants born to B mothers.

Another instance of the possible operation of selection on the blood groups is the finding of a statistically significant excess of MN children from marriages involving two MN individuals. The data, as summarized by Wiener (1943), indicate that 58.4 per cent of the offspring of such marriages are MN, rather than the expected 50.0 per cent. Technical difficulties offer an alternative explanation of these findings.

There is, finally, some evidence that selection does not operate independently for the different blood groups but that, rather, there may be significant interrelationships. Levine in 1943 first drew attention to the fact that the mothers of children with hemolytic disease due to Rh antibodies were more often compatibly mated with respect to the A, B, O system than were a random series of women (review in Van Loghem and Spaander, 1948). A compatible mating in this sense is one in which the husband could be a blood donor to his wife as far as the A, B, O groups are concerned. While the explanation of this finding remains in doubt, it is clear that the strength of selection to which the Rh genes are subjected is not independent of the A, B, O gene frequencies.

It is important to recognize that the relative effectiveness of selection varies with the frequency of the gene in question. Consider the Rh situation. In a population where 50 per cent of the genes are r, and the other 50 per cent R (using the latter term to include all the various genes responsible for the Rh-positive reaction), then the death of a child with erythroblastosis fetalis removes one gene (or gene complex) of each kind, and the proportion is unchanged. But when less than 50 per cent of the genes are r in type, selection is against the r gene, whereas when more than 50 per cent of the genes are r, selection is now in favor of this gene. Levine, Vogel, and Rosenfield accept the figure that, in all its forms, Rh hemolytic disease occurs in a population with about 15 per cent Rh-negative individuals (the United States frequency) in 1 out of 150 to 1 in 200 of all full-term pregnancies. This estimate represents a minimum index of the amount of selection involving the Rh-system. Even so, the apparent magnitude of the selection involved is, in terms of the population geneticist, very considerable, suggesting an "unstable" system. Wiener (1942) and Haldane (1942) were quick to recognize this, the latter pointing out three alternative explanations of the findings, namely, that (1) recent environmental changes have altered for the worse the biological consequences of maternal-fetal Rh incompatibility; (2) the selection against the Rh heterozygotes is balanced by a type of counterselection yet to be discovered; or (3) the present genetical situation is of recent origin and probably due to racial mixture.

In the absence of evidence for either 1 or 2, both Haldane and Wiener appear to incline toward the third possibility. While this appears to present a solution of the immediate dilemma, we are then confronted with the even more perplexing problem of the origin of an Rh-positive and an Rh-negative race.

Reference should be made at this point to the phenomenon of *balanced polymorphism*. Consider a gene pair (or complex), *A-a*, with the corresponding genotypes *AA*, *Aa*, and *aa*. A considerable number of cases are now known from experimental genetics where *Aa* individuals enjoy a reproductive superiority over both *AA* and *aa*. It can readily be shown that under these circumstances the proportion of $AA:Aa:aa$ individuals tends to stabilize at some constant value. The relative frequencies at equilibrium of the three genotypes is a function of their relative reproductive efficiencies. Ford (1942) has suggested that herein may be the explanation of some of the blood-group findings. Such a hypothesis is difficult to prove for man.

Another mechanism which must be reckoned with in any consideration of selection in man is reproductive overcompensation, i.e., the possibility that certain parents, some of whose children are dying for genetic reasons, undertake more pregnancies than normal, with the end-result that their family is as large as, or even larger than, the average. The possibility that something of this nature may currently be operative for such diverse inherited diseases as hereditary spherocytosis, thalassemia, and erythroblastosis fetalis has been summarized by Glass (1950). In view of the high probability that the earth's population was stable or increasing only very slowly for long periods of time (it being all the average parent could do to raise two children to maturity), it seems likely that reproductive overcompensation, to the extent that it exists, is a product of relatively recent times, historically speaking, and was of little or no significance in the determination of present-day gene frequencies.

2. MUTATION

A second factor capable of disturbing gene frequencies is mutation. The subject of mutation has already been discussed at some length. There is relatively little known concerning the rate of mutation of the human blood-group genes. Positive proof of such an event depends upon the birth of an AB child to an OO mother, an MM child to an NN mother, an $r'r'$ child to an rr woman, etc. Even if the blood-group genes had a mutation rate in the neighborhood of 1×10^{-4}, it would require an enormous accumulation of family data to demonstrate this fact. Such an accumulation of data simply does not exist at this time. We are therefore in no position to speculate concerning the place of mutation in the present blood-group frequencies.

Even a low mutation rate which was maintained over a period of many generations could slowly bring about significant changes in gene frequencies. Thus, with respect to any locus, a mutation rate of 3×10^{-5}, if unopposed by any counterforces, could bring about an appreciable change in gene frequency over a period of a thousand or so generations. When a "new" gene is first introduced into a population, as a result either of mutation or of migration (see below), it is the period immediately following its appearance which is the most critical for its survival. Chance alone, regardless of whether the gene has favorable or unfavorable effects, may often be expected to lead to its loss from the population. Repetitive mutation, however, may be expected to give the gene (or a similar allele) another opportunity, until ultimately, so to speak, a gene with favorable effects proves its worth and becomes established in the population. Even an unfavorable gene, if it arises sufficiently frequently through mutation, may achieve an appreciable frequency.

Once a favorable gene is established in a population, the rate of its increase can be readily calculated. For instance, in a population where individuals heterozygous or homozygous for a dominant gene, A, have a 1 per cent selective advantage over aa individuals, the rates of change are as shown in the accompanying tabulation (Pätau, 1938).

Change in Frequency of A	No. of Generations Required
0.001–0.010	232
.010– .500	559
.500– .980	5,189
.980– .990	5,070
0.990–0.999	90,231

Two points are important. First, in terms of geological time, the rate of change is rather rapid at first. Second, the frequencies change very much more slowly as the limiting value of 100 per cent (complete fixation) is approached. In passing, it should be noted that a selective force of 1 per cent, which, as stated above, can be very effective in changing gene frequencies, is one difficult to demonstrate in man because of the many other variables which affect mean family size, the ultimate criterion of selective value.

3. GENETIC DRIFT

A third factor to be reckoned with in interpreting gene frequencies is genetic drift. The concept of genetic drift may be illustrated by considering a population of individuals in which there occur alleles A and a, whose frequencies, p and q, equal 0.5. Let us assume that the two genes are equivalent from the standpoint of natural selection and that the population in question is an isolated tribe numbering only 400 persons, among whom there are 80 married couples in the reproductive age. At the assumed gene frequency, if marriage occurs at random with respect to the A-a pair, we would expect the following numbers of marriages of the specified types: $5AA \times AA : 20AA \times Aa : 20Aa \times aa : 20Aa \times Aa : 10AA \times aa : 5aa \times aa$. If each marriage resulted in the same number of children, the proportion of the various genotypes among the children would be the same as in the parental generation. However, from our previous considerations on probability, it is apparent that it is unlikely that the mating types would occur in precisely the proportions expected. It is also unlikely that all marriages would be equally fertile. It follows that it is very unlikely, even in the absence of mutation or selection, that the gene frequencies among the children would be exactly the same as among the parents.

Consider, now, a population of 1,000 individuals in which again, for two "neutral" alleles, $p = q = 0.5$. To simplify the problem, we will assume that the various genotypes occur in exactly the expected numbers, i.e., $250AA$, $500Aa$, and $250aa$. Assume, now, that, in consequence of some catastrophe— drought, war, pestilence—this population is suddenly reduced to 20 individuals. The probability that the genotypes of these individuals will be $5AA:10Aa:5aa$, an exact duplicate in miniature of the original population, is given by the term

$$\frac{20!}{5!10!5!} (p^2)^5 (2pq)^{10} (q^2)^5$$

of the expansion of $[p^2(AA) + 2pq(Aa) + q^2(aa)]^{20}$, and is approximately 1 in 23. There are thus 22 chances in 23 that a different genotype frequency will result from the catastrophe. (In the exact solution of this probability, allowance would have to be made for failure of replacement in a sample of limited size. With a population of 1,000, the present approximate solution is sufficiently accurate.)

Such random departures from an original gene frequency are referred to as "genetic drift." The significance of this as an actual factor in establishing gene frequencies is difficult to assess. Increasingly, anthropologists picture primitive man as living in small groups, the situation under which the effects of genetic drift are potentially most marked. At the present time, we have no completely convincing examples of the operation of genetic drift in man, since in the strict sense this would require observations at widely spaced points in time. However, on theoretical grounds it is certain that genetic drift does occur, and anthropological studies have brought to light a number of situations *suggestive* of the operation of genetic drift. For instance, the Australian aborigines at the time of the colonization of that continent have been estimated to number 250,000–300,000 (Brown, 1930), divided into some 574 tribes, each tribe containing from 100 to 1,500 persons (Tindale, 1940). Birdsell (1950) has reported that the Pitjandjara tribe has a frequency of the blood-type gene A of 48.8 per cent, whereas the adjacent and anthropologically undifferentiated Ngadadjara tribe has an A frequency of 27.7 per cent. The numbers tested were under 100 in each case, and the exact values must be accepted with reservation, but the difference is clearly significant. Laughlin (1950) has described similar differences between Eskimo tribes. It is tempting to suggest that this is a result of genetic drift. The danger in yielding to this temptation lies in the creation of a type of thinking wherein genetic drift functions as a *deus ex machina* for the anthropologist beset by inexplicable differences in two sets of data.

4. MIGRATION

The final cause of change in gene frequencies which should be mentioned is migration. This can be treated quite briefly. There is much to suggest that in his early days, before the advent of agricultural practices, man was quite nomadic. The extent of his wanderings can, of course, never be known. It was inevitable that, in the course of these wanderings, tribe would meet tribe. And unless human nature has changed greatly in the last 5,000 years, it may be surmised that, either peacefully or following appropriate hostilities, there often occurred a substantial exchange of genes between the two groups. Such an exchange would, of course, tend to bring the two groups closer to a common mean. The precise treatment of this problem will be considered later.

15.5. *The validity of distinguishing between "genes of universal distribution" (sec. 15.2) and "genes of restricted distribution" (sec. 15.3).*—Earlier in this chapter a distinction was drawn for purposes of convenience between "genes of universal distribution" and "genes of restricted distribution." In addition to apparent differences in the manner in which they occur throughout the world, the genes falling into the two subdivisions have seemed to differ in the extent to which they are affected by selective forces. Thus the genes of apparently restricted distribution which we have considered seem, in homozygous form, to be the objects of strong natural selection, whereas the genes of wider distribution which were discussed, such as those responsible for the serological traits, have seemed, as a rule, to be more neutral. It now begins to appear that the distinction between the two types of genes may be more apparent than real. Thus we are coming to appreciate that many of the genes responsible for serological reactions, ordinarily thought of as "genes of universal distribution," may be subject to considerable selection. Moreover, some of the genes responsible for serological traits are now being found to have almost as "spotty" a distribution throughout the world as do the "restricted" genes. Furthermore, as we have seen, on the basis of recent work the "restricted" genes appear to have much wider distributions than was formerly thought to be the case. It seems probable that in the complex interplay of forces which are concerned in establishing the frequency of any given gene there is a continuous spectrum in gene frequencies as related to mutation and selection; no sharp classification of genes into two classes is justified.

15.6. *The error involved in estimating gene frequencies from samples of related individuals.*—In addition to the foregoing dynamic causes for possible changes in gene frequencies, we should at this point mention a purely "technical" cause for apparent gene-frequency differences between two or more groups. Many of the gene-frequency estimates in the literature are based upon relatively small samples and have not been calculated with the efficient techniques outlined in chapter 13. Speculation concerning the reason for a difference between two populations as regards the frequency of a particular gene is fruitless if the difference in question is not at the level of significance. But there is a more pertinent criticism to be directed toward certain estimates. The maximum-likelihood methods discussed in the preceding chapters assume that the individuals who compose the sample are unrelated. This is often not the case, anthropologists having tested small settlements or villages in which many of the inhabitants are related to one another. Techniques for estimating the gene frequency and its variance under these circumstances have been proposed by Cotterman (1947) and Finney (1948a, b). Failure to employ these techniques results in a spuriously low estimate of the variance of the gene frequency. This, in turn, would lead to underestimating the variance of the difference in two gene frequencies, with the subsequent effect of suggesting significant differences between two populations when, in fact, such differences do not exist. Thus, in a sample of 307 American Negroes studied by one of the authors with reference to the S serological reaction, treating the sample as composed of unrelated persons resulted in an estimate of the frequency of the S-bearing chromosomes as 0.152, with a variance of 0.000225, whereas, when allowance was made for relationship, although the estimate of chromosome frequency remained similar (0.159), now the estimate of the variance was 0.000672.

15.7. *The complexities of interpreting gene-frequency differences.*—The problems inherent in an evaluation of the roles of selection, mutation, genetic drift, migration, and sampling error in the origin of apparent differences between two populations as regards the frequency of a given gene are formidable. Some idea of the range of the interpretive problems which can arise is provided by a further consideration of the sickling phenomenon. Approximately 15–20 per cent of the estimated 200,000,000 inhabitants of Africa south of the Sahara exhibit the sickling phenomenon. But whereas it may be regarded as established that homozygotes for the gene in question in the United States are seriously ill, there are those who feel that, in view of the relative paucity of the reported cases from Africa, this is not the situation in that country (Raper, 1950). Granting for the moment that homozygotes are as severely handicapped in Africa as in the United States, the loss of "sickling" genes through natural selection should be very rapid. If the population is not at equilibrium with respect to the gene and if the gene is therefore being eliminated, then we must account for the original high frequencies which must have existed. If the population is at equilibrium, the gene frequency can be maintained only through an unprecedentedly high mutation rate or balanced polymorphism. No matter which explanation proves correct, the further problem arises as to the restricted distribution of the gene. As touched upon earlier, did it originate from within or from without Africa? If from without, why has it become so firmly established only in the Negroes? As previously noted, among Bantu tribes the frequency of the sickling trait varies from 2.0 to 45.0 per cent. This would suggest that the term "Bantu" embraces a rather heterogeneous group. On the other hand, studies on the A, B, O gene frequencies of the Bantu have thus far failed to reveal a comparable heterogeneity (Elsdon-Dew, 1939; Shapiro, 1951a, b). How is this difference to be resolved? Very similar questions arise in the case of the thalassemia gene, particularly in view of the above-mentioned report of a high frequency of the disease in Thailand. Here again we have no answers.

15.8. *Conclusions suggested by gene-frequency data.*—Mindful now of the problems involved in the interpretation of gene-frequency studies on populations (which problems, we might add parenthetically, are no worse than the interpretive problems of traditional anthropology), we may consider the chief results of anthropological value to date. Boyd (1950) has extended an earlier proposal of Wiener (1943), to recognize the basic serological groupings shown in Table 15-3 and described as follows:

1. Early European group (hypothetical)—Possessing the highest incidence (over 30%) of the Rh negative type (gene frequency of $rh > 0.6$) and probably no group B. A relatively high incidence of the gene Rh_1 and A_2. Gene N possibly somewhat higher than in present-day Europeans. Represented today by their modern descendants, the Basques.

2. European (Caucasoid) group—Possessing the next highest incidence of rh (the Rh negative gene), and relatively high incidence of the genes Rh_1 and A_2, with moderate frequencies of other blood group genes. "Normal" frequencies of M and N, i.e., M = ca. 30%, MN = ca. 49%, N = ca. 21%. (The italicized symbols stand for the genes, as opposed to the groups.)

3. African (Negroid) group—Possessing a tremendously high incidence of the gene $Rh°$, a moderate frequency of rh, relatively high incidence of genes A_2 and the rare intermediate A ($A_{1, 2}$, etc.) and Rh genes, rather high incidence of gene B. Probably normal M and N.

4. Asiatic (Mongoloid) group—Possessing high frequencies of genes A_1 and B, and the highest known incidence of the rare gene Rh^z, but little if any of the genes A_2 and rh (the Rh negative gene). Normal M and N. (It is possible that the inhabitants of India will prove to belong to an Asiatic subrace, or even a separate race, serologically, but information is still sadly lacking.)

5. American Indian group—Possessing varying (sometimes high, sometimes zero) incidence of gene A_1, no A_2, and probably no B or rh. Low incidence of gene N. Possessing Rh^z.

6. Australoid group—Possessing high incidence of gene A_1, no A_2, no rh, high incidence of gene N (and consequently a low incidence of gene M). Possessing Rh^z.

How do these serological classifications agree with the traditional anthropological schemes? There is not a great deal of difference. One inclined to discount the contributions of population genetics could say that we are not much, if any, further ahead than before. This is not entirely true. To begin with, even if we had no anthropological measurements whatever, we would still have a valid scheme of classification every bit as useful as the anthropological one and, as a precise tool, considerably more elegant. Furthermore, this scheme, instead of being at a virtual standstill, is a living, growing collection of data. Great developments are in the offing, bound to result in a system of relationships superior to any that has existed in the past.

TABLE 15-3*

APPROXIMATE GENE FREQUENCIES IN SIX GENERALLY
DEFINED SUBDIVISIONS OF MAN

Gene	Early European (1)	European (Caucasian) (2)	African (Negroid) (3)	Asiatic (Mongoloid) (4)	American (5)	Australian (6)
$A(A_2+A_1)$	ca. 0.25	0.2—0.3	0.1- 0.2	0.15-0.4	0 -0.6	0.1-0.6
Ratio A_2/A_1†	> .5?	.1 - .3	ca. .4	0	0	0
B	< .01?	.05- .20	0.05- .25	.1 - .3	0	0
N	> .5?	.3 - .5	ca. .5	0.4 - .5	0.1-0.2	0.8-1.0
Rh-neg (r)	> .5?	.4	ca. .25	0	....	0
Rh^0 (R^0)	< .1?	ca. .1	ca. .6	ca. 0.1	ca. 0.01	ca. 0.01
ptc‡	ca. 0.5	.55- .7	ca. .45	0	0	0
Nonsecreting§	?	ca. 0.5	>0.6	?	?	?
Other genes‖	?	rh'	$A_{1,2}$	Rh^z	Rh^z	Rh^z

* By permission from Genetics and the races of man, by Dr. William C. Boyd, copyright 1950, Little, Brown and Co.

† For convenience in calculation, the ratio of the two subgroups, A_2 and A_1, and not the ratio of the gene frequencies p_2/p_1 is given.

‡ The recessive gene for *not* tasting phenylthiocarbamide.

§ The recessive gene for *not* secreting water-soluble blood-group substances into the gastric juice, saliva, etc.

‖ Other genes the frequency of which seems to be higher in this population than in other races. Although not in Boyd's table, the sickling gene should be entered in column 3.

Moreover, in some instances the serological data, taken in conjunction with other known facts, have materially aided in clarifying certain relationships. Space permits consideration of only a single example of this. The inhabitants of the extreme southwest of France and northeast of Spain, at the western extreme of the Pyrenees Mountains, and along the coast in the angle of the Bay of Biscay, numbering several hundred thousand persons, speak a rather primitive type of language termed "Basque" or "Euskara." Their language is unlike any other spoken in Europe. Their social customs and music differ from those of other western Europeans. Physically, the Basque-speaking peoples are somewhat mixed but are characterized in the main by brachycephaly, having broad heads at the temples and triangular faces, and in this respect are distinguished from the more dolichocephalic inhabitants of present-day Spain. They are even more brachycephalic than the inhabitants of southern France. They tend to be tall. The relation of the Basques, as these people are known, to the other European stocks has been a matter of lively debate for many years. Inasmuch as the foregoing has not prepared the student to appreciate the fine points of these arguments, based largely on the findings of physical and cultural anthropology, suffice it to say here that, aside from agreement on their Caucasian ancestry, the most divergent views possible have been held.

The blood groups of these people likewise attest to their uniqueness (Chalmers, Ikin, and Mourant, 1949; Van der Heide *et al.*, 1951). The blood-group gene B has the very low frequency of 0.0265. The O frequency is relatively high, 0.7172. The $A_1:A_2$ ratio is about the same as for most European populations. MN frequencies show a slight but not significant excess of N over the usual European figures. But the Rh frequencies differ from those of all other peoples thus far tested, in the relative commonness of cde (r), 0.5316, and the relative rareness of $E[cDE(R^2)] = 0.0707, cdE(r'') = 0.0025]$. These findings have led Chalmers, Ikin, and Mourant (1949) to write: "Thus we see that in nearly every respect in which the blood groups of the Basques differ from those of the other peoples of Western Europe, the tendency in Eastern Europe is in the opposite direction. Qualitatively therefore, we have a very good case for supposing the present population of Western Europe to have arisen from the mixing of a people akin to the Basques with peoples related to those now inhabiting Eastern Europe, though perhaps showing more extreme features than the modern Eastern Europeans." It is these findings which led Boyd (Table 15-3, p. 252) to identify the Basques with an early European race, now reduced to this small remnant. It is of interest that the discovery of the Basque Rh-types supplies the "Rh-negative race" postulated by Haldane and Wiener (sec. 15.4) as a way out of the problem in natural selection posed by the present Rh gene frequencies in European populations.

15.9. *"Isolates."*—The term "isolate" has been applied to a group of people who for several to many generations are more or less cut off, for one reason or the other, from reproductive access to their neighbors. The isolating influence may be geographical or social. The term is, of course, a relative one. At one extreme are complete reproductive isolates. These are rare. More often the term is used in a relative sense, being applied to groups the members of which rarely mate outside the group. As noted earlier, it is becoming increasingly probable that prehistoric man was divided up into numerous such isolates. Birdsell's (1953) studies on the tribal organization of the Australian Bushmen, who represent essentially a Paleolithic culture, are especially illuminating in this respect. This has important consequences for our evaluation of the origin and nature of the physical differences now encountered between groups of people. For, as we have already seen, it is precisely under such conditions that the operation of chance, through genetic drift, is relatively most important. It is therefore a moot point how far attempts to see adaptive significance in the differences between groups of people, such as have been summarized by Coon, Garn, and Birdsell (1950), should be carried.

It is apparent that all over the world reproductive isolates are fast breaking up, sometimes spectacularly, as in the marriages contracted by United States Army occupational forces; sometimes more significantly, as in the intermarriage of natives, Orientals, and Caucasians, notably Portuguese, in Hawaii. The result, of course, will be regression to a common mean.

15.10. *The study of mixed populations.*—Given the knowledge that within relatively recent times there has been an intermingling of two populations, the concepts of population genetics can in some instances enable us to estimate with considerable accuracy the extent to which each of the original populations is represented in a third, derivative population, or, given the original gene frequencies and the amount of admixture, we can calculate the expected final frequency in a derivative population. We will consider the last problem first. The appropriate formula where a single pair of alleles is involved is

$$q_n = \frac{c\,q_c + d\,q_d}{c+d},$$

where

q_n = the final frequency of gene x,

q_c = the frequency of gene x in population C,

q_d = the frequency of gene x in population D,

c = the number of persons from population C entering into the new population, and

d = the number of persons from population D entering into the new population.

In other words, the expected final frequency is merely the weighted mean of the two original population frequencies.

The question more often asked in anthropological work is not the composition of some hypothetical derivative population but, given a gene frequency

q_n, thought to be due to an intermingling of two populations with gene frequencies q_c and q_d, what is the proportion in which the two have intermingled? The appropriate formula, derived from the above by algebraic manipulation (Bernstein, 1931), is

$$\frac{c}{c+d} = \frac{q_n - q_d}{q_c - q_d}.$$

This formula may be applied, for instance, to the question of how much non-Negro ancestry may be attributed to the average American Negro. It is well established that the vast majority of the Negro slaves brought to this country were derived from the Gold Coast and its hinterland, with a lesser contribution, in the later days of slaving, from the Belgian Congo and elsewhere. In a total of six studies the amount of sickling observed among 7,096 Gold Coast natives and 852 from the Belgian Congo was 15.5 per cent. Of a total of 10,858 American Negroes tested by various investigators, 10.6 per cent sickled. These are, with rare exceptions, the heterozygotes. Thus in the American Negro $2q_n(1 - q_n) = 0.106$, from which $q_n = 0.056$. In the African Negro, q_c can be calculated to be 0.085. The value of q_d is, of course, zero; $c/(c + d)$, therefore, is 0.66, or 34 per cent of the ancestry is non-Negro.

A second, although somewhat more uncertain, approach to this same problem is supplied by figures concerning the frequency of the R^o gene. In West Africans this gene has a frequency in the neighborhood of 0.63. In Europeans, on the other hand, the gene frequency is about 0.02. So far as is now known, the frequency of the gene in American Indians is similar to, or somewhat less than, the European figure. In American Negroes the gene frequency is about 0.44 (literature summary in Glass and Li, 1953). If we let the symbol c represent the combined Indian-Caucasian contribution, then $c/(c + d)$ is 0.69. By this approach, then, 31 per cent of the ancestry is non-Negro. The discrepancy between the results of these two approaches is small and will perhaps disappear entirely as more accurate figures concerning the frequencies of both the sickling and the R^o genes become available.

In view of the fact that in each generation a certain number of persons with some Negro ancestry are able, because of genetic recombination, to "cross the color line" and "pass" as whites, the actual amount of racial admixture has been greater than 31-34 per cent. This general figure agrees remarkably well with the social anthropologists' suggestion that approximately one-third of the ancestral background of the American Negro is Caucasian and Indian (Herskovits, 1942; Montagu, 1944; Meier, 1949).

Alvarez (1951) has recently described what appears to be a beautiful application of the principles of serological genetics to ferreting out events obscured by the passage of time and lack of adequate records. This concerns the question of the fate of the Indians inhabiting the area now known as the Dominican Republic, following the arrival of the Spanish Conquistadors some 400 years ago. One point of view has been that the Conquistadors exterminated the Indians so rapidly that there was very little—practically no—crossing between the Conquistadors and the Indians or between the Indians and the Negro slaves who were brought in shortly thereafter. Alvarez has shown that the existing A, B, O, MN, and Rh blood groups cannot be explained on the basis of a simple admixture of Negroes and Spanish, and it is necessary to postulate a third element. Assuming, in the absence of actual information, that the blood groups of the original Indian inhabitants were similar to those of such Indians as the present-day Indian of the southwestern United States, Central and South America (high frequency of group O and M, and low frequency of Rh-negative) (cf. Ottensooser and Pasqualin, 1949), then the existing blood-group frequencies suggest a 15-20 per cent Indian component. The departures from expectation on the basis of a simple Negro-Caucasian admixture are most marked in the mountainous, less accessible regions, a finding which gives further support to the hypothesis of a substantial Indian admixture, since it is in such areas as these that the Indians would be expected to take refuge.

15.11. *The concept of race.*—Thus far in this book we have avoided the use of the term "race." This is because in the past the term has been used in so many senses by so many writers that it has largely lost its usefulness. In particular, the term has become thoroughly tainted with attempts to promulgate and disseminate concepts of the superiority of some groups and the inferiority of others—attempts which have no standing in the courts of present-day science. It is an obvious fact that man is divided into a number of subgroups which may be termed "races." The number of races which one recognizes depends to a large extent on the fineness of the distinctions between groups of people which one recognizes as valid. The significant contribution of population genetics to this problem is the demonstration that in so far as the differences between races have been subject to a precise genetic analysis, these differences are rarely absolute but are due rather to differences in gene frequencies. In other words, the differences between races appear to be predominantly those of degree rather than those of kind. No better definition of "race" has yet been suggested than that of Stern (1949): "a genetically more or less isolated division of mankind possessing a corporate genic content which differs from that of all other similar isolates."

More extended treatments of the contemporary views on race are to be found in Huxley and Haddon (1936), Dunn and Dobzhansky (1946), Dahlberg (1946), Count (1950), Boyd (1950), and the UNESCO statement.

Bibliography

SPECIFIC REFERENCES

ALLEN, F. H.; DIAMOND, L. K.; and NIEDZIELA, B. 1951. A new blood-group antigen, Nature, 167:482.

ALVAREZ, J. J. 1951. Studies in the A-B-O, M-N, and Rh-Hr blood factors in the Dominican Republic, with special reference to the problem of admixture, Am. J. Phys. Anthropol., 9:127–48.

BANTON, A. H. 1951. A genetic study of Mediterranean anaemia in Cyprus, Am. J. Human Genetics, 3:47–64.

BIANCO, I.; MONTALENTI, G.; SILVESTRONI, E.; and SINISCALCO, M. 1952. Further data on genetics of microcythaemia or thalassaemia minor and Cooley's disease or thalassaemia major, Ann. Eugenics, 16:299–315.

BIRDSELL, J. B. 1950. Some implications of the genetical concept of race in terms of spatial analysis, Cold Spring Harbor Symp. Quant. Biol., 15:259–314.

———. 1953. Some environmental and cultural factors influencing the structuring of Australian aboriginal populations, Amer. Naturalist, 87:171–207.

BOYD, W. C. 1939. Blood groups, Tabulae biol., 17:113–240.

BROWN, A. R. R. 1930. Former numbers and distribution of the Australian aborigines, Official Year Book of the Commonwealth of Australia, 23:669–96. Melbourne: Government Printer.

CAMINOPETROS, J. 1952. Sickle cell anomaly as sign of Mediterranean anaemia, Lancet, 1:687–93.

CHALMERS, J. N. M.; IKIN, E. W.; and MOURANT, A. E. 1949. The ABO, MN, and Rh blood groups of the Basque people, Am. J. Phys. Anthropol., n.s., 7:529–44.

CHOREMIS, C.; ZERVOS, N.; CONSTANTINIDES, V.; and ZANNOS, L. 1951. Sickle-cell anaemia in Greece, Lancet, 1:1147–49.

COON, C. S.; GARN, S. M.; and BIRDSELL, J. B. 1950. Races: a study of the problem of race formation in man. Springfield, Ill.: Charles C Thomas.

COTTERMAN, C. W. 1947. A weighting system for the estimation of gene frequencies from family records, Contr. Lab. Vert. Biol., University of Michigan, No. 33, pp. 1–21.

ELSDON-DEW, R. 1939. Blood groups in Africa, Pub. South African Inst. M. Res., 9:29–94.

FINNEY, D. J. 1948a. The estimation of gene frequencies from family records. I. Factors without dominance, Heredity, 2:199–218.

———. 1948b. The estimation of gene frequencies from family records. II. Factors with dominance, ibid., pp. 369–90.

FOX, A. L. 1932. The relationship between chemical constitution and taste, Proc. Nat. Acad. Sc., 18:115–20.

GLASS, B. 1950. The action of selection on the principal Rh alleles, Am. J. Human Genetics, 2:269–78.

GLASS, B., and LI, C. C. 1953. The dynamics of racial intermixture—an analysis based on the American Negro, Am. J. Human Genetics, 5:1–20.

HALDANE, J. B. S. 1942. Selection against heterozygosis in man, Ann. Eugenics, **11**: 333–40.

HERSKOVITS, M. J. 1942. The myth of the Negro past. New York: Harper & Bros.

HIRSZFELD, L., and HIRSZFELD, H. 1919. Serological differences between the blood of different races, Lancet, **197**:675–79.

HUXLEY, J. S., and HADDON, A. C. 1936. We Europeans. London and New York: Harper & Bros.

IKIN, E. W., and MOURANT, A. E. 1952. The frequency of the Kidd blood-group antigens in Africans, Man, **52**:21.

LAUGHLIN, W. S. 1950. Blood groups, morphology and population size of the Eskimos, Cold Spring Harbor Symp. Quant. Biol., **15**:165–73.

LEHMANN, H., and CUTBUSH, M. 1952. Sub-division of some southern Indian communities according to the incidence of sickle-cell trait and blood groups, Tr. Roy. Soc. Trop. Med. & Hyg., **46**:380–83.

LEHMANN, H., and RAPER, A. B. 1949. Distribution of the sickle-cell trait in Uganda and its ethnological significance, Nature, **164**:494.

LEVINE, P. 1943. Serological factors as possible causes in spontaneous abortions, J. Hered., **34**:71–80.

LEVINE, P.; VOGEL, P.; and ROSENFIELD, R. E. 1953. Hemolytic disease of the newborn, Adv. Pediatrics, **6**:97–156.

LUNDMAN, B. 1948. Geography of human blood groups, Evolution, **2**:231–37.

MEIER, A. 1949. A study of the racial ancestry of the Mississippi college Negro, Am. J. Phys. Anthropol., **7**:227–40.

MOLLISON, P. L. 1951. Blood transfusions in clinical medicine. Oxford: Blackwell Scientific Publications.

MONTAGU, A. M. F. 1944. Origins of the American Negro, Psychiatry, **7**:163–74.

MOURANT, A. E. 1950. The blood groups of the peoples of the Mediterranean area, Cold Spring Harbor Symp. Quant. Biol., **15**:221–31.

NEEL, J. V., 1951a. The inheritance of the sickling phenomenon, with particular reference to sickle cell disease, Blood, **6**:389–412.

———. 1951b. The population genetics of two inherited blood dyscrasias in man, Cold Spring Harbor Symp. Quant. Biol., **15**:141–58.

OTTENSOOSER, F., and PASQUALIN, R. 1949. Blood types of Brazilian Indians (Matto Grosso), Am. J. Human Genetics, **1**:141–55.

PÄTAU, K. 1939. Die mathematische Analyse der Evolutionvorgänge, Ztschr. f. Abstamm.- u. Vererbungslehre, **76**:220–28.

PANTIN, A. M., and JANQUEIRA, P. C. 1951. Blood groups of Brazilian Indians, Nature, **167**:998.

RACE, R. R. 1950. The eight blood group systems and their inheritance, Cold Spring Harbor Symp. Quant. Biol., **15**:207–20.

RACE, R. R., and SANGER, R. 1950. Blood groups in man. Oxford: Blackwell Scientific Publications.

RAPER, A. B. 1950. Sickle-cell disease in Africa and America—a comparison, J. Trop. Med., **53**:49–53.

SANGHVI, L. D. 1951. ABO blood groups and sex ratio at birth in man, Nature, 168: 1077.

SHAPIRO, M. 1951a. The ABO, MN, P, and Rh group systems in the South African Bantu, South African M.J., 25:165–70.

————. 1951b. Further evidence of homogeneity of blood group distribution in the South African Bantu, ibid., pp. 406–11.

SILVESTRONI, E., and BIANCO, I. 1949. Microcythemia, constitutional microcytic anemia, and Cooley's anemia, Am. J. Human Genetics, 1:83–93.

TINDALE, N. B. 1940. Distribution of Australian aboriginal tribes: a field survey, Tr. Roy. Soc. South Australia, 64:140–231.

VAN DER HEIDE, H. M.; MAGNÉE, W.; VAN LOGHEM, J. J.; and SOUCHARD, L. 1951. Blood group distribution in Basques, Am. J. Human Genetics, 3:356–61.

VAN LOGHEM, J. J., and SPAANDER, J. 1948. L'influence de l'incompatibilité du système ABO sur l'antagonisme Rh, Rev. d'hémat., 3:276–86.

WATERHOUSE, J. A. H., and HOGBEN, L. 1947. Incompatibility of mother and foetus with respect to the iso-agglutinogen A and its antibody, Brit. J. Social. Med., 1: 1–17.

WIENER, A. S. 1942. The Rh factor and racial origins, Science, 96:407–8.

————. 1948. Blood grouping tests in anthropology, Am. J. Phys. Anthropol., 6:236–37.

GENERAL REFERENCES

BOYD, W. C. 1950. Genetics and the races of man. Oxford: Blackwell Scientific Publications.

COUNT, E. W. (ed.). 1950. This is race. New York: Henry Schuman.

DAHLBERG, G. 1946. Race, reason, and rubbish. London and New York: Columbia University Press.

————. 1947. Mathematical methods for population genetics. New York: S. Karger, Ltd.

DUNN, L. C., and DOBZHANSKY, T. 1946. Heredity, race, and society. New York: Mentor Books.

FORD, E. B. 1942. Genetics for medical students. London: Methuen & Co., Ltd.

LAHOVARY, N. 1953. Bioserological methods of human classification, Science, 117: 259–69.

STERN, C. 1949. Principles of human genetics. San Francisco: W. H. Freeman & Co.

UNESCO. 1952. The race concept. Paris: UNESCO.

WIENER, A. S. 1943. Blood groups and transfusion. 3d ed. Springfield, Ill., and Baltimore: Charles C Thomas.

Problems

1. Among 569 Eskimos, 83.5 per cent were observed to be of blood type MM, 15.6 per cent of MN, and 0.9 per cent of NN. Determine the gene frequencies and their standard errors.

2. A survey of 707 West Georgians living in Tiflis disclosed the following distribution of the A, B, O blood groups: 418 (group O), 243 (group A), 43 (group B), and 3 (group AB). Compute the frequencies of the three genes. What are the standard errors associated with these three estimates?

3. A serological study based on 972 Finns revealed the following frequencies: 34.0 (group O), 42.4 (group A), 17.1 (group B), and 6.5 (group AB). Do Finns differ significantly from West Georgians in the frequency of the gene responsible for the O antigen?

4. Are the frequencies of the M gene significantly different among the Australian aborigines, Fijians, and Papuans (use data of Table 15-2)?

5. In population I, 70 per cent of the persons react positively when tested for antigen X (i.e., are XX or Xx). In population II, 15 per cent of those tested react positively for the X antigen. Finally, in population III, thought to be derived from the intermingling of populations I and II, 55 per cent of persons possess the X antigen. If the assumption of the origin of population III is correct, in what proportions must populations I and II have been combined?

CHAPTER 16

Twins

ONE IN approximately every eighty-six pregnancies in the United States terminates in the birth of more than one infant. This phenomenon of multiple births has provided the human geneticist with a useful tool for the appraisal of the nature-nurture interaction in the determination of a given phenotype. It is our intention in this chapter to explore the phenomenon of twinning and the use of twins in genetic research.

16.1. *The origin and types of twins.*—One may postulate at least five distinctly different biological situations which could give rise to twins, namely: the proliferation and fertilization of (1) two ova; (2) only one ovum, if this fertilized egg subsequently divides at some early stage in development; (3) a binucleate egg which subsequently divides; (4) an egg and a large polar body; and (5) the fission products of an egg which has divided prior to fertilization. While direct cytological evidence is lacking to demonstrate that all these possibilities are realized in man, there is a considerable body of circumstantial evidence, largely serologic and somatologic, indicating that at least the first two of the above phenomena do occur and with some degree of regularity. We shall limit our discussion to these more common situations.

Binovular twins, that is, twins arising from two fertilized eggs, may be alike or unlike in sex and are called "dizygous" (DZ) or "fraternal" twins. Monovular twins, that is, those twins arising from just one fertilized egg are, on the other hand, always alike in sex and are termed "monozygous" (MZ) or "identical" twins. As a general rule, these two types of twins differ (1) in their placentation, that is, in the manner in which the placenta, the organ of communication between the fetus and the mother, is attached to the uterus; and (2) in the membranous structures which surround the fetuses during development. In the case of DZ twins, implantation of the embryos is usually such that two separate placentas develop, and each fetus is ultimately enveloped by two membranous sacs, the innermost being termed the "amnion" and the outermost the "chorion" (see Fig. 16-1). Dizygous twins are, therefore, dichorionic, that is, a separate chorion exists for each twin. In identical twins the placentation and the number of embryonic sacs associated with each twin is a function of the time at which division of the pre-

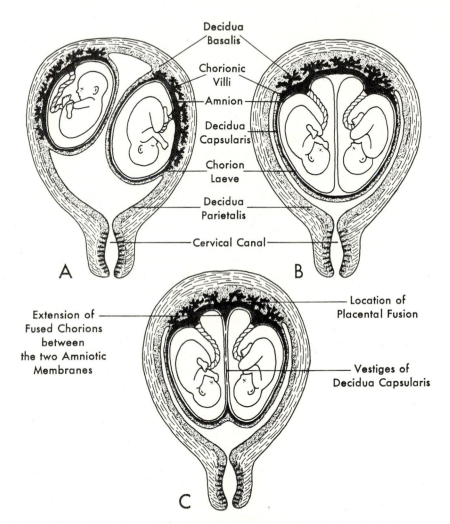

Decidua
Basalis

Chorionic
Villi

Amnion

Decidua
Capsularis

Chorion
Laeve

Decidua
Parietalis

Cervical Canal

A

B

Extension of
Fused Chorions
between
the two Amniotic
Membranes

Location of
Placental Fusion

Vestiges of
Decidua Capsularis

C

Fig. 16-1.—A diagrammatic representation of three of the different relationships between placentas and fetal membranes which may exist. *A*, Binovular twins with totally independent placentas and fetal membranes; *B*, monovular twins with but one placenta and a common chorion; *C*, binovular twins with secondary fusion of the placentas and chorions. (By permission from Human embryology, by Dr. Bradley M. Patten, copyright 1946, Blakiston Co.)

twin embryo occurs. If division occurs prior to implantation, the division products can be implanted at a distance compatible with the formation of two placentas and two chorions. On the other hand, when division occurs during the course of or following implantation, varying degrees of placental fusion will result, and the twins will be enveloped by a common chorion. Thus MZ twins may be either mono- or dichorionic. That both types do occur has been demonstrated by Steiner (1935) in a study of the placentas associated with 132 pairs of like-sexed twins. Steiner found that in all instances (32) where only one chorion was demonstrable, the twins were MZ, but, in addition, in 24 of the 100 dichorionic cases the twins were also MZ. While it would seem that MZ twins are more frequently monochorionic, it is also apparent that the presence of two chorions does not constitute prima facie evidence of dizygosity.

16.2. *The frequency of twin births.*—Both the over-all frequency of twinning and the frequency of each of the two types of twins may vary considerably from one country to the next. Variations in the over-all frequency ranging from once in every 35 terminations in southern Rhodesia (Ross, 1952) to once in every 1,000 terminations among the Annamese of Cochin China (Newman, 1940) have been reported. Whether the actual variation in the frequency of twinning is as great as this is open to some question. The former estimate was based on hospital terminations in a country where hospitalization for the birth of a child is the exception rather than the rule. Such a series is liable to be biased in such a way as to overestimate the frequency. The latter estimate may, on the other hand, underestimate the frequency, because of the high infant mortality in Asiatic countries coupled with the tendency to report only the living members of a multiple birth (Komai and Fukuoka, 1936). Despite objections to these particular estimates, there exist ample and reasonably reliable data attesting to variations in twin frequency ranging from approximately 1 in 70 terminations (American Negroes, the Scandinavians, and the Belgians) to 1 in 145 terminations among the Japanese.

Such differences as are observed in the over-all frequency of twinning could be due to variations in the frequency of either MZ or DZ twins or both. The frequencies of these two types of twins can be approximately determined as follows (Weinberg, 1901): If twin-pairs are divided into groups depending upon whether they are alike or unlike in sex, then three categories are recognizable—pairs in which both are females, pairs both of which are males, and pairs consisting of one male and one female. If we assume that fertilization is a random process and that p and q are the frequencies of male and female births in the total population, respectively, then the frequencies of DZ twins with respect to sex are given by

$$p^2\,(\male\male) + 2pq\,(\male\female) + q^2\,(\female\female) = 1 .$$

Of these three groups of DZ twins, only those which are of unlike sex can be differentiated from MZ pairs. We note, however, that

$$\frac{\text{Unlike-sexed DZ}}{\text{all DZ}} = \frac{2pq}{p^2 + 2pq + q^2} = \frac{2pq}{1} ;$$

whence

$$\text{All DZ} = \frac{\text{unlike-sexed DZ}}{2pq} \quad \text{and} \quad \text{all MZ} = (\text{total twins} - \text{all DZ}) .$$

A simpler approach and one which constitutes a satisfactory approximation to the proportion of MZ twins in most instances may be calculated by mere-

TABLE 16-1*

FREQUENCY OF MONOZYGOTIC AND DIZYGOTIC TWINS AMONG
JAPANESE AND "WHITE" AND "COLORED"
POPULATIONS OF UNITED STATES

PERIOD	POPULATION	MEAN NO. OF TWINS OF KNOWN SEX PER YEAR	MEAN NO. OF SETS BY SEX			FREQUENCY OF MALE BIRTHS IN TOTAL POPULATION	FREQUENCY OF TWINS PER 100 BIRTHS	
			$\male\male$	$\male\female$	$\female\female$		MZ	DZ
1922–36..	"White" (U.S.)	21,014	7,251	6,911	6,852	51.60	0.385	0.744
	"Colored" (U.S.)	3,363	1,091	1,195	1,077	51.32	.408	1.005
1926–31..	Japanese (selected rural areas)	1,420	598	275	547	52.20	0.425	0.272

* Data from Komai and Fukuoka (1936) and Strandskov and Edelen (1946).

ly subtracting from the total number of like-sexed twins the observed number of unlike-sexed twins. The difference between these latter numbers is attributed to MZ twins. This procedure amounts to assuming that the sexes are equally frequent at birth, and, although this generally does not obtain, the departure from 1:1 is sufficiently small to justify this as an approximation procedure.

In Table 16-1 are given the frequencies of MZ and DZ twins among the "white" and "colored" populations of the United States and among the Japanese as determined by the more "exact" method. These data indicate almost a fourfold variation in the frequency of DZ twinning among these three populations; the variation in the over-all frequency of twinning in these populations would seem to be due largely to variation in the frequency of DZ twins.

16.3. *The frequency of multiple births of various types.*—While by far the most common type of plural birth is that which terminates in the presentation of two infants, a single human pregnancy may give rise to three, four, and even more infants. As would be expected, the frequencies of these latter

TABLE 16-2*

FREQUENCY OF PLURAL BIRTHS AMONG JAPANESE AND "WHITE" AND "COLORED" POPULATIONS OF UNITED STATES

PERIOD	POPULATION	MEAN NO. OF BIRTHS PER YEAR	TWINS		TRIPLETS		QUADRUPLETS	
			Mean No. per Year	Per Cent of All Births	Mean No. per Year	Per Cent of All Births	Mean No. per Year	Per Cent of All Births
1922–36..	"White" (U.S.)	1,862,641	21,023	1.129	202.87	0.01088	3.27	0.000177
	"Colored" (U.S.)	237,897	3,365	1.413	46.87	.01939	1.00	0.000397
1926–31..	Japanese (selected rural areas)	204,351	1,425	0.697	9.67	0.00473	......	....

* Data from Strandskov (1945) and Komai and Fukuoka (1936).

events are inversely related to the number of infants. The frequencies with which the higher orders of plural births occur among the "white" and "colored" populations of the United States and among the Japanese are indicated in Table 16-2. We note that among the "white" population approximately 1 in every 10,000 terminations gives rise to triplets, whereas quadruplets occur approximately once in every 600,000 births. Furthermore, we observe that these types of plural births are more common among the "colored" than among the "white" populations and are least common among the Japanese.

With respect to higher orders of plural births, there exists an interesting approximation to the frequencies known as "Hellin's law." Hellin in 1895 observed that the frequencies of triplets, quadruplets, etc., were approximately described by $(f_t)^2$, $(f_t)^3$, etc., where (f_t) is the frequency of twin births in the population. Hellin's approximation would suggest a frequency of $(\frac{1}{89})^2$ for triplets in the "white" population of the United States, whereas $(\frac{1}{100})^2$ are observed. Similarly for the "colored" and Japanese populations, frequencies of $(\frac{1}{71})^2$ and $(\frac{1}{145})^2$ are expected on the basis of Hellin's law, and $(\frac{1}{71})^2$ and $(\frac{1}{141})^2$ are observed.

16.4. *The identification of the zygosity of twins.*—Frequently it is of interest to know whether a specific pair of twins is MZ or DZ. In reaching a judgment in the past, reliance has usually been placed on (1) a careful evaluation of the reported nature of the birth membranes and (2) the degree of physical resemblance between the twins. We have already seen how unreliable a guide the nature of the birth membranes is. In evaluating the degree of physical resemblance, particular attention has been devoted to such traits as sex, height, weight, hair and eye color, hair form, skin pigmentation, facial conformation, etc. Although a correct diagnosis as to zygosity can usually be made on the basis of these morphological criteria, there is, unfortunately, an unavoidable subjective element in the evaluation of many of these traits, and those interested in the study of twins have long felt the need of a more objective approach to the problem of zygosity.

In general, if twins differ in sex or any other known inherited characteristic, they cannot be identical twins. The obverse, however, is not true; that is to say, if twins are alike in sex and other inherited characteristics, they need not be identical. But, given a number of simply inherited and widely distributed traits, a probability statement as to the zygosity of the twins may be made in the latter instance. The recent developments in the field of serological genetics provide inherited traits which satisfy the dual criteria of simple inheritance and universal distribution and so afford a reliable basis for estimating the zygosity of twins. The exact use made of the blood groups in determining the probability that a given pair of twins is either DZ or MZ depends on whether or not the blood types of the parents of the twins are known. We will consider first the procedure when the parental types are known and then the procedure when they are unknown.

1. PARENTAL BLOOD TYPES KNOWN

Since the different systems of blood antigens are not known to be linked, with the possible exception of the Lewis and Lutheran systems, the genes favoring the production of these antigens will be distributed independently within a family. The probability, therefore, of DZ twins both having the A and B antigens and the M antigen, say, would be the product of the probability of having the A and B antigens *times* the probability of having the M antigen, with both these latter probabilities being determined for this specific family. In general, if $p(d_i)$ is the probability that DZ twins will have the same antigen in the ith system, based on the mating type, then the probability that DZ twins will have the same antigens in all n systems is

$$\prod_{i=1}^{n} p(d_i),$$

where

$$\prod_{i=1}^{n} p(d_i) = p(d_1) p(d_2) \ldots p(d_n). \qquad (16.4.1)$$

Similarly, the probability that MZ twins will be alike for all n systems is

$$\prod_{i=1}^{n} p(m_i),$$

where

$$\prod_{i=1}^{n} p(m_i) = p(m_1) p(m_2) \ldots p(m_n). \qquad (16.4.2)$$

The probability that a given pair of twins alike in all n systems is DZ will, therefore, be

$$P[DZ] = \frac{\displaystyle\prod_{i=1}^{n} p(d)_i}{\displaystyle\prod_{i=1}^{n} p(d)_i + \prod_{i=1}^{n} p(m)_i}, \qquad (16.4.3)$$

and, for MZ twins,

$$P[MZ] = 1 - P[DZ]. \qquad (16.4.4)$$

We may incorporate into these combined probabilities any inherited condition to which a specific probability may be assigned, such as sex, the ability to taste PTC, etc. If we apply this method, which is due to L. S. Penrose, to the example in Figure 16-2, we would obtain the following results: With respect to the MN system, for example, we note that two types of children, MN or NN, may be produced and with equal probability. Now the genotypes of fraternal twins are independent, and so the a priori probability that they will be alike in the MN antigens is the sum of the probabilities that two children will have the same phenotype for the various ways in which they could be alike. In this case, fraternal twins could both be NN or MN. The probability of the former event is $(\frac{1}{2})^2$, and of the latter event $(\frac{1}{2})^2$; hence the a priori probability that fraternal twins will be alike in their MN phenotype in this family is $(\frac{1}{2})^2 + (\frac{1}{2})^2$, or 0.50. The probability that fraternal twins will

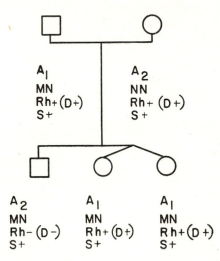

FIG. 16-2.—A pedigree of the serological responses in a family containing a pair of twins.

be alike in their A, B, O phenotypes, etc., is derived in the same fashion (see accompanying tabulation).

Probability of:	DZ	MZ
Twins at birth...	0.66	0.34
Twins being of the same sex......................	0.50	1.00
Twins in this family having the same antigens of the A, B, O series...............................	0.50	1.00
Twins in this family having the same antigens of the MN series..	0.50	1.00
Twins in this family having the same secretor response	1.00	1.00
Twins in this family having the same antigens of the Rh series..	0.25	1.00
Combined probability.......................	0.0206	0.3400

We calculate from (16.4.3)–(16.4.4) that the probability that this specific pair of twins is DZ is 0.0571, or that the pair is MZ is 0.9429. It should be noted that testing the sibling of the twins revealed that the father was heterozygous A_1A_2 and that both parents were heterozygous with respect to the Rh series. Had the brother not been tested, the problem would have taken the form given in the accompanying tabulation. The probability that

Probability of:	DZ	MZ
Twins at birth..................................	0.66	0.34
Twins being of the same sex......................	0.50	1.00
Twins in this family having the same antigens of the A, B, O series................................	1.00	1.00
Twins in this family having the same antigens of the MN series...................................	0.50	1.00
Twins in this family having the same secretor response.	1.00	1.00
Twins in this family having the same antigens of the Rh series...................................	1.00	1.00
Combined probability.......................	0.1650	0.34

the twins are DZ is now 0.3267, the difference between the two estimates, amply demonstrating the advisability of testing as many siblings of twins as possible, as well as their parents, when attempting to reach a conclusion as to zygosity.

2. PARENTAL BLOOD TYPES UNKNOWN

If the parental blood types are unknown, we may still employ the preceding method with but a slight modification. When the parental types are known, the probability of similar blood antigens may be calculated from the mating type. Here, since the parental types are unknown, we must resort to gene-frequency estimates to determine the probability of a pair of DZ twins having the same blood antigens for all the systems considered. To illustrate how this is accomplished, consider a system of antigens such as the MN antigens. If m is the frequency of the antigen M and n is the frequency of the antigen N, then the frequencies of the various possible mating types are those given in the accompanying table [column headed "$P(x)$"]. Now for any given

Mating	$P(x)$	$P(dz)$	$P(x)P(dz)$
MM×MM........	m^4	1	m^4
MM×MN........	$4m^3n$	$\frac{1}{2}$	$2m^3n$
MM×NN........	$2m^2n^2$	1	$2m^2n^2$
MN×MN........	$4m^2n^2$	$\frac{3}{8}$	$3m^2n^2/2$
NN×MN........	$4mn^3$	$\frac{1}{2}$	$2mn^3$
NN×NN........	n^4	1	n^4
....	......		$m^4+2m^3n+7m^2n^2/2+2mn^3+n^4$

mating type we can calculate the probability that DZ twins will be alike with respect to the MN antigens. For example, consider the mating MM × MM:

this mating can produce only type MM offspring; and hence the probability, $P(dz)$, that DZ twins from such a mating will be alike is precisely 1. The mating MN $\times$ MN, on the other hand, results in three phenotypes (MM, MN, and NN) with probabilities $\frac{1}{4}$, $\frac{1}{2}$, and $\frac{1}{4}$. The total probability that DZ twins from such a mating will be alike is

$$p\,(2\text{MM}) + p\,(2\text{MN}) + p(2\text{NN}) = (\tfrac{1}{4})^2 + (\tfrac{1}{2})^2 + (\tfrac{1}{4})^2 = \tfrac{3}{8} \, .$$

In general, the probability that a pair of DZ twins drawn at random will be alike in the MN blood antigens will be the sum of the product of $P(x)$ and $P(dz)$ for all types of mating involving the MN factors. We may say, then, that

$$p\,(d)_{\text{MN}} = \sum_i P\,(x)\,_iP\,(d\,z)\,_i \, .$$

In the specific case of the MN antigens,

$$p\,(d)_{\text{MN}} = m^4 + 2\,m^3 n + \frac{7\,m^2 n^2}{2} + 2\,m\,n^3 + n^4$$

$$= 1 - [\,(\tfrac{1}{2})\,m\,n\,(4 - 3\,m\,n)\,] \, . \tag{16.4.5}$$

The extension of this method to the case of a single pair of genes with dominance is straightforward; and we would find that the probability that a pair of DZ twins drawn at random will be alike with respect to such a system is

$$p(d) = 1 - \tfrac{1}{2}pq^2(3 + q) \, , \tag{16.4.6}$$

where p and q are the frequencies of the two alleles. But the analysis of systems of multiple alleles or closely linked genes is considerably more laborious; the general method has been dealt with by Wiener and Leff (1940), Fisher (1951), and Wiener (1952). In the case of the A, B, O series, when testing with anti-A and anti-B, Wiener and Leff find that the probability that dizygous twins will be alike with respect to this series is

$$p\,(d)_{\text{ABO}} = 1 - [\tfrac{1}{2}r^2\,(1 - r)\,(3 + r) + \tfrac{1}{2}p\,q\,(4 - 3\,p\,q + 6\,r)\,]\,, \tag{16.4.7}$$

where p, q, and r are the frequencies of the genes I^A, I^B, and I^O, respectively. Wiener and Leff (1940) have also shown that when the two principal subgroups of A are differentiable, then

$$p(d)_{A_1A_2\,BO} = 1 - [p\,(d)_{\text{ABO}} + p_1 p_2$$
$$\times\,(2p_2 + 3\,r + r^2 + p_1 q + 3\,q^2 + p_2 r - \tfrac{1}{2}p_1 p_2)\,]\,, \tag{16.4.8}$$

where p_1 and p_2 are the frequencies of the genes I^{A_1} and I^{A_2}, and $p(d)_{\text{ABO}}$ is obtained from (16.4.7), where $p = p_1 + p_2$.

A general formula useful for the Rh series is not so readily arrived at, since the probability is a function of the number of antisera available for testing and the availability of the Rh antisera varies considerably. For the simplest case, where differentiation is on the basis of Rh+ or Rh−(cf. Sec. 8.6, p. 89), equation (16.4.6) is applicable.

If we return to the example in Figure 16-1 and assume that the parental types were unknown, then the probability that these twins are MZ or DZ could be determined with the aid of the following gene-frequency estimates: $A^{M} = 0.524$, $A^{rh-} = 0.383$, $I^{O} = 0.687$, $I^{A_1} = 0.208$, $I^{A_2} = 0.053$, and $I^{B} = 0.052$. We find $P(MZ) = 0.7608$ and $P(DZ) = 0.2391$. It is apparent that, in general, effort expended in blood-typing the parents and siblings of twins is more than repaid by the greater accuracy in the diagnosis of mono- and dizygosity.

16.5. *The inheritance of twinning.*—The importance of heredity in the phenomenon of twinning is by no means the least controversial aspect of the general problem of twinning and the use of twins in genetic research. Nevertheless, certain facts associated with twin births seem to be reasonably well established: (1) twin births tend to cluster in specific families, and (2) the frequency of twinning varies with maternal age and to some extent with parity. With reference to these latter observations, first made by Duncan in 1865, it appears that the frequency of monovular twins increases steadily with increasing maternal age and that, for young mothers for a given age, there is also a tendency for the frequency to increase with birth order (McArthur, 1953). Similarly, in binovular twins the frequency increases with maternal age, reaching a maximum between the ages of thirty-five to thirty-nine, and at each age level the frequency appears to increase after the second parturition. These observations are certainly not consistent with the behavior to be expected of a single gene difference, nor, for that matter, is the pattern of clustering observed within families that to be expected from the simpler modes of inheritance.

At the present time, the best approach to the problem of the inheritance of twinning would appear to be a purely empirical one. Dahlberg (1952), intentionally ignoring the effects of age and parity, has estimated that a mother who has given birth to one set of twins will at her next delivery repeat with another twin birth approximately 3.6 per cent of the time. In other words, the probability of a repeat performance is approximately three times greater than the probability associated with a twin birth in the general population. He has further subdivided this risk on the basis of whether the first twin-pair was MZ or DZ and finds that if the first pair was DZ, then the probability that the mother will repeat with another set of twins is 4.55 per cent. Conversely, if the first pair was MZ, then the risk of a second set of twins is only 1.43 per cent or not much higher than the probability of a twin birth in the general population.

The use of these empirical estimates should be tempered by two considerations, namely, (1) these estimates were derived from the population of Sweden, and, in view of the known differences between populations in the frequency of twinning, it is uncertain to what extent it is legitimate to extrapolate directly to other populations; and (2) if twins may arise either as accidents in development or because of some genetic mechanism, then these estimates, like all empirical risks, are of limited applicability, since they overestimate the probability of a repeat when the first set arose as a consequence of an accident and underestimate the risk when there is a genetic mechanism underlying the origin of the first pair.

16.6. *The argument for the use of twins in genetic research.*—The reason for the great interest of the geneticist in twins is that, since MZ twins have identical genotypes, any dissimilarity which may exist between pairs must be due to the action of environmental agencies either postnatally or *in utero*. The interest in dizygous twins, on the other hand, stems from the fact that, while differing genetically, such twins enjoy certain environmental similarities, such as birth rank and maternal age, not enjoyed by single-born offspring, and thus afford a measure of environmental effects not otherwise possible. If we accept the validity of these arguments, then twins occupy a unique position in human genetic research. J. Späth (1860) and Francis Galton (1875) were the first to call attention to this uniqueness on the part of twins and to indicate their usefulness in the appraisal of the nature-nurture problem. The fullest measure of the contribution to human heredity which twins can make is realized, however, only when the two types of twins are studied. Under such circumstances there is provided a means of appraising either the effect of different environments on the same genotype or the expression of different genotypes under the same environment. We will now consider in some detail the two conventional ways that twin data are utilized.

1. COMPARISON OF MZ WITH DZ TWINS

If, under a given environment, both members of a twin-pair develop the same phenotype, the twins are said to be "concordant"; if they have differing phenotypes, they are said to be "discordant." Consider the case of a single gene difference, where the environment does not appreciably alter the expression of the gene. When this is the case, we should expect identical twins to be always concordant; fraternal twins, however, since they are no more similar genetically than single-born siblings, may be either concordant or discordant. We may use this expected difference in concordance as a measure of the percentage of phenotype variance attributable to "heredity." One such measure which has been used is

$$H = \frac{CMZ - CDZ}{100 - CDZ},$$ (16.6.1)

where CMZ and CDZ are the percentages of concordant MZ and DZ twins, respectively. In the event that the environment exercises little effect upon the genotype, then the expected concordance in MZ twins will be 100 per cent and some lesser value in DZ twins, the exact value depending upon the frequencies of the genes under study. Similarly, if the environment is the principal determiner, as may be the case in an infectious disease where there is no genetically conditioned resistance, then we should expect MZ twins to be

TABLE 16-3*

CONCORDANCE WITH RESPECT TO THREE OF THE MORE COMMON
CONGENITAL ABNORMALITIES OBSERVED AMONG
MONOZYGOTIC AND DIZYGOTIC TWINS

MALFORMATION	TYPE OF TWIN	NO. OF PAIRS	CONCORDANT		H
			No.	Per Cent	
Spina bifida..	MZ	18	13	72.22	0.583
	DZ	36	12	33.33	
Mongolism...	MZ	18	16	88.89	.881
	DZ	60	4	6.67	
Clubfoot.....	MZ	35	8	22.86	0.211
	DZ	133	3	2.26	

* Data from Gedda (1951).

concordant no more frequently than DZ twins. It is evident, then, that, in the limit, H will vary between 0 and 1 and that these values may be used as estimates of the relative contributions of heredity and environment to a given phenotype. In Table 16-3 are given heritability estimates for three of the more common developmental anomalies. These data suggest that heredity plays a more important role in the etiology of Mongolism than in either spina bifida or clubfoot. It is quite likely, however, that this impression is false because of bias in the data introduced by the preferential reporting of concordant monozygous twins. Furthermore, we observe that the empirical risk of repetition for these three anomalies, once an affected child has been born (cf. Table 18-1, p. 314) is approximately the same. If the twin data are accepted at face value, then one must conclude from the latter evidence that if heredity is not the important factor, then the environmental situations which give rise to clubfoot, for example, are preferentially distributed. Does this conclusion seem justified? And, if so, to what preferentially distributed environmental agencies are we to attribute these defects? In addition, how is one to adjust the findings from twin data in the light of the known increase in frequency of both monozygous twinning and Mongolism which accompanies increasing maternal age? This is but a small demonstration of the problems posed in drawing inferences from twin data.

Concordance or discordance on the part of twins implies a distinct discontinuity in phenotypes. Such a discontinuity does not exist in the case of the quantitative characteristics, such as intelligence, stature, etc. We may, though, extend the preceding argument to include these characteristics by the use of the intra-class correlation coefficient. Assume that we have available a series of intelligence quotients. If these scores were obtained for both members of, say, n sets of MZ twins, then we should have a sample of paired measurements $x_1y_1, x_2y_2, \ldots, x_ny_n$, where the x's indicate the performance of one twin and the corresponding y the performance of the other. We have seen in section 9.5 (p. 105) that the relationship between two variables may be measured by the interclass correlation coefficient, when there is some sensible biological criterion by which measurements may be assigned to x or y. There arise situations, however, in which there exists no sensible criterion. In Table 9-3 (p. 109) this problem did not exist, because the measurements were readily divisible, and relevantly so, on the basis of father and son. In twin data, on the other hand, there generally exists no relevant distinction, such as age, parentage, etc. It is customary in this situation to derive the correlation coefficient from a mean and a standard deviation based upon all measurements. Such a correlation coefficient, in contrast to the previous case, is called the "intra-class correlation coefficient" and is, by definition,

$$r = \frac{\sum_i (x_i - a)(y_i - a)}{n s^2} = \frac{\sum_i x_i y_i - n a^2}{n s^2}, \qquad (16.6.2)$$

where a is the mean common to all measurements, n is the number of pairs of measurements, and s^2 is the variance about the mean a. This definition of the intra-class correlation coefficient may be altered as follows: Subtracting 1 from each side of the equality, substituting for ns^2 in the numerator, and reducing, we obtain

$$1 - r_{\text{MZ}} = \frac{n s^2 - \Sigma x y + n a^2}{n s^2} = \left[\frac{\Sigma (x - y)^2}{2 n} \right] \left[\frac{1}{s^2} \right].$$

Now the left term on the right side of the equality is the mean-square deviation (difference) between members of a twin-pair, and s^2 is the variance of all the measurements, irrespective of twinning. If we designate the mean-square deviation for MZ twins as V_{MZ} and the total variance as V, then

$$1 - r_{MZ} = \frac{V_{MZ}}{V}.$$

The intra-class correlation coefficient is equivalent, therefore, to a linear function of the ratio of two variances—the variance between members of a twin-pair and the variance over the whole set of observations. Similarly, for the case of DZ twins, we obtain

$$1 - r_{DZ} = \frac{V_{DZ}}{V}.$$

It is apparent from the two preceding formulas that as V_{DZ} or V_{MZ} tend toward zero, that is, as intra-pair differences diminish, r tends toward 1.

Holzinger (1929) has suggested that the best comparison to appraise the nature-nurture interaction for a quantitative characteristic is a comparison of the intra-class correlation coefficients for MZ with similar coefficients for like-sexed DZ twins, on the supposition that such a comparison minimizes variation due to birth rank, age of mother, and sex. Holzinger has proposed, as the measure of heritability,

$$H = \frac{r_{MZ} - r_{DZ}}{1 - r_{DZ}}, \qquad (16.6.3)$$

where heritability is a general term for the relative contribution of heredity (as versus environment) to the variation observed with respect to a particular trait. This may also be written as

$$H = \frac{V_{DZ} - V_{MZ}}{V_{DZ}}. \qquad (16.6.4)$$

And, again, we note that, in the limit, H is bounded by 0 and 1.

Newman, Freeman, and Holzinger (1937) have applied this method to test differences in intelligence quotients among twin-pairs. They found that the correlation between scores for the Binet I.Q. was 0.881 for MZ twins, 0.631 for DZ twins, from which H was computed to be 0.678. This finding could be interpreted as meaning that, under the circumstances of their study, the relative contribution of heredity was approximately twice that of environment in the determination of mental ability as measured by the Binet I.Q. That this may be a hazardous interpretation will be subsequently indicated.

2. COMPARISON OF IDENTICAL TWINS REARED APART

The comparison of MZ with DZ twins yields, as we have seen, information on the percentage of phenotype variance ascribable to "heredity." The existence of MZ twins who have been reared apart permits an extension of this study of nature-nurture interaction to the appraisal of the effects of two different environments on the same genotype. The conventional approach, which is a logical extension of Holzinger's thesis, consists of comparing identical twins reared apart (MZA) with identical twins reared together (MZT). Such comparisons are more commonly made with respect to the graded characteristics. An estimate of the percentage of phenotype variation ascribable to "environment," that is, to the effect of differing environments on the same genotype, is given by

$$E = \frac{r_{MZT} - r_{MZA}}{1 - r_{MZA}}, \qquad (16.6.5)$$

where r_{MZA} and r_{MZT} are the intra-class correlations. Here, too, the scale of E is continuous but with end-points 0 and 1. It should be noted, however, that, when dealing with small samples, sampling variations may give rise to values of E (or H) which exceed these end-points.

TABLE 16-4*

DEGREE OF SIMILARITY BETWEEN MONOZYGOTIC
TWINS, REARED APART OR TOGETHER,
AND DIZYGOTIC TWINS

CHARAC-TERISTIC	CORRELATION			E	H
	MZA	MZT	DZ		
Height....	.969	.932	.645	−0.544	0.808
Weight...	.886	.917	.631	+ .272	.775
I.Q.......	.670	.881	.631	+0.639	0.678

* Data of Newman, Freeman, and Holzinger (1937).

Widespread application of this method of analysis has been seriously handicapped by the difficulty in obtaining identical twins who have been reared apart. Nonetheless, Newman, Freeman, and Holzinger (1937) have reported on a series of nineteen monozygotic twin-pairs who were reared apart, the largest group studied to date. Their attention was focused, among other things, on the effects of different environments on height, weight, and I.Q. Their results are reproduced in Table 16-4. Of these three characteristics, it would seem that the environmentally most plastic characteristic, as judged by the magnitude of E, is mental ability when measured by the I.Q. In this respect, the most extreme difference in I.Q. (twenty-four points) reported by these workers involved a set of female twins, one of whom had had only two years of primary schooling, while her co-twin had had a college education and had herself taught school. Despite this seemingly greater plasticity, it is worth noting that none of the identical twins differed by more than twenty-four points, whereas over 17 per cent of siblings will differ by this amount.

16.7. *Limitations of the twin method.*—The inferences which may be drawn from twin data are subject to certain reservations, statistical as well as biological. An assumption basic to all the analytical methods applicable to twin data is that the probability of ascertaining "affected" twins is independent of their type. That this assumption is often not valid is indicated by Macklin's data (1940) on tumors in twins (cf. sec. 17.6, p. 299). This unfortunately is not an isolated case; almost all the twin studies which have relied upon data available in the medical literature are subject to this criticism. The effects of differing levels of ascertainment will vary, depending upon what class or classes of twins are involved. For example, if a tendency exists to report discordant monozygous twins more frequently than the frequency with which they actually occur in the general population, we would be led to underestimate the genetic component and, by the same token, to overestimate environmental effects.

In addition to biases which may arise from unequal probabilities of ascertainment, there exist biological biases of both pre- and postnatal origin.

1. BIASES WITH PRENATAL OR NATAL ONSET

Price (1950) has divided the biases with prenatal onset, the so-called "primary" biases, into natal factors, lateral inversions, and the effects of mutual circulation. These biases affect both types of twins but to varying degrees.

Natal factors: The natal factors which may lead to congenital or subsequent dissimilarities are the position of the fetuses in the uterus, *in utero* crowding, the special conditions of implantation, and the order of, and manner through which, delivery occurs. That these factors may exert considerable influence may be demonstrated by the following extreme case. If we may assume that the order of delivery is faithfully recorded, then the evidence indicates that if one of a pair of twins is stillborn, the risk is several times greater that the stillborn twin will be the second rather than the first in the order of birth. The cause of this may be attributed to the premature separation of the placenta, with subsequent anoxia on the part of the second-born twin. If these data are a reliable index of the effects that delivery order alone may have, then it seems likely that collectively these natal factors may exert a considerable influence.

Lateral inversions: In comparing members of monozygous twin-pairs, we may observe three general classes of structural similarities. We observe (1) bilateral similarities, i.e., marked similarity of the right and left **side of** one twin of a pair; (2) homolateral similarities, i.e., cases in which the bilateral similarity within a twin is less than the similarity between comparable sides of the twins collectively; and (3) heterolateral similarities or mirror imaging, wherein the right side of twin A_1 is more similar to the left side of twin A_2 then to the left side of twin A_1. These latter similarities are known as the "lateral inversions." All gradations of lateral inversions are known to occur in MZ twins, from complete situs inversus viscerum to such minor reversals as the direction of hair whorls. It has long been believed that the degree of lateral inversion is directly related to the time of fission of the pretwin embryo. This belief receives some support from the evidence from conjoined or Siamese twins. Conjoined twins are generally regarded as representing attempts at fission late in embryonic development and are much less similar than MZ twins who develop separately, in the sense that conjoined twins show much greater evidences of mirror imaging and tend to resemble one another less than their "normal" MZ counterparts. Furthermore, conjoined twins show the most marked lateral inversions. It is a moot question as to whether or not lateral inversions may influence the physical and mental development of twins; but in the absence of more concrete evidence it seems hazardous to ignore the possibility that they may exert an influence on development.

Effects of mutual circulation: A frequent finding in the case of twins whose placentas are fused is the existence of a mutual circulatory system which may vary from the anastomosis of a few of the smaller vessels to what is tantamount to one circulatory system. The effects of this mutual circulation, if it arose during the development of identical twins, might well be negligible, since such twins have identical heredities and presumably the circulating chemical and cellular constituents of their bloods would be identical. In the case of fraternal twins, the potential effects of mutual circulation are difficult to appraise, but that the mutual circulation may exert an effect seems indicated by three findings with respect to twin cattle. First, it has been known for many years that among unlike-sexed cattle twins the female member of the pair is often a sexual intergrade, a so-called "freemartin." This finding is readily explicable when one recalls that the male begins to elaborate sex hormones before the female and that a mutual circulation would provide a means for these hormones to gain access to the female at a critical stage in development. Second, Owen (1945) finds that, in cattle, fraternal twins are identical in their blood types more frequently than one would expect on the basis of chance alone. The explanation advanced for this finding is that, as a consequence of vascular anastomosis, there is an interchange of the embryonal cells which are ancestral to the erythrocytes of the animal. It has actually been demonstrated in certain cattle twins that the blood of each is a mixture of the two types of corpuscles. Such twins are said to be "red-blood-cell mosaics." Last, recent work by Anderson *et al.* (1951) suggests that the effects of the mutual circulation may be even still more far-reaching. It has been recognized for some time that skin grafts between unrelated or related individuals are not successful, that is, they "do not take." Exceptions to this are transplants between identical twins or members of highly inbred strains. One consequence of these findings has been the belief that skin grafting could be used as a measure of genetic relationship and as a means of determining zygosity in twins. Anderson and his co-workers find, however, that DZ twin cattle will frequently tolerate skin grafts from one to another but will not tolerate grafts from unrelated individuals or siblings. They suggest that the mutual circulation during development may lead to an antigenic balance, as it were, which permits one twin to tolerate a graft which normally would be expected to fail because of an antigen-antibody reaction. Stormont and Woolsey (see Stormont, 1954), using DZ bovine twins known to be red-blood-cell mosaics, have confirmed the findings of Anderson and his co-workers but failed to establish grafts between cattle not twins but known to have the same immunological blood types. Stormont suggests that "the anomalous tolerance of DZ bovine twins to grafts of each other's skin may involve the exchange and establishment *in utero* of cells other than or in addition to erythrocyte-precursors." Whatever the exact explanation of these three phenomena may be, it is apparent that the net effect of a mutual circulatory system will be to make fraternal twins more alike than should be the case on the basis of their heredities.

To what extent these factors are operative in human twins is unknown. On the one hand, we do know that freemartinism is an uncommon occurrence among human twins, if it occurs at all. But then need we expect freemartinism to be a sequella of mutual circulation in human twins, in view of the slower rate of maturation of human beings as opposed to cattle? On the other hand, an instance of red-blood-cell mosaicism has recently been reported in England involving a pair of unlike-sexed twins (Dunsford *et al.*, 1953). As yet, no concerted attempt has been made to determine how successful skin grafts between fraternal twins might be.

In addition to the three classes of primary biases listed, there also arise prenatal factors which are less readily classifiable. For example, Penrose (1937) has reported a monovular pair of twins in which only one had congenital syphilis. Such a finding, since congenital lues generally involves the placental transfer of the organism responsible for syphilis, is not readily assignable to natal factors or to the other two classes of primary biases.

Despite the dissimilarities which may result from the primary biases, it is important to remember in evaluating studies of MZ twins reared apart that such twins have enjoyed the same prenatal environmental similarities, such as birth rank, gestation, and maternal age. Sweeping generalizations seem to be uncalled for in view of the possible effects of comparable prenatal environments. For a documentation of the effects of one set of recognized prenatal factors, namely, manifest maternal disease during pregnancy, the reader is referred to the review article of Bass (1952).

2. BIASES WITH POSTNATAL ONSET

In the comparison of MZ with DZ twins the assumption is made that the individual members of a twin-pair enjoy comparable or equivalent environments. It has been clearly shown that rarely is this the case. Identical twins are treated more similarly than are fraternal twins by both their families and their acquaintances. It is unlikely that this could influence physical characteristics, such as eye color; but it is conceivable that variation in mental development or in exposure to infection could occur. Differences in postnatal exogenous factors can occasionally produce appreciable differences in monozygous twins. Komai and Fukuoka (1934) have reported a pair of male twins, purportedly monozygous, who until five years of age were strikingly similar. But at fifteen years of age, diabetes insipidus had developed in one, and they differed markedly in stature. Siemens (1927) has described an MZ pair, one twin of which was markedly scoliotic, apparently as a consequence of rickets. Finally, Lewis (1936) has reported a dissimilar pair of MZ twins in which, possibly in relation to a head injury, one twin developed an atypical acromegaly in later years. A history of some of these exogenous factors, particularly those with social implication, would be difficult to obtain; yet uncorrected data could lead to erroneous conclusions.

16.8. *General remarks on the twin method.*—An objective evaluation of the contribution which twin studies have thus far made to human heredity is a matter of some complexity. That the studies have not made the contributions which Galton envisaged is certain. In part this is due to the fact that many such studies have been approached with more perseverance than perspicacity. But more important is the inherent difficulty in arriving at sound conclusions in the face of the obscuring effects of the biological biases to which twin data are subject. Twin researchers may point, and justifiably, to the scientific "hunches" which twins have helped to crystallize. However, when one considers the quantification of the components of variation ascribable to heredity and environment—the avowed purpose of twin studies—twins have contributed little which may be extrapolated to other genetic situations. This, again, is in no small part the result of lack of information on the effects which biases may produce. As Price (1950) and others have suggested, evidence on the magnitude of these biases must be obtained if we are to determine whether the twin method can be properly implemented as a tool for the appraisal of nature-nurture interaction. In its present context, the twin method has not vindicated the time spent in the collection of such data.

Bibliography

SPECIFIC REFERENCES

ANDERSON, D.; BILLINGHAM, R. E.; LAMPKIN, G. H.; and MEDAWAR, P. B. 1951. The use of skin grafting to distinguish between monozygotic and dizygotic twins in cattle, Heredity, 5:379–97.

DAHLBERG, G. 1952. Die Tendenz zu Zwillingsgeburten, Acta genet. med. et gemellologiae, 1:80–88.

DUNCAN, J. M. 1865a On some laws of the production of twins, Edinburgh M.J., 10:767–81.

———. 1865b. On the comparative frequency of twin-bearing in different pregnancies, ibid., pp. 928–29.

DUNSFORD, I.; BOWLEY, C. C.; HUTCHISON, A. M.; THOMPSON, J. S.; SANGER, R.; and RACE, R. R. 1953. A human blood-group chimera, Brit. M. J., No. 4827, p. 81.

FISHER, R. A. 1951. Standard calculations for evaluating a blood group system, Heredity, 5:95–102.

GALTON, F. 1875. The history of twins as a criterion of the relative powers of nature and nurture, Fraser's Mag., 12:566–76.

HOLZINGER, K. 1929. The relative effect of nature and nurture influences on twin differences, J. Educ. Psychol., 20:241–48.

KOMAI, T., and FUKUOKA, G. 1934. Post-natal growth disparity in monozygotic twins, J. Hered., 25:423–30.

———. 1936. Frequency of multiple births among the Japanese and related peoples, Am. J. Phys. Anthropol., 21:433–47.

LEWIS, A. J. 1936. A case of apparent dissimilarity of monozygotic twins, Ann. Eugenics, 7:58–64.

MCARTHUR, N. 1953. The frequency of monovular and binovular twin births in Italy, 1949–1950, Acta genet. med. et gemellologiae, 2:11–17.

MACKLIN, M. T. 1940. An analysis of tumors in monozygous and dizygous twins, J. Hered., 31:277–90.

OWEN, R. D. 1945. Immunogenetic consequences of vascular anastomoses in bovine twins, Science, 102:400–401.

PATTEN, B. M. 1946. Human embryology. Philadelphia: Blakiston Co.

PENROSE, L. S. 1937. Congenital syphilis in a monovular twin, Lancet, 1:322.

ROSS, W. F. 1952. Twin pregnancy in the African, Brit. M. J., 2:1336–37.

SIEMENS, H. W. 1927. The diagnosis of identity in twins, J. Hered., 18:201–9.

SPÄTH, J. 1860. Studien über Zwillinge, Ztschr. d. Wien. Gesellsch. d. Ärzte zu Wien, 16:225–41.

STEINER, F. 1935. Nachgeburtsbefunde bei Mehrlingen und Ähnlichkeitsdiagnose, Arch. Gynec. Berlin, 159:509–23.

Stormont, C. 1954. Research with cattle twins. *In:* Kempthorne, O., *et. al.*, Statistics and mathematics in biology, pp. 407–18. Ames: Iowa State College Press.

Strandskov, H. H. 1945. Plural birth frequencies in the total, the "white," and the "colored" U.S. populations, Am. J. Phys. Anthropol., n.s., 3:49–55.

Strandskov, H. H., and Edelen, E. W. 1946. Monozygotic and dizygotic twin birth frequencies in the total, the "white," and the "colored" U.S. populations, Genetics, 31:438–46.

Weinberg, W. 1901. Beiträge zur Physiologie und Pathologie der Mehrlingsgeburten beim Menschen, Arch. f. d. ges. Physiol., 88:346–430.

Wiener, A. S. 1935. Heredity of the agglutinogen M and N of Landsteiner and Levine. IV. Additional theoretico-statistical considerations, Human Biol., 7: 229–39.

———. 1952. Heredity of the M-N-S blood types: theoretico-statistical considerations, Am. J. Human Genetics, 4:37–53.

Wiener, A. S., and Leff, I. L. 1940. Chances of establishing non-identity of binovular twins, with special reference to individuality tests of the blood, Genetics, 25:187–96.

GENERAL REFERENCES

Bass, M. H. 1952. Diseases of the pregnant woman affecting the offspring, Adv. Int. Med., 5:15–58.

Gedda, L. 1951. Studio dei gemelli. Rome: Edizioni Orizzonte medico.

Newman, H. H. 1940. Multiple human births: twins, triplets, quadruplets, and quintuplets. New York: Doubleday, Doran & Co.

Newman, H. H.; Freeman, F. N.; and Holzinger, K. 1937. Twins: a study of heredity and environment. Chicago: University of Chicago Press.

Price, B. 1950. Primary biases in twin studies: a review of prenatal and natal difference-producing factors in monozygotic pairs, Am. J. Human Genetics, 2:293–352.

von Verschuer, O. 1939. Twin research from the time of Galton to the present day, Proc. Roy. Soc., London, B, 128:62–81.

CHAPTER 17

Genetics and Epidemiology

17.1. *The epidemiological approach.*—Epidemiology has been defined as the study of the circumstances responsible for the development of specific diseases in specific groups of people. As such, it is a very broad field indeed. With reference to any particular disease, it is convenient to divide the possible etiological factors which must be considered into three groups, namely: (1) the causative agent (virus, bacterium, parasitic worm, toxic chemical, etc.), (2) the environmental factors responsible for, or favoring, the development and spread of the disease, and (3) the host factors, this last term as usually employed being more or less synonymous with genetic factors but including also such features as nutrition, fatigue, immunity status, body temperature, etc.

The relative importance of these three groups of factors varies widely from one condition to the next. There are diseases, such as the consequences of being struck by lightning, which are entirely of environmental origin. At the other extreme, there are, as we have seen, conditions which may be considered entirely genetic in origin, on the assumption of an environment compatible with life. However, in the causation of most of the diseases which form the subject matter of medicine, we are concerned with a complex interaction between two or three of the groups of factors enumerated above. In the so-called "infectious" or "contagious" diseases, we have to take into account all three of these groups of factors. In the metabolic, endocrine, and "degenerative" diseases, on the other hand, it appears that we are concerned to a much greater extent with the "host factor."

In general, the epidemiologist has in the past tended to concentrate his attention on studies of the viral, bacterial, protozoal, etc., agents of disease and the environmental factors that are related to their distribution and spread. The reasons are chiefly twofold: (1) In the practice of medicine, the physician is concerned with healing the sick patient and preventing the well patient from becoming ill. The most immediate results can be obtained through steps which eliminate the causative agent of a disease or control the environment in such a manner as to limit its distribution. By contrast, much less "can be done about" the host factor. The individual physician treating the individual patient is, of necessity, far more concerned with disposing of the problem at hand and preventing, where possible, the spread of disease to other members of the family and community than with a detailed analysis of the constitutional reasons *why* that particular patient was singled out by the disease. (2) A second reason for the character of the subject matter of epidemiology in past years is that the interests of the epidemiologist have tended to reflect those of the medical profession as a whole, i.e., a more pressing concern with the infectious and contagious diseases.

This concentration of medical attention upon the causative agents of disease and their control through environmental measures, as well as the general

TABLE 17-1

COMPARISON OF TEN LEADING CAUSES OF DEATH IN
THE UNITED STATES IN 1900 AND 1949

1900		1949	
Cause of Death	Deaths per 100,000	Cause of Death	Deaths per 100,000
Cardiovascular-renal disease...	345.2	Cardiovascular-renal disease.	483.5
Influenza and pneumonia.....	202.2	Malignant neoplasms.......	137.3
Tuberculosis................	194.4	Accidents.................	63.7
Gastrointestinal disease.......	142.7	Influenza and pneumonia....	34.3
"Senility".................	117.5	Diabetes mellitus..........	29.7
Accidents..................	72.3	Tuberculosis..............	27.4
Malignant neoplasms.........	64.0	"Senility".................	15.9
Diphtheria.................	40.3	Congenital malformations...	13.0
Typhoid fever...............	31.3	Suicide...................	11.4
Complications of pregnancy...	13.4	Cirrhosis of liver..........	11.3
All causes..............	1,719.1	All causes.............	971.7

* Data taken from Vital statistics of the United States (1949), Table 14, Part 1.

improvement in the standard of living throughout the world, has yielded spectacular dividends (Table 17-1). For instance, in the United States between 1900 and 1949 the total death rate per 100,000 inhabitants dropped from 1,719.1 to 971.7. Three infectious diseases which in 1900 were on the list of the "first ten" are in 1949 no longer found there. These are the gastrointestinal diseases (largely composed of the diarrheas), diphtheria, and typhoid fever. Deaths from pneumonia, influenza, and tuberculosis have dropped from 396.6 to 61.7 per 100,000. There is a truly dramatic trend toward the control of the infectious and contagious disease entities. Continuing discoveries regarding antibiotics and chemotherapy make it apparent that the full impact of this trend is yet to be felt. As a result of these advances, an increasing proportion of medical attention is being forcibly directed toward those diseases in which the "host factor" appears to be relatively more important: certain types of heart disease and cancer, diabetes, kidney disease, and congenital malformations. The epidemiologist and the medical investigator of today are now expending upon the study of man himself much of the energy they have previously poured into studies of the agents of disease. In this chapter we propose to analyze the manner in which the genetic approach is of value to the present-day epidemiologist, using the term "epidemiologist" in the broadest sense. The concept of an "epidemiological team" is well ingrained in medicine. We shall attempt to show how the geneticist can and should function, not as a competitor, but as a member of that team. In passing, it might be pointed out that although we have used the term "epidemiologist" with special reference to the study of disease, because this is the usual sense of the word, the epidemiological approach can obviously be extended to the study of nondiseased states—"genius," "laziness," or "muscular strength," to mention only three examples.

17.2. *Characteristics of genetically determined disease.*—The medical literature contains numerous examples of the problems that arise in deciding whether the appearance of a given disease is or is not due in whole or in part to genetic factors. For instance, for years it was known that syphilitic women tended to have syphilitic babies, and syphilis was spoken of as "hereditary." Once the causative organism of syphilis was recognized, it was soon demonstrated that the parent-child resemblance with respect to syphilis was due to the fact that the child acquired the disease while still in the uterus. The chromosomal mechanisms of inheritance are in no way involved. Tuberculosis has also been referred to as "hereditary." While there is almost certainly a greater predisposition on the part of some people to acquire the disease than on the part of others, a major factor in certain "pedigrees" of tuberculosis is probably the exposure factor. The presence in a family of one or several older persons with tuberculosis means that any children in this family will have an increased exposure. A more recent illustration of the care to be exercised in referring to a disease as inherited may be drawn from the field of hematology. Some individuals, for obscure reasons, develop a greatly diminished blood platelet count and, because of the role of the platelets in blood clotting, are subject to serious hemorrhages. There are observations in the literature of the occurrence of the disease in a mother and her newborn child (summary in Robson and Walker, 1951). This seemed to point to a genetic etiology. However, recent studies indicate that in patients with this disease, which is known as "idiopathic thrombocytopenic purpura," there is a substance in the blood plasma which is capable of lowering the platelet number (Harrington *et al.*, 1951; Sprague *et al.*, 1952). The lowered platelet counts sometimes observed among the children of such mothers may be due to the fact that the substance responsible for a low platelet count in the mothers can cross the "placental barrier" and exert a similar effect upon the child as long as it persists in the circulation.

Sufficient has been said to indicate the need for caution in concluding from the occurrence of a disease in successive generations that it is inherited in the genetic sense. What, then, are the criteria from which the influence of genetic factors in the etiology of a given trait, disease susceptibility, etc., may be detected? There are four principal criteria, some of which we have already considered in detail, some of which are introduced here for the first time.

1. THE OCCURRENCE OF THE DISEASE IN DEFINITE NUMERICAL PROPORTIONS AMONG INDIVIDUALS RELATED BY DESCENT

As has been discussed in preceding chapters, where a trait has a simple genetic basis, in families where the trait occurs there are definite numerical relationships between affected and nonaffected persons, *although these relationships may not become apparent until the proper corrections have been made for mode of ascertainment, age, etc.* With present-day genetical techniques, little difficulty should be encountered in deciding whether a given trait is due to a single, dominant or recessive, sex-linked or autosomal, gene with complete penetrance. Traits due to the interaction of several genes, or to a gene of incomplete penetrance, present greater difficulties. It must be admitted, for instance, that the precise genetic analysis of a human trait due to three different, nonallelic genes presents almost insuperable problems at present.

2. FAILURE OF THE DISEASE TO "SPREAD" TO NONRELATED INDIVIDUALS

It is characteristic of a disease with a specific causative agent whose transmission is intimately related to various features of the environment that in the course of a single generation it spreads in a "horizontal fashion," without regard to blood relationship. In genetically determined illness, on the other hand, there is a "vertical" distribution pattern. In dominant heredity this vertical pattern is usually obvious. In recessive inheritance this is less so, one of the chief clues to vertical transmission being the excess of consanguineous marriages seen among the parents of children with rare recessive traits.

3. ONSET OF DISEASE AT A CHARACTERISTIC AGE WITHOUT A KNOWN PRECIPITATING EVENT

A gradual onset without a clear cause is characteristic of many of the "degenerative" diseases in which genetic factors play a role. However, such an onset could also be seen in diseases with definite but undiscovered bacterial, viral, or traumatic etiologies as well. In a sense, evidence of this nature is negative evidence—"the trait must be genetically determined because we can't recognize any other cause"—and negative evidence is never so satisfactory as positive.

4. GREATER CONCORDANCE OF THE DISEASE IN IDENTICAL THAN IN FRATERNAL TWINS

The uses (and abuses) of the twin method in genetic investigation have already been discussed; suffice it to reiterate here that, before the greater concordance of a given trait in identical than in fraternal twins is acceptable as evidence of a genetic factor, the possible environmental sources of bias enumerated earlier must be considered one by one.

17.3. *The varying roles of heredity and environment with respect to a given trait.* —We have repeatedly stressed the fact that a given trait may be inherited in two different ways in two different families. It is equally important to recognize that the balance between the roles of heredity and environment in the causation of a given trait may vary from one family to the next. Otherwise stated, a given trait may sometimes be genetically determined, sometimes environmentally (including in the latter term uterine influence). The term "phenocopy" has been applied to environmentally produced abnormalities which mimic traits known to be genetically determined in man. There are a number of human traits which may be genetically determined but may also occur as phenocopies. These probably include such diverse findings as harelip, clubfoot, depth of skin pigmentation, or an increased amount of blood cholesterol. Unfortunately, the unequivocal demonstration of "phenocopies" requires an experimental approach not feasible in human material. The case is somewhat clearer in Drosophila and mice. There are mouse strains in which 20–30 per cent of the newborn regularly show harelip, undoubtedly because of the genetic constitution of the strain (Reed, 1936). On the other hand, this same defect may occur as a phenocopy, either because of treatment of the pregnant mouse with certain hormones at an early stage of her pregnancy (Fraser and Fainstat, 1951) or because of a reduction in the oxygen content of the air she breathes at certain critical stages in gestation (Ingalls, Curley, and Prindle, 1950). With respect to Drosophila, numerous investigators have found that treatment of larvae or early pupae with relatively high temperatures for periods of 1–2 hours results in a wide variety of abnormalities in the adult fly, many of the abnormalities mimicking defects which in some strains are determined by single genes (Goldschmidt, 1945, 1949).

17.4. *Experimental studies on the host factor in susceptibility to infectious disease, as illustrated by mouse typhoid.*—A classical piece of investigation on the manner in which heredity, environment, and an agent of disease may interact has been carried out on mice by Webster (1932). This investigator has studied in detail the factors which regulate the spread of mouse typhoid through a population of mice. It has been shown that the following must be taken into account: (1) the initial infecting dose of mouse typhoid, (2) the number of mice infected initially, (3) the strain of bacteria used in the experiment, (4) the strain of animals used, and (5) the diet of the mice. Variations in any one of these factors significantly influenced the spread of the disease through the population. It was possible by selective breeding to establish strains of mice relatively resistant to the disease, as well as relatively susceptible strains. In one series of experiments, after the initial mouse population had been established and the infection introduced, two mice were added to the population each day. When the mice added were from a susceptible strain, explosive outbreaks of the disease were observed. When the mice were from the resistant strain, only sporadic cases were seen. In another series of experiments, "herds" composed of both susceptible and resistant animals were established, and infection was then introduced into the herd. Under these circumstances, 70 per cent of the susceptible animals but only 12 per cent of the resistant ones died.

The analysis of the genetic basis for resistance to mouse typhoid has been carried farthest by Gowen and his students (1948, 1950, 1953). Resistance appears to be due to a complex of many genes. An effort has been made to define the physiological mechanisms through which these genes work. Resistant and susceptible strains arising through selection from a common stock have been observed to differ with respect to the greater number of leucocytes normally found in the blood stream of the resistant animals, the apparent greater ability of the leucocytes and also the macrophages of resistant strains to digest engulfed bacteria, and the higher pH of the blood of resistant strains. The exact significance of these findings to the defense mechanisms of the mouse against infection is not clear (cf. Weir, Cooper, and Clark, 1953). In the event that infection does occur, resistant strains may show extensive lesions of the liver and survive, whereas the susceptible will die without any apparent damage to that organ. In contrast, resistant strains show but little damage to the spleen even when severe typhoid exists, whereas moderate lesions are found in the susceptible animals. These observations have been interpreted to mean that in the resistant animals the liver acts as an effective first line of defense against the toxin released by the typhoid organism. It is apparent from this that the genetic control of disease resistance in the mouse involves a number of complex mechanisms. It is certain that the situation is no less complex in man.

17.5. *The genetic factor in resistance to infectious diseases in man.*—The manner in which the geneticist participates in the study of the epidemiology of disease varies somewhat with the disease at hand. In this and the following section we shall deal with two illustrative categories of disease, namely, the contagious diseases and the neoplastic diseases. Other categories, such as congenital defects or endocrine disorders, might equally well have been selected. The contagious diseases will be discussed under four headings: "childhood diseases," poliomyelitis, leprosy, and tuberculosis.

1. THE "CHILDHOOD DISEASES"

Studies on disease resistance in man have produced evidence of inherited differences between individuals with respect to susceptibility to some infections but not to others. Table 17-2 summarizes the results of twin studies

TABLE 17-2*

COMPARISON OF BEHAVIOR OF IDENTICAL AND FRATERNAL TWINS IN DEVELOP-
MENT DURING CHILDHOOD OF VARIOUS ACUTE DISEASES, FOR TWIN-
PAIRS IN WHICH AT LEAST ONE TWIN DEVELOPED THE DISEASE

DISEASE	IDENTICAL TWINS		FRATERNAL TWINS		PERCENTAGE OF DISCORDANCE	
	Concordant	Discordant	Concordant	Discordant	Identicals	Fraternals
Measles.............	866	19	762	47	2.1±0.5	5.8±0.8
Scarlet fever.........	84	59	69	91	41.3±4.1	56.9±3.9
Diphtheria..........	79	79	57	95	50.0±4.0	62.5±3.9
Mumps..............	95	21	89	32	18.1±3.6	26.4±4.0
Whooping cough......	524	15	447	34	2.8±0.7	7.1±1.2

* Table compiled from data published by Von Verschuer, Gebbing, Camerer, and Schleicher, and summarized by Degkwitz and Kirchmair (1940).

with respect to five of the infectious diseases commonly encountered in childhood. It is apparent that for each disease the identical twins are more frequently concordant than the fraternals. However, the differences, while undoubtedly significant, are not striking. Because of the greater environmental similarities of identical twins—which would be especially important in exposure to a contagious disease—these differences must be interpreted with caution. In passing, it might be pointed out that if for any reason the attack rate in one type of twin is higher than in the other, this fact alone can alter the concordance-discordance ratio.

2. PARALYTIC POLIOMYELITIS

The findings with respect to paralytic poliomyelitis, leprosy, and tuberculosis warrant a somewhat more extended treatment. Poliomyelitis with paralysis is a rather uncommon disease. However, it is now established that infection with the poliomyelitis virus is rather common but that only a small fraction of persons who are infected—in the neighborhood of 1 per cent or less—become paralyzed. The virus appears to be spread both through droplet infection and through contamination of food and water. A number of antigenically different strains of the virus have been identified. Thus in understanding any epidemic of poliomyelitis, we have to reckon with the strain of virus involved, the environmental factors which favor its dispersion, and differences in the reactions of people to their infection, i.e., the development of paralysis in some but not in others.

It has long been observed that in epidemics the distribution of paralytic poliomyelitis does not appear to be at random in human populations, the frequency of multiple cases within family groups appearing to be greater than would be expected on the basis of chance alone (Aycock, 1942). As pointed out above, this does not necessarily indicate a constitutional susceptibility to the disease, since it might mean only that certain families have a particularly heavy exposure to the virus or, because of dietary peculiarities, are especially susceptible to its ravages. When, however, one finds histories of different individuals from a family becoming paralyzed in epidemics many years apart, the evidence for a "constitutional factor" is strengthened. Further evidence is derived from a careful study of the physical characteristics of affected children. For over a hundred years physicians have commented on the tendency of paralytic poliomyelitis to attack strong and healthy-appearing children rather than the weak and "rundown." This clinical impression was finally put to precise anthropometric tests (Aycock, 1941; Draper, Dupertuis, and Caughey, 1944). It has been shown that, prior to and following adolescence, children who have had paralytic poliomyelitis tend to be larger than controls, although the difference disappears during adolescence. The impression is of a different kind of growth curve. There appear to be other, more qualitative, differences. In comparison with a group of controls, the children with paralytic poliomyelitis were found to exhibit a greater tendency to (1) pigmented spots on the body, (2) long and curved eyelashes, (3) large central incisor teeth, with a gap between the upper incisors, (4) relaxation of the joints, and (5) an oblique fold of skin at the inner aspect of the upper eyelid, the so-called "internal eyefold." None of these differences between the paralyzed and their controls was absolute; rather, it was a matter of certain "tendencies." The genetic interpretation of these findings would be that susceptible persons often, but not always, have received certain genes with well-defined physical effects, with the familial constellations of the disease explained by the genetic similarities between related individuals.

The most precise evidence concerning the importance of this genetic factor comes from studies by Herndon and Jennings on twins (1951). An attempt was made to locate all families in the state of North Carolina in which paralytic poliomyelitis had occurred in one or both members of a pair of twins during the years 1940 through 1948. Attention was restricted to those families in which the twins had been living together at the time of the illness of one or both. Data were also obtained on the siblings and parents of the twins. The results are shown in Table 17-3. Two facts are apparent: (1) both members of a pair of monozygotic twins are more often attacked than both members of a pair of dizygotic twins, and (2) the similarities between dizygotic twins are essentially the same as between siblings; both groups appear to contain more affected individuals than would be expected on the basis of chance alone. This, however, could as well be attributed to environmental as to genetic influences. Although conclusions concerning the genetic mecha-

TABLE 17-3*

PERCENTAGE OF PARALYTIC POLIOMYELITIS OCCURRING
AMONG RELATIVES OF TWIN INDEX CASES

Relationship	Total	Para-lytic Polio-myelitis	Percentage± Standard Error
Monozygotic co-twins......	14	5	35.71±12.81
Dizygotic co-twins........	33	2	6.06± 4.15
Exposed siblings..........	101	7	6.93± 2.53
Parents.................	87	0	0.0

* Used by permission of Dr. C. N. Herndon and the American Journal of Human Genetics.

nisms involved appear premature, the data are consistent with the hypothesis that the development of paralysis following an attack of poliomyelitis is due to a recessive gene with about 35 per cent penetrance when homozygous.

3. LEPROSY

Studies on the possible genetic factor in the development of leprosy are of a somewhat different nature. This disease is due to a bacterium closely related to that responsible for tuberculosis. It is transmitted through contact with an infected person, but apparently the contact must be of a longer duration than is true of most contagious diseases. There are clearly delimited, persistent "centers of distribution" of the disease in North America, the two best known being in New Brunswick (Canada) and Louisiana. Many of the inhabitants of these two areas have a common origin. In 1775 the Norman immigrants who had settled in the Acadian region of Canada were forcibly uprooted by the English because of the possible imminence of war with France and the uncertain loyalties and future conduct of the Acadians. Some of these people then settled in New Brunswick, others were transported to Louisiana. These events form the basis for the poem *Evangeline*, so well known to school children.

The disease is not known to have been present in the original group from which these two subgroups are derived; at least, there can be found no official notice of the condition. Aycock (1938, 1941) has pointed out that the disease does not appear to be distributed at random among the descendants of these Acadian emigrants but to exhibit a predilection for certain families. Two "explanations" are possible. One can, on the one hand, think in terms of a hereditary susceptibility to the bacterium, which susceptibility can, of course, manifest itself only when the disease has been introduced into a community. One can, on the other hand, postulate a uniform susceptibility but postulate further that the conditions of prolonged and intimate contact necessary to infection are realized only in certain families in which *by chance* the disease has become established.

The development of leprosy in the spouse of an infected individual (conjugal leprosy) is very much rarer than in the children of a leper. This finding has been given a genetic interpretation; but proponents of the other point of view call attention to the fact that the finding can also be explained if children are more susceptible to the disease than adults. This argument, in turn, is countered with the statement that many studies of leprosy among the relatives of lepers reveal an unusual concentration of cases in distant relatives not living under conditions of "prolonged and intimate exposure." The available data do not permit a clear decision, but the possibility of different inherent susceptibilities to the disease is strong.

4. TUBERCULOSIS

The problem of the extent to which genetic factors enter into susceptibility to tuberculosis is one of the oldest in human genetics. Thus in 1883 Hirsch wrote: "That phthisis propagates itself in many families from generation to generation is so much a matter of daily experience, that the severest sceptic can hardly venture to deny a hereditary element in the case: even if we be unable for the present to decide whether it consists in the transmission of a specific poison, something like that of syphilis, or, in other words, whether it be hereditary in the narrower sense; or whether it do not depend, as seems to me more probable and more consonant with some things soon to be mentioned, upon a congenital disposition towards the disease, a disposition that has to be looked for, naturally, in the organization of the respiratory system."

Because the factor of familial exposure is so important in the development of this disease, the pedigree method of approach is of very limited value, and here again the most critical data are derived from twin studies. Table 17-4 summarizes the findings in the most comprehensive study to date, that of Kallmann and Reisner (1943). Over a five-year period an attempt was made to discover as many individuals as possible in the state of New York whose tuberculosis was diagnosed during that time and who were twins. Data were then collected concerning the presence or absence of tuberculosis in the other twin and in members of the immediate family. The concordance between identical twins is striking. On the other hand, the higher rate in spouses as compared with the general population underlines the importance of the factor of proximity to an infected individual and raises the question of how much

TABLE 17-4*

CONCORDANCE IN DEVELOPMENT OF TUBERCULOSIS AT SOME TIME DURING
LIFE IN IDENTICAL AND FRATERNAL TWINS AND THEIR
SIBLINGS, PARENTS, AND SPOUSES

	GENERAL POPULATION OVER FOURTEEN YEARS OF AGE	RELATION TO TUBERCULOSIS INDEX CASES					
		Spouses	Parent	Half-sibling	Full sibling	Dizygotic Co-twin	Monozygotic Co-twin
No. of such individuals studied	....	226	688	42†	720†	230	78
No. with tuberculosis.........	....	14	114	4	136	42	48
Uncorrected morbidity rate...	1.1	6.2	16.6	9.5	18.9	18.3	61.5
Morbidity rate corrected for differences in age between groups..................	1.4	7.1	16.9	11.9	25.5	25.6	87.3

* Used by permission of Dr. Franz Kallmann and the Journal of Heredity.
† Individuals below the age of fourteen have not been included in the study.

the greater tendency of identical twins to have the same playmates and close associations both within and without the family enters into the picture. It is noteworthy that identical twins tended to be alike not only with respect to the presence or absence of the disease but also in the clinical course which the disease pursued.

The studies of Lurie and his collaborators (1952, 1953) on susceptibility to tuberculosis in the rabbit have demonstrated that various strains differ widely in their susceptibility to the disease. As might be expected, resistance appears to be a complex phenomenon, depending on such possibly interrelated and interdependent factors as ability to localize the disease at the portal of entry, high or low skin permeability, rapid or slow development of allergic sensitivity, and intense or more limited production of antibodies. Each of these steps might be under independent genetic control. The twin studies supply some insight into the total heritability of a complex trait but tell us little about the mechanisms involved and the control of each of these.

Studies by Meyer and Jensen (1951) constitute a step toward defining the nature of the resistance mechanisms under genetic control in man. Following an infection with the tubercle bacillus, there develops a sensitivity to the products of the bacillus, this forming the basis for the skin test commonly used in the diagnosis of tuberculosis. This same sensitivity develops following vaccination with the attenuated BCG strain of the tubercle bacillus. The carefully controlled studies of Meyer and Jensen demonstrate that the degree of the sensitivity which develops following vaccination, as judged by the intensity of the skin reaction to an extract of the bacillus, varies significantly from the children of one family to the next.

It is apparent that despite considerable work we are still far from a complete understanding of the infectious agent-host-environment reaction with regard to *any* of the contagious diseases of man. As geneticists, the authors cannot resist the temptation to suggest—and not entirely facetiously—that the time to study these interrelationships is now; if the present trends with respect to the contagious diseases continue, it may soon be impossible for the investigator to amass sufficient material for a critical study of many of the contagious diseases. Poliomyelitis and leprosy were chosen for discussion in this chapter because they appear to represent two extremes—the one (including nonparalytic and subclinical infections) a relatively common disease with a short incubation period, the other a rare disease with a long incubation period. They thus illustrate the opposite ends of the epidemiological range in which the geneticist may be called upon to work.

17.6. *The genetic factor in the development of tumors.*—In our foregoing consideration of the role of heredity in susceptibility to a variety of contagious diseases, one of the chief problems lay in determining to what extent intrafamily similarities in disease patterns were due to a *common heredity* and to what extent to a *common exposure* to a communicable agent of disease. In evaluating the role of heredity in the development of tumors, as well as the various "degenerative diseases," the complication of "common exposure" apparently need not be reckoned with, since, so far as is known, these are not contagious; but now a different sort of problem arises. Because tumors and the "degenerative diseases," as the name implies, are primarily diseases of the older age groups, it is not uncommon to find that only one affected generation is alive at any one time. The inevitable consequence is difficulty in documenting the occurrence and extent of the disease in past generations.

As an illustration of the genetic problems which arise in the study of some common diseases of late onset, we shall examine the development of tumors, i.e., the neoplastic diseases. There are many different types of tumors, some of which grow slowly and remain localized throughout the life of the affected individual, others of which grow rapidly and spread to distant parts of the body. The former are known as "benign" tumors, the latter as "malignant." The most common type of malignant tumor is a cancer. The one thing that all tumors, benign or malignant, have in common is an abnormal proliferation of cells.

As a result of a great deal of research on the subject, many factors which can contribute to the development of tumors are now recognized. Heredity is definitely one of these "contributing factors." It is important at this point to recognize the dangers involved in making sweeping statements about heredity in the susceptibility to tumor formation. In any discussion of heredity in tumor development, one must be quite careful to specify the type of tumor. For in some cases, as we shall now see, heredity may be quite important, whereas in others the genetic factor is very minor, if, indeed, it exists at all.

The medical literature contains many case reports of families in which there was a high incidence of some particular tumor. The significance of these reports is difficult to evaluate. Tumors of all types are so common—approximately one out of each seven deaths in the United States each year being due to a tumor—that by chance we must expect some "familial concentrations." Although some of these concentrations certainly have a genetic implication, it is much safer, in arriving at an opinion concerning the role of heredity in the development of a particular type of tumor, to survey the family history in an unselected group of cases of that particular tumor type.

Such surveys have indicated that there are a number of rare tumors whose development depends on the presence of a single dominant gene. For instance, in the condition known as multiple neurofibromatosis, numerous tumors develop in association with the peripheral nerves (Fig. 17-1). There are also areas of brownish discoloration scattered over the body, known as "café-au-lait" spots, from their resemblance to the color of a half-coffee half-milk mixture. These tumors are usually benign, causing difficulty chiefly because of their unsightly appearance and the pressure they may exert on surrounding structures. Occasionally, however, one of these tumors becomes malignant and causes the death of the affected person. The development of the tumors often, if not always, depends on a single dominant gene, a typical pedigree being shown in Figure 17-2. As another illustration of the inheritance of a precancerous lesion, the dominantly inherited multiple polyposis of the colon should be mentioned. In this disease the large intestine is literally studded with small protuberances, termed "polyps." These polyps have a tendency to develop into malignant tumors in middle life. This is a particularly insidious condition, since the presence of the tumors is not usually appreciated until symptoms due to the development of a cancer appear. Figure 17-3 summarizes the findings in an unusually large kindred in which this disease was present.

In contrast to some of these rare tumors in which a genetic etiology is so clearly apparent, for many of the more common tumors the genetic factor, if

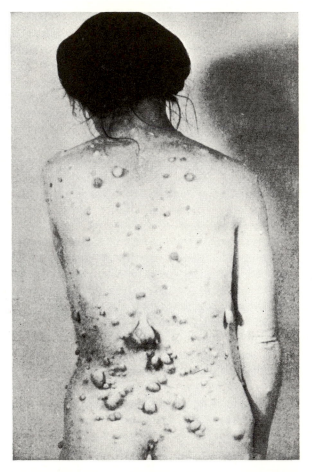

FIG. 17-1.—Photograph of an individual with multiple neurofibromatosis

any, can be demonstrated only by large-scale statistical studies. There are three of the more common types of cancer which have been carefully and extensively studied from the genetic viewpoint. These are cancer of the breast, cancer of the cervix and body of the uterus, and leukemia. We will briefly consider some of the more important studies on each of these.

The subject of familial factors in breast cancer has been scrutinized by a number of different investigators. In Denmark, Jacobsen (1947) conducted family studies on the relatives of 200 women who developed breast cancer. An attempt was made to determine the number of times this cancer had occurred in the mothers, sisters, mothers' sisters, and fathers' sisters of the index cases. In order to have a suitable basis for comparison, he selected 200 individuals of the same ages as his patients who did *not* have breast cancer

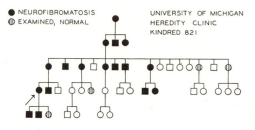

FIG. 17-2.—Pedigree of multiple neurofibromatosis

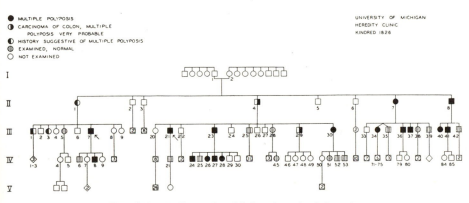

FIG. 17-3.—Pedigree of multiple polyposis of the colon

and studied the occurrence of this condition in the same categories of relatives enumerated above. The results are shown in Table 17-5. The frequency of mammary cancer appeared to be increased in each of the four critical categories of relatives of cancer patients studied, the increase being five- to tenfold. Moreover, the frequency of various other types of cancer appeared to be somewhat increased among the relatives of the patients with cancer.

Penrose, MacKenzie, and Karn (1948) have studied in England the occurrence of mammary cancer among the *deceased* mothers and sisters of patients with this disease. For controls they used English mortality data. They observed an approximately threefold increase in this type of cancer in the relatives of their patients. In this connection they point out that there seem to be fewer affected persons in Jacobsen's controls than one might expect. Penrose, MacKenzie, and Karn failed to observe an increase in other types of cancer among the relatives of affected persons.

The subject of familial factors in cancer of the uterus has been most thoroughly investigated by Brøbeck (1949) in Denmark and Murphy (1952) in

TABLE 17-5

STUDIES ON FAMILIAL INCIDENCE OF CANCER

			PROBAND FAMILIES			CONTROL FAMILIES		
TYPE OF CANCER	INVESTIGATOR	TYPE OF RELATIVE	No. of Such Relatives	No. with Same Type of Cancer	Per Cent	No. of Such Relatives	No. with Same Type of Cancer	Per Cent
Cancer of breast	Jacobsen (1946)	Mothers	200	21	10.5	200	2	1.0
		Sisters	381	13	3.4	433	2	0.5
		Mothers' sisters	316	17	5.4	312	. . . . .	0.0
		Fathers' sisters	224	12	5.4	223	2	0.9
	Penrose, Mac-Kenzie, and Karn (1948)	Mothers*	406	25	6.2	. . . . .	11.1†	2.7
		Sisters*	307	23	7.5	. . . . .	7.0†	2.3
Cancer of cervix of uterus	Brøbeck (1949)	Mothers	200	6	3.0	200	6	3.0
		Sisters	488	10	2.0	449	1	0.2
		Mothers' sisters	314	3	1.0	336	5	1.5
		Fathers' sisters	284	6	2.1	306	1	0.3
Cancer of corpus of uterus	Brøbeck (1949)	Mothers	90	6	6.7	90	3	3.3
		Sisters	217	6	2.8	233	1	0.4
		Mothers' sisters	155	4	2.6	152	4	2.6
		Fathers' sisters	129	1	0.8	118	0	0.0
Cancer of cervix and corpus uteri	Murphy (1952)	Mothers	200	6‡	3.0	214	2	0.9
		Sisters	360	5‡	1.4	341	4	1.2
		Mothers' sisters	252	9‡	3.6	327	5	1.5
		Fathers' sisters	198	6‡	3.0	272	4	1.5

* Refers only to deceased mothers and sisters.
† Expectation calculated on the basis of mortality statistics.
‡ No distinction between cancer of the cervix uteri and corpus uteri.

the United States. Brøbeck distinguished between cancer of the cervix of the uterus as opposed to cancer of the corpus or body of the uterus. For both types of uterine cancer there appeared to be a slight, approximately twofold, increase in the frequency of uterine cancer among the four most pertinent categories of female relatives. The same tendency is apparent in Murphy's data, where no attempt was made to distinguish between cancer of the cervix and that of the corpus. Brøbeck found a slight increase in the frequency of other types of cancer among the relatives of the cancerous patients whom he studied—this finding was not confirmed by Murphy.

The importance of familial factors in leukemia, now usually classed as a type of cancer of the blood-forming tissues, has been studied by Videbaek (1947) in Denmark. The medical histories of 4,041 relatives of 209 patients with leukemia were compared with the histories of 3,641 relatives of 200 control individuals. Presumably because of the greater rarity of leukemia, the findings are not presented in terms of frequency of this disease in specific categories of relatives. Among all the relatives of the patients with leukemia who were investigated, 17 were found to have developed the same disorder, whereas in the control material there was only 1 case of leukemia. A somewhat higher incidence of all types of cancer was reported among the relatives of leukemic patients.

These various studies all agree in suggesting that certain categories of relatives of patients with these kinds of cancer have a slight, but significantly increased, risk of developing the same type of cancer. There is less agreement as to whether there is also a slightly increased susceptibility to cancer of all types. Any attempt at a strictly genetic interpretation of these findings is hazardous, resting largely, as it must, on analogy with the findings in experimental material. In this connection the difficulties involved in the selection of a suitable control material must be emphasized. By improper selection of controls, one can accentuate or minimize the familial factor.

It is in such a situation as this that twin studies can be of especial value. Macklin (1940; see also Gorer, 1938) was able to find, in the medical literature, reports of tumors involving one or both members of fifty-three pairs of monozygous twins and thirty-five pairs of dizygous twins. To these, on the basis of correspondence and personal observation, she was able to add nine additional pairs of dizygous and eight of monozygous. Because of sex differences in frequency and type of tumor, in a comparison of tumor incidence one must eliminate the unlike-sexed dizygous twins and compare only like-sexed dizygous twins with monozygous twins. This is done in Table 17-6. As was pointed out earlier (sec. 16.7, p. 277), one of the biases in data on concordance-discordance in twin-pairs is that concordant twins, because of their greater interest, are more likely to be reported in the medical literature than are discordant. This source of bias is minimized when all the observations are methodically collected by the same individual or group, as in the studies of Herndon and Jennings, and Kallmann and Reisner, referred to earlier, and maximized when a rare event is concerned, with various observers reporting a case or two, as is true of the tumor data. In the data compiled by Macklin, the ratio of identical to like-sexed dizygous twins is approximately 2:1, whereas the true ratio in the population is more like 1:1. As Macklin points out, this indicates a biased sample. Under these circumstances, although there is undoubtedly a greater concordance of identical than fraternal twins, the formulas of section 16.6 (p. 272) cannot be applied to estimate "heritability." In other words, in this instance much of the "usefulness" of the twin approach has been lost because of biased reporting. Before leaving Table 17-6, it should be pointed out that additional evidence tending to implicate the etiological significance of heredity comes from the greater similarity in type of tumor and age of onset reported for monozygous twins.

It seems reasonable to interpret all these findings as indicating that the development of many tumors is due at least in part to genetic factors. However, any attempt to specify the nature of the genetic mechanisms involved

TABLE 17-6*

COMPARISON OF CONCORDANCE-DISCORDANCE IN TUMORS AMONG
MONOZYGOUS AND LIKE-SEXED DIZYGOUS TWINS

Twins	No. of Pairs	No. of Pairs, Both Affected	Concord-ance (Per Cent)	No. of Pairs, Both Affected with Similar Tumors	Concordance in Tumor Type in "Both Af-fected" Twin-Pairs (Per Cent)	Average Difference in Age of Onset, Both Affected (Years)
Monozygous.........	62	38	61.3	36	94.7	1.5
Like-sexed dizygous..	27	12	44.4	7	58.3	9.5

* Used by permission of Dr. Madge T. Macklin and the Journal of Heredity.

seems premature. One can "explain" the findings either on the basis of dominant or recessive genes with poor penetrance or on the basis of the interaction of several genes, again with poor penetrance. It seems quite possible that several different genetic mechanisms may be at work in the inheritance of a tendency to develop a particular type of neoplasm. Thus, as seen earlier, genetic factors are usually not prominent in the appearance of leukemia. However, Anderson (1951) has reported the occurrence of leukemia in five of the eight children of normal parents, a finding compatible with simple recessive heredity (Fig. 17-4, a). We ourselves have observed the occurrence of leukemia in two children who were first cousins through the paternal line and second cousins on the maternal side, again a situation suggesting recessive inheritance (Fig. 17-4, b). Such cases, although rare, indicate the need for caution not only in making sweeping statements about heredity in cancer but also in reaching conclusions even as regards a particular type.

The nature of the basic event which transforms a normal cell into one no longer responsible to the usual checks on unbridled proliferation—in short, into a cancer cell—is unknown. From time to time a variety of theories has been suggested. Chief among these has been the cell-mutation theory, which suggests that a cell becomes capable of giving rise to a cancer because of a mutation in that cell. In view of the known frequency of germinal mutation in man, it seems probable, if the frequency of somatic mutation parallels that of germinal, that many of the somatic cells do contain mutant genes. The recent finding that many of the chemical agents capable of inducing malignant tumors in animals can also produce mutations provides indirect support to the mutation hypothesis (Tatum, 1947; Demerec, 1949, 1951).

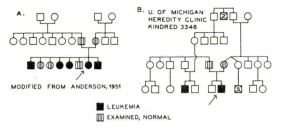

Fig. 17-4.—(a) The appearance of leukemia in five siblings. (By permission of Dr. Ray C. Anderson and the American Journal of Diseases of Children.) (b) The appearance of leukemia in double cousins.

17.7. Population genetics and epidemiology.—As pointed out in an earlier chapter, population geneticists view the inherited traits of man in terms of a dynamic balance between mutation and selection. But in the analysis of factors affecting both mutation and selection, the population geneticist must work with population characteristics which are the stock-in-trade of the epidemiologist. For instance, Falls and Neel (1951), in a study of the rate of mutation to the dominant gene responsible for the type of cancer of the eye in children known as retinoblastoma, located forty-nine families in the state of Michigan in which a presumably mutant individual had appeared between 1936 and 1945. An attempt was made to apply the approach of the epidemiologist to an analysis of why these forty-nine families had been singled out for the appearance of this particular mutation. Figure 17-5 is a map of the state of Michigan showing the distribution of families in which a child exhibiting the results of a mutation to the retinoblastoma gene was born during 1936–45. This map has been compared with the population map for the state—no clear tendency of mutation to favor either rural or urban populations was apparent. Figure 17-6 compares the ages of the parents of children with sporadic retinoblastoma with the ages of the parents of all children born in the state during that same period. No difference is apparent. There was no year or season of the year in which an increased number of children who later developed retinoblastoma were born. The parents of mutant children, in one of whom a mutation may be presumed to have occurred, were not observed to exhibit any distinguishing physical characteristics. In short, none of the obvious epidemiological approaches applicable to this problem yielded a clue as to why mutation occurs when and where it does. Further studies of this question along these and still other lines are highly desirable.

17.8. *The possible complexity of epidemiological genetics.*—We have attempted to present in relatively simple terms the manner in which genetic concepts

Fig. 17-5.—A map of the state of Michigan, showing the distribution of children who, between 1936 and 1945, as a result of mutation, received a retinoblastoma gene from one or the other of their parents. The figure *11* in a circle represents cases occurring in the city of Detroit. (Falls and Neel, 1951, reproduced by permission from Archives of Ophthalmology.)

must be an integral part of the armamentarium of the modern epidemiologist. In concluding this chapter, an example of the potential complexity of the interactions between an agent of disease, heredity, and environment will be considered. For this example we have selected the much-studied mammary cancer of mice. Through selection, lines of mice varying greatly in the frequency with which they spontaneously develop mammary cancer have been established. One of the widely used low-cancer-incidence lines is known as C_{57}, while two high-incidence lines are C_3H and A. Shortly after these lines

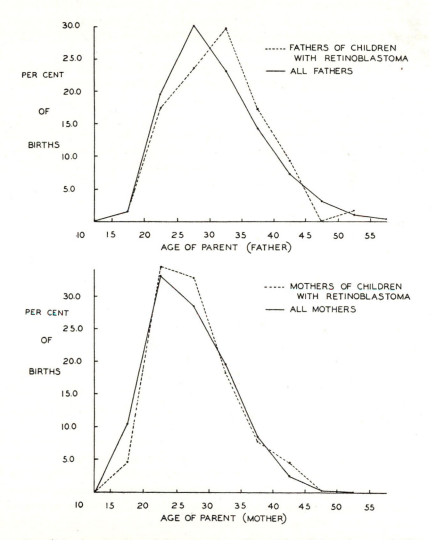

Fig. 17-6.—A comparison of the age at time of conception of the parents of children with sporadic retinoblastoma and the age at conception of the parents of all births occurring in the state of Michigan between 1936 and 1945. (Falls and Neel, 1951, reproduced by permission from Archives of Ophthalmology.)

had been established, it was noted, in 1933, that the results of crosses between low-incidence and high-incidence strains were greatly influenced by which strain supplied the female parent. If the female parent was from a high-incidence line, then the offspring developed very significantly more mammary cancer than if the female parent was from a low-incidence line. It was then demonstrated that if mice from a high-incidence strain were prevented from nursing their mothers but were foster-nursed by C_{57} mice, the frequency of breast cancer in these offspring was greatly reduced, although still elevated over the frequency observed in the C_{57} strain. The milk-factor agent which was thus demonstrated has been shown to have many of the properties of a virus. At this stage of the analysis, then, it was apparent that the frequency of development of mammary cancer in mice was determined by the interaction between a basic predisposition to cancer and a virus-like agent. The next step in the analysis was the demonstration that the ability of the milk factor to survive in a strain was related to the genetic constitution of the strain. Thus when female offspring from the cross of ♀ (high-tumor)C_3H × ♂ (low-tumor)C_{57} were back-crossed to both the C_3H and C_{57} males, definite evidence of genetic differences in the ability to propagate and transmit the agent was revealed. There is, then, genetic control of susceptibility to the milk-factor agent. The final complication in this story is supplied by the fact that in the A strain usually only mice that have reproduced develop the mammary cancer, whereas in the C_3H line mice that have not reproduced develop as many tumors as those that have. An analysis of this situation revealed hormonal differences between the two strains which were presumed to be responsible for the observed differences. These hormonal differences were also shown to be under genetic control. In the final analysis, then, the development of mammary cancer in certain strains of mice appears to depend upon an interaction between at least three sets of factors: (1) the basic genetic constitution, (2) the presence of the milk-factor agent, and (3) the hormonal level, with both factors 2 and 3 strongly influenced by the genetic constitution of the mouse. The interested student will find reviews and references to the literature in the papers of Heston (1946, 1948, 1953) and Bittner (1950). The possibility exists that the epidemiological implications of mouse leukemia may be no less complex (Gross, 1952, 1953). The implications of this for human breast cancer and leukemia are not now clear.

The clarification of this situation has been due to the existence of mouse strains with very different susceptibilities to mammary carcinoma, the short generation time of the mouse, and the combined efforts of many investigators. The difficulties inherent in demonstrating unequivocally a similar situation in man, should such a situation exist, are apparent.

Bibliography

SPECIFIC REFERENCES

ANDERSON, R. C. 1951. Familial leukemia, Am. J. Dis. Child., **81**:313–22.

AYCOCK, W. L. 1941a. Familial susceptibility to leprosy, Am. J. M. Sc., **201**:450–66.

———. 1941b. Constitutional types and susceptibility to paralysis in poliomyelitis, *ibid.*, **202**:456–68.

———. 1942. Familial aggregation in poliomyelitis, *ibid.*, **203**:452–65.

AYCOCK, W. L., and MCKINLEY, E. B. 1938. The roles of familial susceptibility and contagion in the epidemiology of leprosy, Internat. J. Leprosy, **6**:169–84.

BITTNER, J. J. 1950. Genetic aspect of cancer research, Am. J. Med., **8**:218–28.

BRØBECK, O. 1949. Heredity in cancer uteri. ("Opera ex domo biologiae hereditariae humanae Universitatis Hafniensis," Vol. **21**.) Copenhagen: E. Munksgaard.

DEGKWITZ, R., and KIRCHMAIR, H. 1940. Vererbung und Disposition bei Infektionskrankheiten, Handb. der Erbbiol. des Mensch., **4**:1042–78.

DEMEREC, M. 1949. Chemical mutagens, Proc. Eighth Internat. Cong. Genetics, Hereditas suppl., pp. 201–9.

DEMEREC, M.; BERTANI, G.; and FLINT, J. 1951. A survey of chemicals for mutagenic action on E. coli, Am. Naturalist, **85**:119–36.

DRAPER, G.; DUPERTUIS, C. W.; and CAUGHEY, J. L. 1944. Human constitution in clinical medicine. New York: Paul B. Hoeber, Inc.

FALLS, H. F., and NEEL, J. V. 1951. Genetics of retinoblastoma, Arch. Ophth., **46**: 367–89.

FRASER, F. C., and FAINSTAT, T. D. 1951. Production of congenital defects in offspring of pregnant mice treated with cortisone: progress report, Pediatrics, **8**: 527–33.

GOLDSCHMIDT, R. B. 1945. Additional data on phenocopies and genic action, J. Exper. Zoöl., **100**:193–201.

———. 1949. Phenocopies, Scient. Am., **181**:46–49.

GORER, P. A. 1938. The genetic interpretation of studies on cancer in twins, Ann. Eugenics, **8**:219–32.

GOWEN, J. W. 1948. Inheritance of immunity in animals, Ann. Rev. Microbiol., **2**: 215–54.

———. 1950. Significance of genetics to the typhoid disease syndromes in the mouse and fowl, Scientia, **85**:145–50.

———. 1953. Humoral and cellular elements in natural and acquired resistance to typhoid, Am. J. Human Genetics, **4**:285–302.

GROSS, L. 1952. Mouse leukemia, Ann. New York Acad. Sc., **54**:1184–96.

———. 1953. Biological properties of the mouse leukemia agent, Cancer, **6**:153–58.

HARRINGTON, W. J.; MINNICH, V.; HOLLINGSWORTH, J. W.; and MOORE, C. V. 1951. Demonstration of a thrombocytopenic factor in the blood of patients with thrombocytopenic purpura, J. Lab. & Clin. Med., **38**:1–10.

HERNDON, C. N., and JENNINGS, R. G. 1951. A twin-family study of susceptibility to poliomyelitis, Am. J. Human Genetics, **3**:17–46.

HESTON, W. E. 1946. Path of gene action in mammary-tumor development in mice, J. Nat. Cancer Inst., 7:79–86.

———. 1948. Role of genes and their relationship to extrachromosomal factors in the development of mammary gland tumors in mice, Brit. J. Cancer, 2:87–90.

———. 1953. The bearing of mouse genetics on our understanding of human cancer, Am. J. Human Genetics, 4:314–31.

INGALLS, T. H.; CURLEY, F. J.; and PRINDLE, R. A. 1950. Anoxia as cause of fetal death and congenital defect in mouse, Am. J. Dis. Child., 80:34–45.

JACOBSEN, O. 1947. Heredity in breast cancer. ("Opera ex domo biologiae hereditariae humanae Universitatis Hafniensis," Vol. 11.) Copenhagen: E. Munksgaard.

KALLMANN, F. J., and REISNER, D. 1943. Twin studies on genetic variations in resistance to tuberculosis, J. Hered., 34:269–76, 293–301.

LURIE, M. B. 1953. On the mechanism of genetic resistance to tuberculosis and its mode of inheritance, Am. J. Human Genetics, 4:302–14.

LURIE, M. B.; ABRAMSON, S.; and HEPPLESTON, A. G. 1952. On the response of genetically resistant and susceptible rabbits to quantitative inhalation of human type tubercle bacilli and the nature of the resistance to tuberculosis, J. Exper. Med., 95:119–34.

MACKLIN, M. T. 1940. An analysis of tumors in monozygous and dizygous twins, J. Hered., 31:277–90.

MEYER, S. N., and JENSEN, C. M. 1951. Significance of familial factors in the development of tuberculin allergy, Am. J. Human Genetics, 3:325–31.

MURPHY, D. P. 1952. Heredity in uterine cancer. Cambridge: Harvard University Press.

PENROSE, L. S.; MACKENZIE, H. J.; and KARN, M. N. 1948. A genetical study of human mammary cancer, Ann. Eugenics, 14:234–66.

REED, S. C. 1936. Harelip in the house mouse. I. Effects of the external and internal environments. II. Mendelian units concerned with harelip and application of the data to the human harelip problem, Genetics, 21:339–74.

ROBSON, H. N., and WALKER, C. H. M. 1951. Congenital and neonatal thrombocytopenic purpura, Arch. Dis. Child., 26:175–83.

SPRAGUE, C. C.; HARRINGTON, W. J.; LANGE, R. D.; and SHAPLEIGH, J. D. 1952. Platelet transfusions and the pathogenesis of idiopathic thrombocytopenic purpura, J.A.M.A., 150:1193–98.

TATUM, E. L. 1947. Chemically induced mutations and their bearing on carcinogenesis, Ann. New York Acad. Sc., 49:87–97.

VIDEBAEK, A. 1947. Heredity in human leukemia and its relation to cancer. ("Opera ex domo biologiae hereditariae humanae Universitatis Hafniensis," Vol. 13.) Copenhagen: E. Munksgaard.

WEBSTER, L. T. 1932. Experimental epidemiology, Harvey Lect., 27:154–78.

WEIR, J. A. 1949. Blood pH as a factor in genetic resistance to mouse typhoid, J. Infect. Dis., 84:252–74.

WEIR, J. A.; COOPER, R. H.; and CLARK, R. D. 1953. The nature of genetic resistance to infection in mice, Science, 117:328–30.

CHAPTER 18

The Applications of Genetic Knowledge to Man
I. Counseling

THERE ARE three chief "areas" in which the principles of human heredity which have been developed in the preceding chapters find application. These "areas" are (1) genetic counseling, (2) medicolegal genetics, and (3) eugenics. In the present chapter we shall consider the first of these applications, in the form of some representative illustrations of how a knowledge of genetic principles enables one to meet a variety of questions arising in the practice of medicine, social work, marriage counseling, placement for adoption, etc. The list of problems to be considered is by no means exhaustive but is sufficient to illustrate some of the more commonly encountered situations. The cases to be treated are not hypothetical but are based upon the actual experience of the staff members of the Heredity Clinic of the University of Michigan.

18.1. *The philosophy of genetic counseling.*—This chapter will attempt to give the student a "feeling" for the present clinical applications of genetics. It cannot be too strongly emphasized at this point that in the applications of genetic knowledge to the problems of man there are no "cookbook" answers. Each problem and the family concerned must be treated as an entity. To each statement about the "usual" mode of heredity of a given trait there are exceptions. Just as in the practice of medicine the textbook case is conspicuous by its absence, so in the practice of genetic counseling the type of pedigree which graces the genetic literature is not often seen. Any individual rendering genetic advice must be prepared to analyze each pedigree on its own merits and—more important—to recognize the limitations of his analysis.

Many individuals who consult a physician or a geneticist with a genetic problem, once they have the answer, then proceed to ask what, under the circumstances, their future reproductive behavior should be. Unwilling to reach a decision which may serve as a bone of contention within the family or with a background which fails to qualify them for an intelligent decision, they seek guidance. It is appropriate at this point to draw a sharp distinction between genetic counseling, which as defined in this book is concerned solely with the tendering of a genetic prognosis in a specific situation, and eugenics, to be considered in chapter 20, concerned with recommendations as to the advisability of reproduction on the part of individuals or groups of individuals. In the practice of genetic counseling, it is our policy to inform a responsible member of any family with a counseling problem of all the facts at our disposal bearing on the issue. However, with rare exceptions, we do *not* attempt to pass a judgment as to the advisability of parenthood. This is a decision to be reached by the family concerned. In this connection it might be noted that it is often the more responsible and able members of society who are most impressed by the occurrence of abnormality in a child of theirs and most receptive to the possibility of altering their reproductive behavior accordingly. Part of the obligation of the individual engaged in genetic counseling is to attempt to aid those who consult him to arrive at a rounded picture of the problem, with the favorable and unfavorable aspects in proper perspective. These are matters requiring balance and objectivity, both in the family involved and in the geneticist or physician.

One of the important functions of genetic counseling today lies in dispelling superstition and folklore—some of the latter even having crept into the medical literature. We have been repeatedly impressed by the misinformation people have at their fingertips when they contact the Heredity Clinic. In the field of congenital malformation, for instance, this varies from the belief that the next child is almost certain to be affected to the equally strong belief that lightning never strikes twice. Some of this misinformation is at a level where it creates serious psychological problems, which the physician with a knowledge of genetics is in a position to dispel.

18.2. *A consideration of actual cases.*—With this as a background, we now turn to some specific problems. It will be noted that for many of these problems it has not been possible to give clear-cut answers. Although it would have been feasible to select from our files in certain instances less ambiguous situations, this would tend to mislead the reader as to the experience of one rendering genetic advice. The simple fact is that more often than not the geneticist is unable to give definite and simple answers to the questions with which he is confronted.

It is assumed that, by now, the student is adequately equipped to handle questions involving simple dominant and simple recessive inheritance, and no uncomplicated problems of that nature will be presented here. Rather, an attempt has been made to select experiences which afford insight into the potential complexity and the philosophy of counseling.

1. In any chronic disease for which there is no known cure, although there may be palliative therapy, it is often important to the individuals involved to

have the most accurate possible statement concerning the rate of progression of their disease. The patient faced with the necessity of adjusting to progressive ophthalmologic or neuromuscular disease needs some estimate of the time available to him for making this adjustment. By the same token, a young, genetically handicapped person just entering the productive period of his life can profit by as clear a definition as possible of what his physical (or mental) handicap will be. Thus Mrs. D was seen at age thirty-four because of a progressive loss of joint mobility accompanied by moderate contractures. She had first become aware of this during adolescence. Although, in consequence of her disease, her movements were awkward and clumsy, she was leading an active and very useful life. A family history revealed that her brother, maternal half-brother, and mother were all affected. Her chief concern was for her two sons. Would they be affected, and, if so, to what extent would they be physically incapacitated?

It was possible to conduct physical examinations on all the affected members of the family, as well as the patient's two sons. There was considerable variability in the expression of the condition. The most severely affected was the patient's mother, but at age fifty-eight, while unable to perform many of the finer movements of hands and the rest of her body, she was extremely active. A characteristic finding in all affected persons was a broad thumb. It was felt that one of the patient's sons showed the very early stages of the disease, including the broad thumb. On the basis of the experience with the family as a whole, it was predicted that while this son would be greatly handicapped in any trade or profession requiring manual dexterity, he would in all probability not, on the other hand, be physically incapacitated by his disease to the extent seen, for instance, in some cases of rheumatoid arthritis.

This family also illustrates the point that a genetic prognosis can be made even when the precise diagnosis is in doubt. We were for some time uncertain as to what to term this disorder. Not until over a year after the family had been seen was it established that the disease in question was a very rare disorder of the connective tissue and joint capsules known as "Léri's pleonosteosis."

2. Increasingly, agencies concerned with placing children for adoption are realizing that one component of a good pre-adoption evaluation is an inquiry into the family background from the standpoint of potential genetic risks. A single example will suffice: Sue C was a six-month-old illegitimate child whose family background, as far as it could be reconstructed, is shown in Figure 18-1. Her maternal grandfather, two maternal aunts, and one maternal half-brother all had developed marked impairment of hearing during the first two decades of life. Otologic studies revealed that this was a "nerve-type deafness." This trait is often inherited as if due to a dominant gene. If

this is the case in this family, Sue C's mother, because of the appearance of the trait in one of her children, must be presumed to be heterozygous for the gene, even though she herself has normal hearing, i.e., the gene is not completely penetrant. Sue C's mother is schizophrenic. Little is known concerning the paternal history. The agency of which Sue C became a ward wished a statement concerning the probability that Sue might develop either of the two diseases known to be present in the family background.

If the deafness is due to a dominant gene with approximately 50 per cent penetrance, Sue C has a one-in-four chance of becoming deaf. With respect to the schizophrenia, Kallmann's figures (1946) suggest that where one parent is schizophrenic, there is a one-in-six chance that any given child will be affected. However, this figure is based on children reared with a certain degree of contact with the affected parent. There are no figures for children sepa-

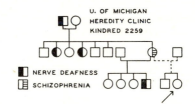

Fig. 18-1.—The pedigree of Sue C. Explanation in text

rated at an early age from a schizophrenic parent; but it may be presumed that, to the extent that environment enters into the development of schizophrenia, separation would tend to decrease the risk involved. The probability that the child will develop neither of the two diseases is therefore at least $\frac{3}{4} \times \frac{5}{6} = \frac{15}{24}$, or 62 in 100, and may be somewhat greater.

This is the type of problem where it is doubtful whether any two geneticists would derive precisely the same answer. On the other hand, there would undoubtedly be agreement concerning the order of magnitude of the risks involved. Other things being equal, Sue C, from the genetic standpoint, is a poorer risk than the average child.

3. A problem illustrating some of the ramifications of recessive inheritance is the following: Dr. B stated that he was the middle of five children. Each of his two older sisters had a child with infantile amaurotic idiocy, a serious disease of the central nervous system which results in death at an early age. The two younger sisters were unmarried. He himself had one infant son. The pedigree is given in Figure 18-2. Dr. B raised four questions: (1) What was the probability that any future child of his two older sisters would develop amaurotic idiocy? (2) What was the probability that his own son or any other children he might have would develop the disease? (3) What should he

advise his two unmarried sisters concerning the probability of the occurrence of amaurotic idiocy in such children as they might have? And, finally, (4) assuming that his son escaped the disease, what was the chance that it would appear in the son's offspring?

Infantile amaurotic idiocy has been shown to be a recessively inherited trait which appears to have complete penetrance. The pattern of affected individuals in this family is in keeping with that hypothesis, although it is unusual for two sisters who are carriers both to marry carrier males. Accordingly, the probability that any further child of the two older sisters will also be affected is one in four. The probability that the physician's own son or any further children will develop the disease is calculated as follows: probability that the physician is a carrier × probability that his wife is a carrier × probability that an affected child will be born to carrier parents. The first prob-

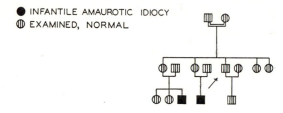

● INFANTILE AMAUROTIC IDIOCY
◐ EXAMINED, NORMAL

FIG. 18-2.—A counseling problem in amaurotic idiocy. Explanation in text

ability is most likely $\frac{1}{2}$, although it could be $\frac{2}{3}$ if both the consanguineous parents of the physician were carriers of the gene. The second probability is in the neighborhood of $\frac{1}{50}$ to $\frac{1}{100}$, this figure derived from the square root of the frequency of amaurotic idiocy in persons of Jewish extraction, this being the background of this family. The third probability is, of course, $\frac{1}{4}$. The compound probability is between one in four hundred and one in eight hundred. This probability also applies to the chance that any future child of the unmarried sisters will be affected. Finally, the chance that the physician's son is a carrier is one in four, and the probability that the son's offspring will be affected is in the neighborhood of $\frac{1}{4} \times \frac{1}{50} \times \frac{1}{4}$ to $\frac{1}{4} \times \frac{1}{100} \times \frac{1}{4}$ or $\frac{1}{800}$ to $\frac{1}{1600}$.

4. X's fiancée had her first epileptic seizure at age three. This and subsequent seizures had been satisfactorily controlled by anticonvulsants. Both were desirous of children, but X was concerned as to the probability that they would be epileptic. The epilepsy seemed to be of the idiopathic type—the family history was negative. Despite a large amount of work, the genetic basis of epilepsy is unclear. However, it *is* well established that, whereas the frequency of epilepsy in the population as a whole is approximately 0.5 per

cent, the frequency among the parents, siblings, and children of epileptics is approximately 3.0 per cent (Lennox, 1951). It appears that the earlier the age of onset, the greater the chance that the disease will appear in a child. In this particular case, with onset at age three, it was felt that the probability of the disease's appearing in a child was in the neighborhood of 4.0 per cent on the basis of the fiancée's history alone. X had a negative family history. Under these circumstances, it is frequently suggested that electroencephalograms be carried out on the apparently normal marital partner. In view of the not inconsiderable incidence of electroencephalographic abnormality in the general population, the exact interpretation of an abnormal EEG in this context is not clear, and such an examination was not made in this case. The final impression, then, was of a one-in-twenty-five probability of epilepsy in a child.

5. Mrs. M's first child, born when Mrs. M was twenty-five, was normal. Her second child, born when Mrs. M was twenty-six, was a Mongolian idiot, a condition characterized by mental retardation, a characteristic appearance, and hypermobility of the joints. There was a negative family history. The M's were anxious to rear a large family but were considerably shaken by their experience with the second child. We were consulted as to the likelihood of a recurrence. The etiology of Mongolism is unknown. However, from Böök and Reed's (1950) compilation of the figures on Mongolism, it appears that, once a child with this condition has been born in a family, the over-all chance of a recurrence in subsequent pregnancies is about one in twenty-five. The occurrence of Mongolism shows a striking relationship to maternal age, the frequency of this condition being some twenty times as great among the children born after the mother is forty than among children born before the mother is thirty. The relative youth of Mrs. M was interpreted as a factor in her favor, and she was advised that should she have another child within the next several years, the probability of a recurrence was approximately one in fifty to one in one hundred. It is worth pointing out that the pediatrician whom this couple had consulted had advised a "rest period" of three or four years. There is no evidence that "maternal reproductive exhaustion" has any relationship to congenital abnormality. Because of the maternal age factor in Mongolism, a "rest period" is, if anything, contraindicated.

6. The first child of Dr. and Mrs. M had a complex malformation of the hands and feet, characterized by a bilateral mid-tarsal amputation of the feet and bilateral hypoplasia of some digits and the absence of others. The family history was completely negative for this or other congenital defect. The parents were not related. What was the probability of a recurrence? Three etiologies were to be considered: (1) that the trait was due to a de-

velopmental accident not likely to be repeated, (2) that the trait was the result of a dominant mutation, and (3) that the trait was due to homozygosity for a rare recessive gene. In view of the striking effect upon all four extremities, a developmental accident seemed unlikely. However, no decision between possibilities 2, with a negligible chance of recurrence, and 3 with a one-in-four probability of recurrence, was possible.

In passing, a word should be said about the growing recognition of the importance of mutation in the etiology of isolated cases of various diseases. With respect to many diseases which commonly follow a dominant pattern of heredity, it is now recognized that the isolated case of today is the starting point for the pedigree of tomorrow. A decision in any individual case is largely dependent on the exact disease involved. There are conditions which are so regularly inherited as if due to a dominant gene—osteogenesis imperfecta, aniridia, neurofibromatosis, Marfan's syndrome, achondroplasia, to mention only a few—that an isolated case has a high degree of probability of being the result of mutation. There are other conditions which only rarely follow a pattern of dominant heredity, where the isolated case has quite a different significance. A judgment in any case can be reached only on the basis of familiarity with the particular disease involved.

Problems such as the above, involving congenital malformation, are among those most frequently encountered in genetic counseling. A few of the congenital malformations have a simple genetic basis. The etiology of most of the common malformations is, however, quite obscure. It has, however, been established that congenital malformations are not distributed at random in the population but show a statistically significant tendency to cluster in certain families. From a study of the subsequent reproductive histories of women who have borne a malformed child, it is possible to derive tables of use in making empirical predictions concerning the probability of recurrence of particular malformations. Table 18-1 summarizes the empirical risk figures for some of the more commonly seen malformations. These empirical risk figures, while useful generalizations, can never be a substitute for a careful family history. Such risk figures are the average of many different situations, ranging from developmental accident to dominant heredity with irregular penetrance. In the one situation, the chance of a recurrence is essentially zero; in the other, very appreciable. Two specific examples will serve to underscore this point. The empirical risk figures given in Table 18-1 would indicate that when parents one of whom has a harelip have had a child with a harelip, the chance of a repetition of this abnormality in the next child is about one in seven. However, there is a type of harelip variably associated with cleft palate, congenital fistulas of the lower lip, and missing teeth, which seems to depend on an irregularly dominant gene. A pedigree of a family in which this syndrome is present is shown in Figure 18-3. Occasionally, also, simple harelip and cleft palate, unassociated with dental anomalies (other than those due to the cleft palate) or with congenital fistulas of the lower lips, may appear to depend on a single dominant gene with high penetrance. A pedigree of this type is shown in Figure 18-4. While both these situations are rare, they do serve to emphasize the wide range of situations encompassed in empirical risk figures.

7. Both the B's two children showed deformities of the extremities of the "lobster-claw" type (Fig. 18-5). The family history was negative, the parents

TABLE 18-1*

PROBABILITY OF RECURRENCE OF SPECIFIC CONGENITAL MALFORMATIONS
IN SUBSEQUENT CHILDREN, IN FAMILIES IN WHICH A CHILD
WITH THIS MALFORMATION HAS BEEN BORN

Type of Malformation	Anen-cephaly	Spina Bifida	Hydro-cephaly	Harelip with or without Cleft Palate	Cleft Palate without Harelip	Pyloric Stenosis	Club-foot	Con-genital Dislo-cation of Hip
Frequency at birth in general population.........	0.2	0.3	0.2	0.1	0.04	0.3	0.1	0.1
Chance of recurrence, both parents normal....	2.1	3.9	1.6	4.0	1.8	4.0	5.0	5.0
Chance of recurrence, one parent affected....	....	....	....	14.0	17.0	....	10.0	10–15

* Where feasible, two sets of probabilities are given, depending on whether or not one of the parents exhibits the same malformation. All risks are given as percentages. Data of Record and McKeown (1949–50); Fogh-Andersen (1943); McKeown, MacMahon, and Record (1951); Bauer and Bode (1940).

unrelated. The "lobster-claw" deformity is usually inherited as a simple dominant trait. In this case the "negative family history" included accurate data on some fifty persons. Simple dominant heredity was excluded. Two possibilities seemed to stand out. One was of a dominant mutation at an early stage in the development of one of the parental gonads, so that a sector of the gonad was affected, or even a somatic mutation involving a portion of the body. If this is so, then the probability of recurrence is something less than one in two, but how much less than one in two cannot be stated. The other possibility was, of course, recessive inheritance, with its one-in-four probability of recurrence of the trait. In any event, the parents could be told that their chances of producing another affected child were quite appreciable, although probably not greater than one in four.

8. Families which present two or more distinct and different problems in genetic counseling are uncommon but have a frequency disproportionate to their occurrence in the general population in the experience of anyone engaged in genetic counseling, because these families are more likely to seek advice than are those with only one problem. The following will serve as an example of the complexity which counseling problems may assume. Both children of the R's had died at an early age, the first with congenital heart

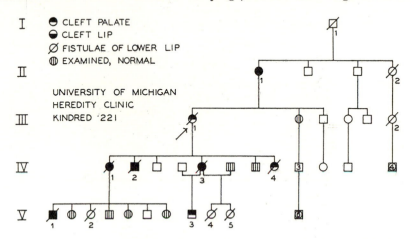

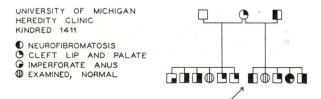

FIG. 18-3.—A pedigree of harelip and cleft palate associated with fistulas of the lower lip.

FIG. 18-4.—A pedigree of simple harelip and cleft palate, imperforate anus, and neurofibromatosis.

disease (anomalous venous return to heart, both the systemic and the pulmonary circulation draining into the right auricle; persistent ductus arteriosus; persistent foramen ovale), the second child with cystic fibrosis of the pancreas. In addition, Mrs. R was Rh-negative and her husband Rh-positive. During the second pregnancy, Mrs. R was found to have Rh antibodies in significant titer in the latter months of gestation, although the child, when born, showed no signs of erythroblastosis foetalis.

There are no figures available for the probability of recurrence of this specific cardiac malformation; but, on the basis of the figures derived for other malformations, it would seem likely that there is an increased risk, perhaps in the neighborhood of one in twenty. Cystic fibrosis of the pancreas has been shown to be distributed in families as if due to a recessive gene. The probability that a future child would *not* show either of these malformations is thus approximately $\frac{19}{20} \times \frac{3}{4}$ = approximately $\frac{71}{100}$. Rh tests showed that Mr. R's most probable genotype was $CDe/cDE;$ it was thus very likely that any future child of this couple would be of a genotype which would render it susceptible to erythroblastosis foetalis, a disease which, despite recent thera-

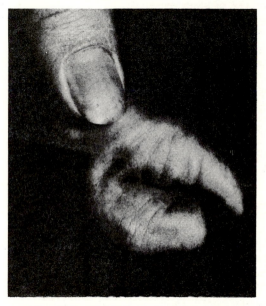

FIG. 18-5.—The "lobster-claw" deformity

peutic advances, still carries a significant mortality. The compound probability that a normal infant will be born to this couple is probably less than one in two.

18.3. *Balance.*—The foregoing examples could easily be multiplied several times. Sufficient has been said, however, to sketch in the main outlines of genetic counseling. In presenting these examples, we have tended to limit the discussion to the particular problem which led some individual from the family concerned to seek genetic advice. It is well to re-emphasize at this point that good genetic counseling requires that more than the immediate problem be taken into account. Let us consider a hypothetical situation concerned with simple recessive inheritance. Consider two families, in one of which the parents are, by all standards, superior and in which the only child has amaurotic idiocy (cf. example 3), and another in which the parents are of low intelligence and physical vigor, and there are three children, one of whom has the condition. Although in each family the probability that another child will be affected is one in four, most geneticists would agree that, from the standpoint of the advisability of future reproduction, the two families present quite different problems.

It may be argued that the distinction we attempt to maintain between genetic counseling and eugenics is acceptable in theory but breaks down in practice. It must be admitted that it is difficult for the counselor to avoid influencing a patient's behavior in one way or the other. This is particularly the case when it comes to presenting the "other side" of the picture. But the fact that the distinction is difficult to maintain in practice can scarcely be taken as a reason for failure to draw a line, particularly in view of the very different problems raised by eugenic considerations.

Bibliography

SPECIFIC REFERENCES

BAUER, K. H., and BODE, W. 1940. Erbpathologie der Stützgewebe beim Menschen, Handb. der Erbbiol. des Mensch., 3:105–334. Berlin: Julius Springer.

BÖÖK, J. A., and REED, S. C. 1950. Empiric risk figures in Mongolism, J.A.M.A., 143:730–32.

FOGH-ANDERSEN, P. 1943. Inheritance of harelip and cleft palate, pp. 1–266. ("Opera ex domo biologiae hereditaria humanae Universitatis Hafniensis," Vol. 4.)

KALLMANN, F. J. 1946. The genetic theory of schizophrenia, Am. J. Psychiat., 103: 309–22.

LENNOX, W. G. 1951. The heredity of epilepsy as told by relatives and twins, J.A.M.A., 146:529–36.

McKEOWN, T.; MACMAHON, B.; and RECORD, R. G. 1951. The familial incidence of congenital pyloric stenosis, Ann. Eugenics, 16:260–81.

RECORD, R. G., and McKEOWN, T. 1949–50. Congenital malformations of the central nervous system. I, II, and III, Brit. J. Social. Med., 3:183–219, 4:26–50, 217–20.

GENERAL REFERENCES

ANDERSON, R. C. 1951. Genetics and medical practice, Journal-Lancet, 71:49–52.

DICE, L. R. 1952. Heredity clinics: their value for public service and for research, Am. J. Human Genetics, 4:1–13.

NEEL, J. V. 1951. Some applications of the principles of genetics to the practice of medicine, M. Clin. North America, 35:519–33.

REED, S. C. 1949, 1951. Counseling in human genetics, Dight Inst. Bull., Nos. 6 and 7.

STERN, C. 1949. Problems of genetic advisers. In: STERN, C., Principles of human genetics, chap. 8, pp. 113–21. San Francisco: W. H. Freeman & Co.

CHAPTER 19

The Applications of Genetic Knowledge to Man
II. Medicolegal Aspects

THE MOST important medicolegal application of genetics—and the only application to be discussed in this chapter—has to do with questions of parentage. Such questions may present themselves in any one of three ways, as follows:

1. Can a man, A, be the father of child a?
2. Can a woman, A', be the mother of child a?
3. Can a man and woman, A and A', be the joint parents of child a?

The first question arises repeatedly in legal cases involving disputed paternity. The second question arises much more rarely, and chiefly when a woman, for reasons at the time best known to herself, has represented a child as being her offspring when such was not the case. The third question arises in a number of forms. A couple, A and A', may attempt to claim a child, a, as their long-lost, strayed, or kidnapped child. Or individual a, perhaps motivated by the prospect of a legal inheritance, may claim A and A' as his parents. Again, A and A' may suspect that their child has been interchanged with another shortly after birth in a hospital. Finally, A' may accuse A of being the father of a child who has actually no biological relationship to either.

19.1. *Objectives of medicolegal genetics.*—In legal cases concerning questions of parentage, one attempts to demonstrate that the child in question possesses one or more inherited traits which cannot possibly have been derived from the parent (or parents) in dispute. In other words, one attempts to determine whether a particular individual can, on genetic grounds, be excluded as the parent of a child. It is important to recognize that *one can never prove parentage;* one can only say that the genotype of this person is such that he *could be* the parent of a particular child.

There are several different criteria which an inherited character must meet if it is to be of use in solving legal problems. The character must be simply inherited, with its mode of inheritance a matter of general agreement. It must be present at birth or soon thereafter and remain relatively constant

318

throughout life. Although rare traits can be utilized in the exceptional family where they happen to be present, to be of general usefulness the trait must be relatively common (but not too common). Finally, the character itself must be one concerning whose presence or absence all competent observers can agree. Of all the known traits of man, certain of the blood groups and the secretor factor best meet these requirements and hence have been widely employed in legal questions of parentage. The use of the blood groups in this context involves no new principle but is merely a technical outgrowth of genetic studies. It is suggested that at this point the student review sections 8.4 through 8.8 (pp. 83–91).

19.2. *Procedures in excluding paternity.*—We will consider, first, the exclusion of paternity, assuming the mother of the child to be well established. The

TABLE 19-1

RESULTS OF VARIOUS TYPES OF MATINGS INVOLVING
A, B, O BLOOD GROUPS

No.	Blood Groups of Parents	Possible Blood Groups in Children	Blood Groups Not Possible in Children
1........	O×O	O	A, B, AB
2........	O×A	O, A	B, AB
3........	A×A	O, A	B, AB
4........	O×B	O, B	A, AB
5........	B×B	O, B	A, AB
6........	A×B	O, A, B, AB	None
7........	O×AB	A, B	O, AB
8........	A×AB	A, B, AB	O
9........	B×AB	A, B, AB	O
10........	AB×AB	A, B, AB	O

law is a conservative institution. There is thus a considerable lag between the recognition of a new inherited serological trait and the admissibility of data on this trait as evidence. At the present time, the results of A, B, O, MN, and Rh typing are being accepted in many courts. However, the results of blood testing are not binding, the court having the right to ignore the results of blood testing if other evidence seems to justify this course of action.

The applications of the A, B, O and MN blood groups to paternity problems are illustrated in Tables 19-1 and 19-2. As indicated earlier, two varieties of the A antigen, termed A_1 and A_2, are commonly recognized. Theoretically, these subgroups could also be applied in problems of disputed parentage. In actual practice, however, technical difficulties, especially in subgrouping blood of newborn infants, interfere with the tests, and so the subgroups are not generally utilized.

Because of the greater number of alleles (or linked gene combinations) at the Rh locus, the use of the Rh genes in paternity problems is more complicated but similar in principle to the use of the A, B, O and MN blood groups. Using the four most common Rh antisera [anti-C (rh'), anti-c (hr'), anti-D (Rh$_o$), and anti-E (rh'')], one can define twelve Rh phenotypes, most of which include one relatively common genotype and several rarer ones. This may occasionally lead to ambiguity in the interpretation of the results of paternity tests. Much of the ambiguity disappears if we think in terms of response to specific Rh antisera. A child's red blood cells cannot be agglutinated by one of these antisera unless one or both parents' cells are also agglutinated. Sometimes, as in the fourth example to be discussed in this

TABLE 19-2

RESULTS OF VARIOUS TYPES OF MATINGS
INVOLVING MN BLOOD TYPES

No.	Blood Groups of Parents	Possible Blood Groups in Children	Blood Groups Not Possible in Children
1......	M×M	M	N, MN
2......	N×N	N	M, MN
3......	M×N	MN	M, N
4......	MN×M	M, MN	N
5......	MN×N	N, MN	M
6......	MN×MN	M, N, MN	None

chapter, the uncertainty as to exact parental genotype may be dispelled by tests of the other children of these parents, the results of these tests throwing light on parental genotypes.

Thus far, the newer blood types have not been extensively used in problems of this nature, both because of the recency of their discovery and because of the scarcity of the necessary antisera. Theoretically, however, the Lutheran, Kell, Lewis, Duffy, and Kidd antigens all have potential value in paternity cases, as does the secretor-nonsecretor test.

The findings in two paternity problems referred to Dr. C. W. Cotterman, of the University of Michigan Heredity Clinic, are shown in Table 19-3. In the first example, an unmarried woman alleged that a certain man was the father of her child. The serological studies included the A, B, O, MN, and Rh-reactions. The child exhibits no serological reactions not accounted for by the genotypes of the mother and alleged father. Accordingly, there is no basis for a paternity exclusion in this case.

In the second example, a man had reason to suspect that he was not the biological father of the two youngest children in his family of nine. Blood tests were performed on the husband, wife, and two youngest children. In this case there is a paternity exclusion, the Rh-types providing the crucial evidence. The serological reactions indicate that the most probable genotype of the husband is CDe/CDe (R^1R^1) and the wife cDE/cde (R^2r). The genotype of the eighth child is probably cDE/cde (R^2r) and the ninth CDe/cDE (R^1R^2). Since the father is homozygous for the CDe chromosome, this chromosome must be present in all his biological children. Conversely, the absence of the chromosome in child No. 8 proves that he is not the offspring of the legal father.

19.3. Usefulness of the various blood types.—The usefulness of any particular blood-group system in paternity problems depends upon the number of the

TABLE 19-3*

BLOOD FINDINGS IN TWO DISPUTED PATERNITY PROBLEMS

BLOOD OF:	A-B Reactions		A, B, O Genotype	M-N Reactions		MN Genotype	Rh-Reactions				Most Probable Rh Genotype
	A	B		M	N		C	D	E	c	
Family 1:											
Man.......	+	−	A-	+	−	MM	+	+	−	−	CDe/CDe
Woman....	−	+	B-	+	−	MM	+	+	+	+	CDe/cDE
Child......	+	−	AB	+	−	MM	+	+	−	−	CDe/CDe
Family 2:											
Husband...	+	−	A-	+	+	MN	+	+	−	−	CDe/CDe
Wife.......	+	−	A-	+	+	MN	−	+	+	+	cDE/cde
Child 1....	+	−	A-	+	+	MN	−	+	+	+	cDE/cde
Child 2....	+	−	A-	+	−	MM	+	+	+	+	CDe/cDE

* Data of Cotterman (unpublished).

genes in the system, the dominance relationships between the genes, and the relative frequencies of the various genes in the system. The simplest genetic system of use in paternity exclusion involves two alleles, one completely dominant to the other. Let p = the frequency of the dominant gene, D, and $q = 1 - p$ the frequency of the recessive gene, d. The frequency of the three possible genotypes DD, Dd, and dd is, of course, $p^2(DD)$, $2pq(Dd)$, and $q^2(dd)$, the sum of the frequencies of the three genotypes equaling 1. Now in a genetic system such as this, the only situation wherein paternity can be excluded is when a dd man is falsely accused by a dd woman of being the father of a D–child. The probability of exclusion when a man is accused at random with respect to the D–d system ($P_{D,\,d}$) is therefore (probability of *dd* woman) $\times$ (probability of such a woman giving birth to a D-child) $\times$ (probability that putative father is *dd*), or

$$P_{D,d} = q^2 p q^2 = p q^4 . \qquad (19.3.1)$$

The problem is slightly more involved in the case of two genes without dominance. Here we may use the actual case of the MN types (disregarding the related S factor), setting the frequency of the M gene at p and of the N gene at q, with $p + q$ again equaling 1. To determine the probability of excluding the putative parent, we reason as follows: Nine genetic combinations may be visualized for the "avowed" parent and the child, namely, parent MM, child MM, or MN, or NN; parent MN, child MM, or MN, or NN; and parent NN, child MM, or MN, or NN. Each of these combinations will occur with some assignable frequency. For example, the combination parent MM-child MM can arise as a consequence of either of two matings, $MM \times MM$, or $MM \times MN$. The former, which occurs with frequency p^4, produces all MM children; the latter, which occurs with frequency $2p^3q$, produces equal numbers of MM and MN children. It follows that in a randomly mating population the combination of parent MM-child MM arises with frequency $p^4 + p^3q = p^3(p + q) = p^3$. Now to each parent-child combination there may or may not correspond a phenotype incompatible with the combination (e.g., in the case of parent MM-child MM, the other parent could not be NN). If an incompatible phenotype does occur, then the probability of its occurrence can be specified. The accompanying table illustrates, ex-

Parent-Child Combination (1)	Probability of (1) (2)	Putative Parent Cannot Be (3)	Probability of (3) (4)	Compound Probability (5)
MM-MM	p^3	NN	q^2	p^3q^2
MM-MN	p^2q	MM	p^2	p^4q
MM-NN	0	Indeterminate	. . . .	. . . .
MN-MM	p^2q	NN	q^2	p^2q^3
MN-MN	pq	None	0	. . . .
MN-NN	pq^2	MM	p^2	p^3q^2
NN-MM	0	Indeterminate	. . . .	. . . .
NN-MN	pq^2	NN	q^2	pq^4
NN-NN	q^3	MM	p^2	p^2q^3

haustively, the parent-child combinations and the incompatible phenotypes. The total probability, $P_{(D, D')}$, of excluding one parent, either father or mother, assuming that the identity of the other parent is firmly established, equals the sum of these separate compound probabilities, i.e.,

$$P_{(D, D')} = p^4 q + 2p^3 q^2 + 2p^2 q^3 + p q^4$$
$$= p q (1 - p + p^2)$$
$$= p q (1 - p q). \tag{19.3.2}$$

It will be noted that, for all values of p (or q) other than 1, the value of $pq(1 - pq)$ is greater than pq^4. In other words, two alleles without dominance are more effective in paternity exclusion than are two alleles, one of which is dominant to the other. This can be taken as a small demonstration that

TABLE 19-4*

CHANCE THAT AN ENGLISHMAN WILL BE EXONER-
ATED OF FALSE CHARGE OF PATERNITY BROUGHT
BY AN ENGLISHWOMAN, WHEN SIX BLOOD
GROUPS AND SECRETOR FACTOR ARE USED IN
EXCLUSION TESTS

No.	Blood-Group System	Exclusion by Each System	Combined Exclusion
1........	A, B, O	0.1760	0.1760
2........	MNS	.2741	.4019
3........	Rh	.2520	.5526
4........	Kell	.0421	.5714
5........	Lutheran	.0333	.5857
6........	Secretion	.0258	.5964
7........	Duffy	0.0496	0.6164

* By permission from Blood groups in man, by Drs. R. R. Race
and Ruth Sanger, copyright 1950, Blackwell Scientific Publications.

genetic systems without dominance are more useful in paternity exclusion problems than are systems with dominance. It may be further stated that, in general, the more different alleles that are recognized at any locus, the greater the probability of a paternity exclusion based on that locus.

Race and Sanger (1950), using methods similar to the above, have calculated the probability of an Englishman's being exonerated of a false charge of paternity brought by an Englishwoman, when six of the better-known blood-type systems and the secretor factor are used. In obtaining the probability for each system, the actual values of the frequencies in the English population of the various genes involved are substituted in equations similar to those given above but, for the Rh and some other systems, somewhat more complicated. The results are shown in Table 19-4. At the present time, an

Englishman falsely accused of paternity may expect to be exonerated in 62 out of each 100 trials.

Earlier in the chapter the statement was made that, to be of general usefulness in paternity exclusion work, a trait must be relatively common. Now that equations have been derived for the probability of exclusion for two alleles with and without dominance, the truth of this statement may be tested by substituting various values of p and q in the equations and plotting the resulting probabilities of exclusion. In this way curves D' and D'' of Figure 19-1 are derived. For the case of two genes with dominance, Wiener *et al.*

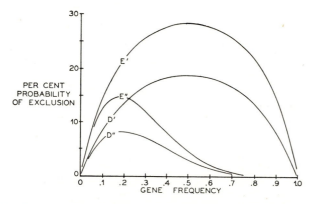

Fig. 19-1.—The probability of exclusion of (1) paternity and (2) joint parentage by the use of a single-gene pair, at different frequencies of the two genes involved. The four curves apply to the following situations: D', exclusion of paternity, two genes lacking dominance; D'', exclusion of paternity, one gene dominant to the other; E', exclusion of parentage, two genes lacking dominance; E'', exclusion of parentage, one gene dominant to the other. (Modified by permission of Dr. C. W. Cotterman and the American Journal of Human Genetics.)

(1930) first pointed out that the gene frequency at which the maximum value for paternity exclusion is obtained may be derived by setting

$$\frac{dP_{D,d}}{dq} = 0 .$$

It will be recalled that $P_{D,d} = pq^4$, or, since $p = 1 - q$, $q^4 - q^5$. Therefore,

$$\frac{dP_{D,d}}{dq} = 4q^3 - 5q^4 = 0 . \tag{19.3.3}$$

Solving for q, we obtain $q = 0.8$. Substituting this value in (19.3.1), we obtain the maximum value of $P_{D,d}$ as 0.08192.

In the same fashion we may find that the maximum probability of exclusion for two alleles without dominance, $P_{D,D'}$, is obtained at a gene frequency derived from setting

$$\frac{dP_{D,D'}}{dq} = 0 .$$

Since

$$P_{D,D'} = p\,q\,(1-p\,q) = q - 2\,q^2 + 2\,q^3 - q^4 ,$$

$$\frac{dP_{D,D'}}{dq} = 1 - 4\,q + 6\,q^2 - 4\,q^3 = 0 . \qquad (19.3.4)$$

This solves at $q = 0.5$. At this value of q, $P_{D,D'} = 0.1875$. In other words, at its most favorable gene frequency of $p = q = 0.5$, a system of two alleles without dominance is more than twice as effective in excluding false accusations of paternity than is a system of two alleles with dominance at its most favorable gene frequency, $p = 0.2$, $q = 0.8$.

19.4. Exclusion of maternity.—This is a very rare problem, such as might arise where a woman was suspected of falsely representing a child as her own. If the father is unknown, exclusion is impossible where reliance is placed on two alleles with one dominant to the other (why?). For two alleles without dominance, as the MN system, exclusion is possible where an M mother presents herself with an N child, or an N mother with an M child. Likewise, in a system of multiple alleles, such as the A, B, O system, maternity could be excluded where an AB mother claimed an O child, or an O mother an AB child. Because of the rarity of this problem, we will not go more deeply into it here.

19.5. Exclusion of joint parentage.—We turn now to the third of the medico-legal questions described in the introduction, that of excluding joint parentage. This is solved along precisely the same lines as a question of paternity or maternity. Two examples from the experience of the Heredity Clinic (Dr. C. W. Cotterman) will make the procedure clear. Example 1 is summarized in Table 19-5. The first child of a couple was kidnapped. Subsequently, they had two more children but continued an active search for the missing child, at length locating a boy, corresponding in age to their missing son, whose background was clouded with such uncertainty that he conceivably could have been the kidnapped child. However, serological studies revealed that, whereas the mother was type N, the alleged kidnapped child was type M. Furthermore, although the most probable Rh genotype of the possible mother was cDE/cde (R^2r) and the possible father CDe/cde (R^1r), the child was CDe/CDe (R^1R^1). The serological studies thus provide double grounds for the exclusion of parentage, the findings being particularly conclusive, because for each of the exclusions it is the mother who is the incompatible parent.

Before the results of blood testing were known in this case, many questions were raised concerning the pertinency of real or fancied physical resemblances between the parents and their alleged child. Physical resemblances, while interesting, cannot serve as critical evidence. It should be emphasized again that parentage cannot be proved in any particular case, but only disproved. In this particular instance, both the possible mother and the child showed a minor degree of webbing of the toes, a fact which seemed of some significance to the lay mind. However, about 8 per cent of the population has some degree of webbing of the toes. Accordingly, the probability that any child selected *at random* would resemble the mother in this case would be one in twelve.

TABLE 19-5*

RESULTS OF BLOOD-TYPING IN A KIDNAPPING CASE

BLOOD OF:	A-B REACTIONS		AB GENOTYPE	M-N REACTIONS		MN GENOTYPE	RH-REACTIONS				MOST PROBABLE RH GENOTYPE
	A	B		M	N		C	D	E	c	
Father.........	−	+	*BO*	+	+	*MN*	+	+	−	+	*CDe/cde*
Mother........	−	+	*BO*	−	+	*NN*	−	+	+	+	*cDE/cde*
Child 1........	−	−	*OO*	−	+	*NN*	+	+	+	+	*CDe/cDE*
Child 2........	−	−	*OO*	−	+	*NN*	+	+	−	+	*CDe/cde*
Alleged kidnapped child	−	+	*B-*	+	−	*MM*	+	+	−	−	*CDe/CDe*

* Unpublished data of Cotterman.

In a second and somewhat more involved problem with which the Clinic has been confronted, two mothers became suspicious that their infants, whom we shall designate as X-1 and X-2, born 4 days apart in the same hospital, had been interchanged. The serological studies employed in an attempt to reach a decision included tests for the following antigens: A, B, M, N, S, C (Rh'), D (Rh$_o$), E (Rh''), c (hr'), P, and Fy^a (Duffy). Cotterman's findings, in the sequence in which they were established, are summarized in Table 19-6. Figure 19-2 is a pedigree of the family; reference to it will aid in keeping the characters involved in mind, Dr. Cotterman, whose report (1951*b*) we paraphrase freely, first saw Mr. and Mrs. W, who brought along the disputed child X-2. Mr. W turned out to be OMN and Mrs. W to be ON, so that X-2, being OMN, was exactly of the phenotypes expected on the assumption of parentage. But Mrs. W's Rh-Hr phenotype was C+D+E+c+, while the child was C−D−E−c+ (*cde/cde;* rh-negative). Therefore, Mrs.

W could be the mother of the child only if she were an example of the rare genotype *CDE/cde*. Among the nine possible genotypes giving the reactions C+D+E+c+ (*CDe/cDE, cDe/CDE, CDe/cdE, cDE/Cde, CDE/cde, CdE/cDe, cDE/CDE, cdE/CDE, CdE/cDE*), *CI E/cde* has a frequency probably not exceeding 2 per cent, so that this evidence was thought to point against Mrs. W's maternity of X-2.

Dr. Cotterman then asked the couple W to return and bring along their older daughter, W-1. Fortunately, this child completed the evidence needed to rule out Mrs. W as the mother of X-2. Since W-1 was C+D+E−c− (genotype *CDe/CDe* or *CDe/Cde*), it was evident that Mrs. W was certainly

TABLE 19-6*

RESULTS OF BLOOD ANTIGEN TESTING ON TWO ALLEGEDLY INTER-
CHANGED CHILDREN AND THEIR ASSIGNED PARENTS AND SIBS

SUBJECT	A-B RE-ACTIONS		AB GENO-TYPE	M-N RE-ACTIONS		MN GENO-TYPE	RH-REACTIONS				MOST PROBABLE RH GENOTYPE
	A	B		M	N		C	D	E	c	
Mr. W........	−	−	O	+	−	MM	+	+	−	+	*CDe/cde*
Mrs. W......	−	−	O	−	+	NN	+	+	+	+	*CDe/cDE*
X-2..........	−	−	O	+	+	MN	−	−	−	+	*cde/cde*
W-1..........	−	−	O	+	+	MN	+	+	−	−	*CDe/CDe*
Mr. Y........	−	−	O	+	+	MN	−	−	−	+	*cde/cde*
X-1..........	−	−	O	+	+	MN	+	+	+	+	*CDe/cDE*
Mrs. Y.......	−	−	O	+	+	MN	−	−	−	+	*cde/cde*
Y-1..........	−	−	O	+	+	MN	−	−	−	+	*cde/cde*
Y-2..........	−	−	O	+	−	MM	−	−	−	+	*cde/cde*
Y-3..........	−	−	O	−	+	NN	−	−	−	+	*cde/cde*
Y-5..........	−	−	O	+	+	MN	−	−	−	+	*cde/cde*

* Used by permission of Dr. C. W. Cotterman and the American Journal of Human Genetics.

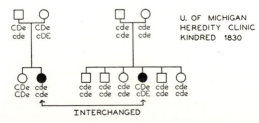

FIG. 19-2.—The genetical findings at the Rh locus in a problem involving the question of interchanged children. (Modified by permission of Dr. C. W. Cotterman and the American Journal of Human Genetics.)

not CDE/cde and must be either CDe/cDE or cDE/Cde. Thus, accepting W-1 as the child of Mrs. W—and this child was said to have been born at home—we have a complete exclusion of Mrs. W as the mother of X-2, and this exclusion is in no way conditional upon the paternity of either X-2 or W-1. The argument does, of course, ignore the possibility of mutation or crossing-over.

Also tested at the second stage were Mr. Y and X-1. Since X-1, like X-2, was found to be OMN, it was clear that no further light could be thrown on the problem by means of A, B, O or MN, if the interchange had occurred in the alleged manner. But Mr. Y was Rh-negative and X-1 was C+D+E+ c+. These findings again make a parent-offspring relationship impossible, for the same reasons as in the case of Mrs. W and X-2.

When later it became possible to test Mrs. Y and four of her five remaining children, all were found to be group O, Rh-negative. It could therefore be concluded that Mr. and Mrs. Y could not jointly have served as parents of X-1, and, even without knowledge of Mr. Y's genotype, it could be said to be quite unlikely that Mrs. Y was the mother of X-1, judging from the Rh-reactions. The final genotypical interpretation given to the serological findings is shown in Figure 19-2. Blood-group evidence in this case provided an exclusion of joint parentage for both assigned couples in respect to the two allegedly interchanged children. Moreover, the exclusions approached the ideal of *unconditional exclusion of maternity* for each woman. If the two children in question are interchanged in the diagram, the serological incompatibilities disappear. It was therefore concluded that there had actually been an exchange of children in the hospital.

The probability that the findings with respect to a given set of alleles will exclude joint parentage, like the probabilities with respect to exclusion of paternity or maternity, vary with the frequencies of the genes composing the allelic system. Cotterman (1951a), using an approach similar to that employed earlier in the chapter, has shown that, for two genes with dominance, the probability of exclusion, $P_{D,d}$, is $pq^4(1 + q)$. The maximum probability of exclusion, 0.14815, is obtained at $p = 0.1835$. For two genes without dominance, the probability of exclusion, $P_{D,D'}$, is $2pq - 5p^2q^2 + 6p^3q^3$; and the maximum probability of exclusion, 0.28125, is again obtained at $p = 0.5$. The respective curves are shown in Figure 19-1.

19.6. *Practical considerations.*—There are many pitfalls to be avoided in testing for blood-cell antigens. The techniques are delicate and easily disturbed by a variety of factors. For these reasons, it is mandatory that serological testing for medicolegal purposes be conducted only by an expert serologist. Even then, it is routine in many laboratories to have the findings, especially if a paternity exclusion is at stake, checked by a second, independent, expert.

In view of the emphasis which we have placed on the subject of mutation, the student has undoubtedly already wondered concerning the extent to which this phenomenon introduces a source of bias into the foregoing considerations. It was stated in chapter 15 that there is at present no estimate of the rate of mutation of the genes responsible for the various blood-cell antigens. We do, however, have enough data to make it very unlikely that mutation occurs sufficiently frequently to interfere with the use of the blood-cell antigens in parentage problems. Andresen (1947) has reported that in a series of 35,300 mother-child MN tests either reported in the literature or carried out in his laboratory, there is no exception (such as could be ascribed to mutation) to the rule that an M mother does not have an N child, or an N mother an M child. In the A, B, O system, of 1,849 AB mothers reported in the literature or tested in his laboratory, 3 are reported to have had O children, while of 15,138 O mothers, 5 are reported to have given birth to AB children. However, *all* of these exceptions were recorded in the early days of serological testing, when the antisera were not so well standardized as today and when, because of confusion concerning the mode of inheritance, the exceptional nature of such children was not recognized and so the crucial retests were not carried out. At present there are only two well-documented instances in which AB mothers have given birth to O children (Kossovitch, 1929, quoted from Wiener, 1943; Haselhorst and Lauer, 1930). In one of these, the child in question was a deaf-mute, nearly completely blind, with the left side of the face deformed. The possibility arises of a chromosomal aberration of some type.

19.7. *Legal status of blood-group determination in problems of parentage.*— Schatkin (1947) has summarized the statistics concerning the use of blood tests in cases of disputed parentage. Because of settlements out of court, it is difficult to compile exact figures, but the blood tests have by now been used in this country in several thousand cases of disputed paternity. They have been even more extensively used in the Scandinavian countries and Germany. The states of New York, Wisconsin, Ohio, New Jersey, Maryland, South Dakota, North Carolina, Maine, and Pennsylvania have passed laws that give the court the power to order blood tests in any case in which the problem of paternity or maternity is relevant to the case. Furthermore, in states where statutes have not been passed, there is nothing to prevent the performance of the tests by mutual consent of the parties involved, and, once significant results have been obtained, they are admissible as evidence. Davidsohn, Levine, and Wiener (1952) suggest that, since the tests can be used only to exclude paternity and cannot be used to prove paternity, in order to avoid misunderstanding, the tests should be admissible as evidence only when they exclude paternity.

Bibliography

SPECIFIC REFERENCES

ANDRESEN, P. H. 1947. Reliability of the exclusion of paternity after the MN and ABO systems as elucidated by 20,000 mother-child examinations and its significance to the medico-legal conclusion, Acta path. Scandinav., 24:545–53.

COTTERMAŃ, C. W. 1951a. A note on the detection of interchanged children, Am. J. Human Genetics, 3:362–75.

———. 1951b. Dermatoglyphic comparisons of two families with probably interchanged children, ibid., pp. 380–92.

HASELHORST, G., and LAUER, A. 1930. Über eine Blutgruppenkombination, Mutter AB und Kind O, Ztschr. f. Konstitutionslehre, 15:205–28.

WIENER, A. S. 1950. Heredity of the Rh blood types. IX. Observations in a series of 526 cases of disputed parentage, Am. J. Human Genetics, 2:177–97.

WIENER, A. S.; LEDERER, M.; and POLAYES, S. H. 1930. Studies in isohemagglutination. IV. On the chances of proving non-paternity; with special reference to blood groups, J. Immunol., 19:259–82.

GENERAL REFERENCES

ANDRESEN, P. H. 1952. The human blood groups utilized in disputed paternity cases and criminal proceedings. Springfield, Ill.: Charles C Thomas.

DAVIDSOHN, I.; LEVINE, P.; and WIENER, A. S. 1952. Medicolegal application of blood grouping tests, J.A.M.A., 149:699–706.

FORBES, G. 1951. Blood groups and disputed paternity, Brit. M. J., 1:227–31.

LOMBARD, J. F. 1952. Adoption, illegitimacy, and blood tests. Boston: Boston Law Book Co.

RACE, R. R., and SANGER, R. 1950. Blood groups in man. Oxford: Blackwell Scientific Publications.

SCHATKIN, S. B. 1947. Disputed paternity proceedings. 2d ed. New York: Matthew Bender & Co.

WIENER, A. S. 1943. Blood groups and transfusions. Springfield, Ill.: Charles C Thomas.

CHAPTER 20

The Applications of Genetic Knowledge to Man
III. Eugenics

THE ANTIQUITY of man is commonly placed at 500,000–1,000,000 years. We possess fairly adequate knowledge concerning the last 5,000 years of his history on earth. Any attempt to reconstruct a picture of how he lived before that is largely a matter of conjecture. Increasingly, however, anthropologists are inclined to think of primitive man as living in small and relatively isolated groups scattered over wide territories. The aggregation of man into larger communities probably awaited the advent of agriculture, perhaps 10,000–15,000 years ago, while cities as we know them today depend upon conditions which have apparently existed for only the last several thousand years, and especially since the Industrial Revolution.

It seems reasonable to postulate that for at least 95 per cent of his time on earth man has been engaged *on a highly individual basis* in a constant struggle with his environment, including in this term his fellow-man. Survival and reproduction must have depended largely upon the individual himself—his ability to find food, secure a mate, defend himself and his family from the wild beasts and his fellow-man. Then, at some point after men began to band together in larger communities, there occurred a subtle change in the nature of the selective factors at work. Much still depended on the individual, but at the same time group factors became important. The individual survived or perished partly on his own merits but partly also on the merits of the society of which he was a member. In any human society there is, as we have seen, a tremendous range of genetic variation. The introduction of civilization as we know it appears to have made it possible for the genetically less endowed to survive and reproduce to a greater extent than would have been possible some thousands of years ago.

20.1. *Dysgenic influences.*—There are at least four distinct ways in which civilization may be adjudged to offset the rigorous natural selection under whose influence man slowly evolved. These "dysgenic influences," as we may term them, are of differing degrees of antiquity. We will proceed to discuss them in the probable order of their age.

1. WAR

Individual strife, based on the survival of the stronger or the craftier, must be regarded as an instrument of natural selection. But strife between groups wherein each group is represented by a picked body of men is quite a different matter. Organized warfare as we know it today and for the past several centuries leads to death the physically and intellectually fittest of the nation. The handicapped stay behind. It may be argued that the most able group tends to predominate, so that there results a kind of mass survival of the fittest. But the brilliant general who leads an army to victory sacrifices the genetic resources of his country in a way that can never be undone. War may be regarded as the oldest of the dysgenic influences at work today. Paradoxically enough, the recent unfortunate developments as regards our ability to exterminate our fellow-man on a really "magnificent" scale, to the extent that they make the front lines almost as safe as the homeland, tend to reverse the previously selective inroads of war.

There is no way of estimating accurately the extent to which, in each generation of the past several thousand years, war has decimated mankind. However, one cannot read the history of any country or area without being impressed by the accounts of merciless wars in which the vanquished army was literally annihilated and the survivors—women and those who for various reasons were not in military service—assimilated. The interested reader is referred to such diverse sources as Jordan (1915), Bodart and Kellogg (1916), Krzywicki (1934), Sorokin (1937), and Montross (1946).

2. PRESERVATION OF THE HANDICAPPED

It is difficult to conceive of the lame, the halt, and the blind reproducing at the same level as the physically fit under the conditions of 100,000 years ago. The practice of infanticide, especially with respect to defective children, was widespread in primitive communities (Krzywicki, 1934). Gradually, however, a compassionate society has evolved an elaborate machinery for the maintenance of the handicapped. We are not concerned here to debate the wisdom of these measures on purely sociological grounds. But the end-result of the support of the crippled beggar of Roman days or the physically handicapped of today, to the extent that the defects are genetically determined, is the same—it provides a mechanism for the perpetuation and, because of mutation, actual relative increase in the frequency of these defects. There is no way at present of specifying the fraction of these defects which is due in each generation to newly arisen mutations which under conditions of natural selection would be rapidly eliminated. However, the studies on human mutation rates which have been described in chapter 11 constitute a small start on this problem. We may choose the case of retinoblastoma as an illustration.

As pointed out earlier, the tendency to develop this particular type of cancer of the eye appears to depend upon a single dominant gene with 90 per cent penetrance, which arises each generation with a frequency of approximately 2×10^{-5}. Without medical care the cancer is close to 100 per cent fatal. With the medical care which has developed in the last half-century, on the other hand, approximately 70 per cent of cases will survive, although they will be blind in one or both eyes. Let us assume that these survivors reproduce at about half the normal rate. The net reproductive expectation at birth of any person with the disease will thus be 35 per cent of normal. This change in reproductive expectation, from approximately zero 50 years ago to the present 35 per cent, will result in a change in the frequency of the responsible gene. The general expression for the frequency of the retinoblastoma gene at any given number of generations from an initial point of departure is

$$p_x = m + mr + mr^2 + mr^3 + mr^4 + \ldots + mr^x$$

or

$$p_x = m(1 + r + r^2 + r^3 + r^4 + \ldots + r^x),$$

where

$p =$ frequency of the retinoblastoma gene at the xth generation,

$m =$ the mutation rate for the gene $= 0.00002$,

$r =$ net fertility of affected persons, and

$x =$ number of generations that have elapsed.

We may, if we wish, write this in another form by designating the net selective disadvantage as s, equal to $(1 - r)$, in which case the equation becomes

$$p_x = m [1 + (1 - s) + (1 - s)^2 + (1 - s)^3 + \ldots + (1 - s)^x].$$

In the limit, as $x \to \infty$,

$$[1 + (1 - s) + (1 - s)^2 + (1 - s)^3 + \ldots + (1 - s)^x] = \frac{1}{s},$$

and the equation becomes an expression well known to the population geneticist, $p = m/s$. In the derivation of this expression, the possibility of back-mutation has been ignored; this would ordinarily be a negligible factor. With any improvement in medical care, then, the frequency of the gene will increase, at first rapidly, then more slowly, until an equilibrium point is reached. For practical purposes, this equilibrium point is closely approximated after eight generations.

Prior to modern ophthalmological care, with $r \cong 0$, the frequency of the gene would equal m, the mutation rate, or 1 in 25,000 persons (1 in 50,000 genes). With the present level of care,

$$\frac{m}{s} = \frac{2 \times 10^{-5}}{0.65} = 3.08 \times 10^{-5},$$

i.e., the disease may be expected to become half again as frequent. If medicine were able to save all persons with this disease and if the reproductive rates of such persons were normal, then, because of constant mutation pressure, retinoblastoma would gradually become the normal characteristic, and persons without the disease who might arise in consequence of back-mutation would become the exceptions.

The study of the population genetics of man is still in its infancy. There is an urgent need for more studies like the above. Only from such investigations can the genetic balance sheets be drawn up which permit an accurate evaluation of the results of the present relaxation of natural selection.

3. DIFFERENTIAL FERTILITY

In primitive society we may assume that the stronger and more intelligent made a disproportionately great numerical contribution to the next generation, by virtue of both the superior heredity of their children and the superior environment into which these children were born. One of the first clear reversals of this tendency is seen in the Middle Ages, when the church encouraged the more intelligent to enter the clergy. Unfortunately, the price of scholarship was celibacy. More recently, there have appeared important differences in the fertility of the various social classes. We have already considered the question of the extent to which the observed differences in intelligence between individuals are inherited. Figure 20-1 illustrates the distribution of Army Alpha Test scores in different occupational groups, the material having been gathered during World War I. The highest average score was made by the persons in the professional group, the lowest by persons in the agricultural group. But there is a very substantial overlap between all groups. The interpretation of this finding has been roundly discussed. We shall adopt the viewpoint which attributes much of this difference to environmental factors, but yet holds that, to an extent which cannot now be specified, these differences also reflect average differences in certain innate abilities.

Table 20-1 contains a representative set of figures on differential fertility in the United States. Even if the genetically determined proportion of the *mean* I.Q. difference between the professional groups, on the one hand, and unskilled laborers, on the other, is only one or two points, there is operative in society today a mechanism which results in a disproportionately small contribution by the intellectually more able to the next generation.

Some investigators have actually attempted to calculate how much of a decline in mean I.Q. should be anticipated each generation if the present trend continues. Such calculations, which must be viewed with certain reservations, take as their starting point, in addition to data such as have been presented in Table 20-1 and Figure 20-1, studies which indicate a negative correlation between mean intelligence test score and size of family (Mehrotra and Maxwell, 1949). These calculations suggest a mean decline in *innate* I.Q. of 1–2 points per generation, although, obviously, no actual decline might be observed for several generations because environmental changes favoring an

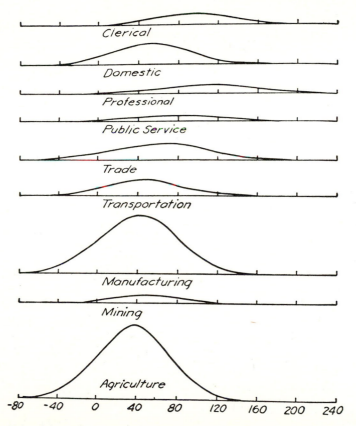

FIG. 20-1.—Schematic curves of the distribution of intelligence of various occupational groups in the United States population on the Army Alpha Test. (By permission of Dr. P. M. Symonds and the Journal of Educational Psychology; also Preface to Eugenics, by Frederick Osborn, rev. ed., copyright 1951, Harper and Bros.)

increased I.Q. could more than offset any genetically determined tendency toward a decline in average I.Q. (cf. Lorimer, 1952, for summarizing article). The concern with which one views this possible trend is in part tempered by how permanent one thinks the above-described differential fertility will be. It is probable that one important reason for the differential fertility seen in the United States is the unequal dissemination of information regarding birth control in a society in a state of rapid evolution. In Sweden, for example, where the practice of birth control is probably somewhat more widespread than in the United States, differential fertility is less apparent than in this country. Lorimer (1952) has summarized the current thought of many on this matter as follows: "Some decline of genetic capacity for intelli-

TABLE 20-1*

REPRODUCTIVE RATIO OF FERTILE AMERICAN COUPLES
BY OCCUPATIONAL GROUPS IN 1928

Occupation of Father	Estimated Reproduction Index
Agriculture, forestry, and animal industry....	1.32
Extraction of minerals....................	1.31
Manufacturing and mechanical industries....	1.08
Transportation.........................	1.04
Domestic and personal service..............	0.96
Public service (not elsewhere classified).....	0.93
Trade.................................	0.87
Clerical occupations.....................	0.78
Professional service.....................	0.76

* A ratio of 1.00 implies replacement in the population. Lower scores indicate that the group is failing to replace itself, while higher scores indicate that the group more than replaces itself. By permission from The dynamics of population, by Dr. F. Lorimer and F. Osborn, copyright 1934, the Macmillan Co.

gence is likely to be an inevitable aspect of the demographic transition in any society from high fertility and high mortality to low fertility and low mortality, but this transition is a necessary condition for sustained economic and social advance. The more rapidly the transition is effected, the smaller will be its possible adverse impact on genetic characteristics."

4. INCREASED MUTATION

Mutation usually results in the appearance of traits which are a handicap to the organism. The frequency of any given deleterious trait is the result of a balance between the frequency of its origin through mutation and the frequency of its elimination through natural selection. We have just reviewed briefly the present tendency toward a relaxation of natural selection. A companion evil must be recognized. The growing intrusion of ionizing radiation into our lives carries with it the threat of increased mutation. We do not know at present precisely how to evaluate this threat.

20.2. *Eugenics.*—The realization of the above-enumerated trends has in the past caused concern among those interested in the future of man. There has resulted the so-called "eugenics movement," which may be broadly defined as a conscious effort on the part of man to arrest the genetic deterioration which is presumed to be taking place. Eugenics has two aspects, a positive and a negative. The positive approach consists in an attempt to encourage reproduction among those whom society would consider useful and valuable citizens. The negative approach consists in an effort to discourage, by various means, the reproduction of those physically and mentally handicapped who owe their defect to heredity.

The eugenics movement has had a lurid and disquieting history. In Nazi Germany the positive aspect was perverted to the doctrine of a race of supermen whose illustrious racial heritage conferred special rights, while the negative aspects were interpreted as permitting the wholesale extermination of elements whom the supermen adjudged undesirable. During that same period there was in the United States a great deal of loose thinking, based primarily upon failure to develop a critical attitude toward both the nature-nurture problem and the precise mathematical consequence of either positive or negative selection. These developments combined to bring discredit upon the entire eugenics movement, an opprobrium from which a sounder eugenics movement is only now emerging. It is imperative that in this renewal of interest the lessons of the past be not forgotten. Every sincere believer in the development of eugenics would do well to refer from time to time to such biased presentations of the problem as Grant's *Passing of the Great Race*, as a reminder of the extremes to which so-called "eugenicists" of other days have gone and the pitfalls to be avoided.

The position of the present-day eugenicist has been ably stated by Frederick Osborn in the most recent edition of his book entitled *Preface to Eugenics*. In brief, it consists of an attempt to encourage the propagation of the "fit" through a system of marriage and child allowances, housing facilities for those with children, etc. On the negative side, it consists of an effort to persuade the "unfit" to limit their family size on a voluntary basis, through birth control, induced abortion, or even voluntary sterilization. There are few who would take exception to the general principles of so reasonable a program. But practical difficulties arise when one considers the machinery of such an undertaking. There are persons whose unfitness as parents—not only on genetic grounds but also because of their inability to care for children— would be generally acknowledged. There are other persons who, by general agreement, should be encouraged to reproduce. But there is also a large group concerning whose desirability as parents there would at the present time be a very considerable debate, with no two competent geneticists taking precisely the same stand.

Even with respect to the group which at first glance would appear eugenically most acceptable, there arise very real problems. Thus man's most precious natural endowment is, of course, his brain. From the standpoint of the survival of the species, physical defect and disability are far less important than mental defect and disability. We know very little concerning the inheritance of mental attributes. There can be no doubt as to the great plasticity of the mind. This has led some to deny altogether the existence of inherited mental differences. As is so often the case, the argument is all the brisker here for want of facts. It has frequently been pointed out that among the research workers, the musicians, the inventors, there are disproportionately many "obsessive-compulsive" personalities—persons constantly driving toward a particular goal. Let us assume for the moment that this is an inherited mental attribute. To what extent should this trait be selected for? One can imagine that life in a society composed entirely of obsessive-compulsive personalities would be strenuous and full of conflicts. One can also imagine that without any such individuals the tempo of our social evolution would be much slower.

The protagonists of eugenics can at this point accuse the authors of setting up a straw man, and with some justification. Nevertheless, there is an important point at issue here. Time and again man, in consequence of an insufficient understanding of the factors involved, has contrived to upset nature's balance. Where this results in the introduction of a new insect pest or the extinction of a once numerous animal, such as the passenger pigeon, the results, although serious economically, can be compensated for. But where the organism involved is man himself, it behooves us to proceed with great caution.

20.3. *The rate of genetic change.*—Earlier in this chapter we have listed the principal possible dysgenic factors at work today. It is important, as the next step in our thinking, to understand the time factor involved in those changes in the frequencies of various inherited traits which result from selection. Let us assume for the moment that certain arbitrarily designated proportions of individuals with various inherited diseases could be prevented from reproduction. What would be the mathematical consequences? This is a question which has attracted many persons (summaries in Hogben, 1946; Li, 1948; Von Hofsten, 1951).

We shall consider, first, the very simple case of selection against a trait determined by a single dominant or recessive gene. The effectiveness of such selection in the case of recessive genes is mathematically related to the initial frequency of the trait, being a function of the relative frequency of the heterozygous carriers (cf. Table 7-4, p. 70). Since the eugenicist has thus far been concerned largely with genetic minorities whose frequency in the population does not exceed 1 per cent, we shall consider the case of complete selection against (1) a dominant trait with a frequency of 1 per cent, (2) a dominant with a frequency of 0.1 per cent, (3) a recessive with a frequency of 1 per cent, and (4) a recessive with a frequency of 0.1 per cent. The re-

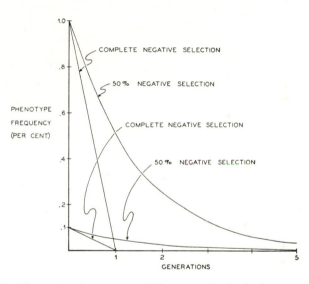

FIG. 20-2.—The results of complete and 50 per cent selection for five generations against a trait determined by a single dominant gene. In the one case the initial frequency of the trait is placed at 1 per cent and in the second case at 0.1 per cent. (Modified by permission of Dr. Nils von Hofsten and Hereditas.)

sults, which can be calculated from simple formulas to be found in the references to Hogben, Li, and Von Hofsten, given above, are summarized in Figures 20-2 and 20-3. It is assumed that there is no mutation while selection is proceeding. Complete selection against a dominantly inherited trait will, of course, eliminate the trait in a single generation. Selection against a recessively inherited trait is much less effective, the frequency after ten generations of complete negative selection being 25 per cent of the original value with an initial frequency of 1 per cent, and 58 per cent of the original value at the lower initial frequency of 0.1 per cent. At recessive-trait frequencies less than 1 in 1,000, the results of selection are, of course, even less striking.

For instance, it requires 50 generations of complete negative selection to decrease the frequency of a recessive trait from 0.04 to 0.01 per cent.

Because of variations in the age of onset of the disease, failures of penetrance, failures of diagnosis, etc., it is unlikely that any program of negative selection against the "common" inherited diseases could be 100 per cent effective. Let us consider, next, the consequences of 50 per cent negative selection. The results are again given in Figures 20-2 and 20-3. Selection against the dominantly inherited trait is still quite effective, but the recessively inherited trait decreases in frequency quite slowly.

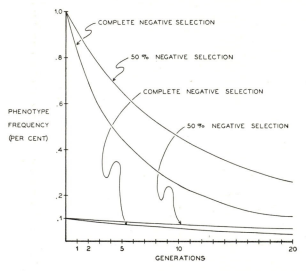

FIG. 20-3.—The results of complete and 50 per cent selection for 20 generations against a trait determined by homozygosity for a single recessive gene. In the one case the initial frequency of the trait is put at 1 per cent and in the second case at 0.1 per cent. (Modified by permission of Dr. Nils von Hofsten and Hereditas.)

When a trait depends upon the interaction of two genes, the effectiveness of negative selection is, in general, decreased by comparison with single-gene inheritance. The shape of the selection curve at a given selection pressure varies considerably with the frequency and dominance relationships of the genes involved. We will not go into the matter here, other than to point out that many undesirable traits are multigenic, and the results of selection against multigenic traits are quite complex—and slow.

The foregoing calculations and figures all neglect the influence of mutation on the rate of decrease. This is a very real factor. Figure 20-4 illustrates how a mutation rate of 1:20,000 to an undesirable recessive gene will impede the effects of selection against that gene.

It must be apparent from the foregoing that it is difficult, if not impossible, to make broad generalizations concerning the effects of negative selection in reducing the frequency of eugenically undesirable traits. However, from what is now known of the mode of inheritance of various pathological conditions, the effectiveness of negative selection under various circumstances, and human mutation rates, it seems safe to state that any "moderate" eugenic program will have only a very small effect on the genetic composition of man for the next several hundred years.

There are two possible reactions to these facts. We can, on the one hand, adopt the stand that any program which appears to impede to any extent

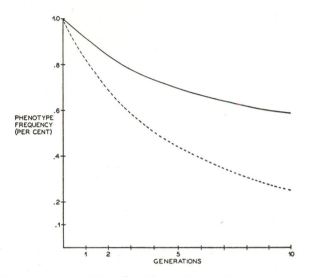

PHENOTYPE
FREQUENCY
(PER CENT)

GENERATIONS

Fig. 20-4.—A comparison of the effect of complete negative selection upon a recessive trait with an initial frequency of 1 per cent (*a*) when there is no mutation (*dashed line*) and (*b*) when the mutation rate is 5×10^{-5} (*solid line*). (Modified by permission of Dr. Nils von Hofsten and Hereditas.)

man's apparent genetic deterioration is desirable. Or we can support the thesis that, in view of the slowness of this presumed deterioration and the relative ineffectiveness of our foreseeable efforts to control it, effort spent now in *improving* man might better be directed toward a more complete *understanding* of his genetics, with the hope that with this understanding will come the wisdom to utilize this knowledge justly.

20.4. *The status of eugenic sterilization.*—No discussion of eugenics would be complete without reference to the present thinking regarding so-called "eugenic sterilization," particularly since the term "eugenics" is unfortunately almost synonymous in the minds of some with the term "sterilization." Some thirty-two states in this country, beginning with Indiana in 1907, have at one time or another passed what are often loosely termed "eugenic laws," although in five states these laws have subsequently been declared unconstitutional. Most of these laws were passed prior to 1930, in what may be termed the "first flush" of eugenic enthusiasm. Switzerland (certain cantons), Canada, Denmark, Mexico, Norway, Sweden, Finland, Japan, and Iceland also have legislation relating to sterilization. The sterilization laws of the United States through 1932 have been summarized by Landman (1932).

Sterilization as usually performed consists of a relatively simple and safe procedure which prevents parenthood by surgically destroying the continuity of the duct through which eggs or sperm, as the case may be, are transported. The operation in no way interferes with normal sexual life, but simply prevents parenthood. Sterilization may be regarded as the ultimate in planned parenthood—a final, irrevocable decision on the part of the individual or those responsible for him that he shall reproduce no more.

The three classes of individuals who are particularly affected by sterilization laws are the feeble-minded, the insane, and the epileptic. It is argued that, inasmuch as these individuals tend to reproduce their kind (see below), it is sound preventive medicine to anticipate the birth of children who run a very significantly increased risk of becoming state charges. It is further argued that the insane and feeble-minded, regardless of the caliber of their children, are poorly equipped to discharge their obligations as parents in a society which accepts the dictum that parenthood carries with it the responsibility of caring for the child. The first may be termed the "genetic argument," the second, "the sociological." These two arguments have not always been kept separated. We will not be concerned here with the sociological argument. Rather, in what follows, we shall examine sterilization from the genetic point of view, with particular reference to the manner in which the laws are enforced and their probable genetic effectiveness.

In passing, it might be pointed out that the "eugenic laws" of seven states provide for the sterilization of habitual criminals; the laws of eight states for the sterilization of moral degenerates; and the laws of nine states for the sterilization of sexual offenders and perverts. The inclusion of such individuals in a eugenics law cannot by any stretch of the imagination be justified on genetic grounds and can be regarded only as an unfortunate carry-over from the early, uncritical days of eugenics.

Sterilization may be on either a voluntary or an involuntary basis. In the first instance the person concerned or his guardian applies for sterilization; in the second instance it is so ordered by the courts. Those states in which **there are** statutes regarding compulsory sterilization have, **by and large,** what may be termed sufficient safeguards to protect the rights of the individual involved. The procedure in the state of Virginia, which may be taken as a model, has been described in a famous legal verdict by Justice Oliver Wendell Holmes, as follows: The first step is for the superintendent of an institution or colony to present a petition to the board of directors of his institution or colony, stating the facts and the grounds for his opinion, verified by affidavit. Notice of the petition and of the time and place of the hearing in the institution is to be served upon the inmate and also upon his guardian. If there is no guardian, the superintendent is to apply to the circuit court of the county to appoint one. If the inmate is a minor, notice is also given to his parents and/or guardian with a copy of the petition. The board is to see to it that the inmate may attend the hearings if desired by him or his guardian. The evidence is all to be reduced to writing, and after the board has made its orders for or against the operation, the superintendent, or the inmate, or his guardian, may appeal to the circuit court of the county. If there is still dissatisfaction with the verdict, the case may be carried to the Supreme Court of Appeals.

Carrie Buck was a feeble-minded white woman who was committed to the Virginia State Colony for the Feeble-minded. She was the daughter of a feeble-minded mother in the same institution and the mother of an illegitimate feeble-minded child. The legality of sterilizing Carrie Buck was carried to the Supreme Court of the United States, where, in 1927, Justice Oliver Wendell Holmes delivered a famous decision which reads in part as follows:

"We have seen more than once that the public welfare may call upon its best citizens for their lives. It would be strange if it could not call upon those who already sap the strength of the State for these lesser sacrifices, often not felt to be such by those concerned, in order to prevent our being swamped with incompetence. It is better for all the world, if instead of waiting to execute degenerate offspring for crime, or to let them starve for their imbecility, society can prevent those who are manifestly unfit from continuing their kind. The principle that sustains compulsory vaccination is broad enough to cover cutting the fallopian tubes. . . . Three generations of imbeciles are enough.

"But, it is said, however it might be if this reasoning were applied generally, it fails when it is confined to the small number who are in the institutions named and is not applied to the multitudes outside. It is the usual last resort of constitutional arguments to point out shortcomings of this sort. But the answer is that the law does all that is needed when it does all that it can, indicates a policy, applies it to all within the lines, and seeks to bring within the lines all similarly situated so far and so fast as its means allow. Of course so far as the operations enable those who otherwise must be kept confined to be returned to the world, and thus open the asylum to others, the equality aimed at will be more nearly reached."

This rousingly phrased verdict on the part of Mr. Justice Holmes has been widely publicized. There are actually two more soberly worded verdicts of state supreme courts which most certainly are more searching. In a verdict rendered in 1929, the Utah State Supreme Court failed to uphold the compulsory sterilization of a young sexual pervert and habitual criminal, on the grounds of insufficient evidence for the hereditary nature of these conditions. Here was a step, unfortunately largely unheeded, toward the clarification of "eugenic laws." The second of the two critical verdicts was rendered by the Supreme Court of Idaho in 1931. In a test case the legality of sterilizing a feeble-minded male was upheld. The evidence for the hereditary nature of this defect was far more critically reviewed than in previous decisions. The import of this decision, also largely unheeded, is to replace the sweeping assertions of early eugenicists, on which so many of these laws are based, with a more critical attitude toward the question of what is and what is not inherited.

Although there appears to be, then, ample legal precedent for *compulsory* sterilization, as well as adequate safeguards of the rights of the individual, in most of our states sterilization is on a *"voluntary"* basis, i.e., an institutionalized individual or his guardian applies for sterilization. Since, however, the act of sterilization often carries with it the possibility of release from the institution, one may question to what extent a type of persuasion is employed.

20.5. *The effectiveness of eugenic sterilization.*—Figures compiled by the Human Betterment Association of America, Inc., reveal that in 1947 there were 1,232 official sterilizations in this country and 2,322 in 1948. These sterilizations involved the insane and the feeble-minded in about equal numbers. During those same two years the number of children born in the United States was 3,699,940 and 3,535,068, respectively. It is conservatively estimated that at least 0.8 per cent of the population is feeble-minded, and at least 1.2 per cent develop the major psychoses which may be used as an indication for sterilization. Accordingly, it may be estimated in very round numbers (neglecting for the moment the lag between time of birth and time of sterilization) that about 3 per cent of the feeble-minded and 2 per cent of the insane are each year permanently barred from parenthood as a result of present sterilization procedures. In many cases, however, these individuals have already produced one or more defective children.

The fact is often lost sight of that the insane and certain types of mentally defective individuals have significantly decreased rates of reproduction. It is difficult to obtain accurate estimates as regards the insane. The situation is more clear cut as concerns the mentally defective. Approximately 20–25 per cent of these unfortunates fall into the idiot and imbecile category. It is uncommon for such individuals to reproduce (Penrose, 1949; Dahlberg, 1951). It may be presumed that this has been the case for centuries. Assuming that these individuals reproduce to no greater extent than those artificially sterilized reproduce prior to their sterilization, it is apparent that "natural" sterilization involves some seven or eight times as many people as the present legislative sterilization. There is evidence that the extreme degrees of mental defect are more often due to single genes than is the borderline type of defect. We have seen earlier that negative selection is far more effective against traits determined by single genes than against traits determined by several genes. It follows that, at present, "natural" sterilization is many more times as effective as legal sterilization. It further follows that present-day sterilization programs are making a very small contribution to controlling the numbers of the insane and mentally defective.

The financial argument has often loomed large in discussions of sterilization programs. It is stated that sterilization is sound economically because it makes it possible to release individuals who would otherwise be institutionalized at public expense and also prevents their (unborn) children from becoming public charges. The first argument is difficult to evaluate, for when such persons leave the institution, they often become public charges in their local communities. The second argument is on somewhat clearer grounds. It has been estimated that in the case of the feeble-minded each sterilization prevents an average of 2.5 births (Tietze and Johnson, 1950). Among the group on which this estimate was based, 36 per cent of the children born prior to sterilization were feeble-minded (Johnson, 1950). Assuming that these birth rates continue, then the sterilization of each 100 mentally deficient individuals results in the prevention of the birth of some 90 feeble-minded children. The advocates of sterilization programs argue that any such decrease is a step in the right direction.

There are groups in this country which are actively engaged in furthering the cause of sterilization. There are undoubtedly individual cases in which sterilization is desirable, as much on sociological as on genetic grounds. It would seem, however, that before any attempt gets under way to persuade large numbers of the population, either here or abroad, to submit to sterilization, it would be well to scrutinize the basic tenets carefully. Thus, as noted above, the three groups most commonly listed in state statutes as subject to sterilization are the feeble-minded, the insane, and the epileptic. The empirical risk figures for these three groups, summarized in Table 20-3, are quite different. Attention is particularly directed toward the disparity between the figures for epilepsy and those for the other diseases included in the table. It must be apparent from the earlier discussion that very little can be expected to result from the sterilization of a small fraction of the epileptics. The present authors cannot help feeling that before man attempts to grasp the reins of his genetic destiny, he should have a far more comprehensive knowledge of human inheritance than is now at his disposal. Few would doubt the wisdom of curbing the reproduction of the mentally unbalanced. But the principle involved, once unleashed, has ramifications which touch upon the lives of all of us. One of the dilemmas posed by these ramifications has recently been examined by Muller (1950). In our discussion thus far we have been thinking for the most part in terms of clear-cut traits determined by one or several genes. Muller has drawn attention to the fact that the aver-

TABLE 20-2

PROBABILITY THAT CHILD OF INDIVIDUAL AFFLICTED WITH ONE OF MENTAL CONDITIONS AGAINST WHICH EUGENIC MEASURES ARE COMMONLY DIRECTED WILL DEVELOP SAME CONDITION

Disease in Parent (Only One Parent Affected)	Empiric Risk for Child (Per Cent)	Authority
Feeble-mindedness	30–35	Report of Departmental Committee on Sterilization (1934); Johnson (1950); Dahlberg (1951)
Manic-depressive psychosis	10–25	Report of Departmental Committee on Sterilization (1934); Röll and Entres (1936); Slater (1938); Stenstedt (1952)
Schizophrenia	10–18	Kallmann (1938, 1952)
Epilepsy	3– 6	Kallmann and Sander (1947); Lennox (1951)

age human is heterozygous for a considerable number of semidominant genes with slight but deleterious effects when heterozygous. It is suggested that the average number of such genes per individual is eight, but may well be double that number. Whatever the average may be, it is certain that some individuals are heterozygous for considerably more than the average, while other more fortunate individuals have less than the average. An individual with a high concentration of such genes just as definitely passes a genetic handicap to his children as does one with a single major inherited pathologic trait due to a completely dominant gene. How in any just scheme are the two to be equated? And when will we have the genetic knowledge to detect these carriers of the many genes with slightly deleterious effects? Eugenic procedures based upon our present limited knowledge cannot help being discriminatory, in the sense that they single out for action the obviously handicapped, while failing to touch those no less handicapped but in less apparent ways.

20.6. *A positive program for human genetics.*—We have indicated some of the problems which arise in the formulation of a program of either positive or negative eugenics. Under these circumstances, what, if any, measures are justified at the present time?

Realism bids recognition of the fact that family planning is practical in large areas of the world today, with every indication that the practice will spread. One element in such planning is information as to the probable outcome of any pregnancy, to the extent that such information exists. We are therefore in favor of as rapid an increase as possible in the genetic counseling services available to interested parties.

Who will carry out the counseling? It must be apparent by now that, although the general principles of genetics are simple enough, their ramifications are complex. An insight into genetic methods and the ability to dispose of simple genetic questions are not difficult to acquire; the ability to treat the more complex issues depends upon an apprenticeship no less rigorous than that to which any medical specialist is subjected. A two-pronged attack on the problem of providing adequate genetic counseling therefore seems indicated. On the one hand, there is the need for such instruction of medical students and graduates as will enable them to recognize genetic problems in their practice and dispose of the simpler ones. There is, at the same time, a need for the training of specialists who, in heredity clinics in connection with major medical centers, can meet the more complex problems.

It is obvious that the feeble-minded and the psychotic will avail themselves little, if at all, of such counseling programs or, if they did, would be poorly equipped mentally to implement personal decisions concerning family size restrictions. The possibility thus emerges of the more responsible conforming to genetic principles which the less responsible ignore. It is felt that, for the present, the institutionalization of the latter on purely sociological grounds must provide the principal legal balance wheel.

From the standpoint of reproduction, it is difficult to see any essential moral difference between voluntary sexual sterilization and the rigid practice of birth control. There seems to be no reason to accept one but not the other. Accordingly, we would feel that there are no contraindications to the voluntary sterilization of certain classes of people, as much on sociological as on eugenic grounds, where the individuals concerned are aware of the consequences of their decision. It is a moot point whether those whom society adjudges insane or mentally deficient (and who numerically constitute one of the largest problems) are in a position to make responsible decisions, including those concerning reproduction. Any sterilization program which encompasses such persons will have difficulty avoiding a large element of "persuasion," if not compulsion.

The day may yet come when man has sufficient accumulated knowledge concerning his heredity and its interaction with his environment that a comprehensive program of self-directed evolution may be undertaken. But for the present the effort which would be expended on a eugenics program might better go into efforts to explore the many gaps in our present fragmentary information.

Bibliography

SPECIFIC REFERENCES

BODART, G., and KELLOGG, V. L. 1916. Losses of life in modern wars. Oxford: Oxford University Press.

DAHLBERG, G. 1951. Mental deficiency, Acta genet. et stat. med., 2:15–29.

JOHNSON, B. S. 1950. A study of sterilized persons from the Laconia State School, Am. J. Ment. Deficiency, 54:404–8.

JORDAN, D. S. 1915. War and the breed. New York: Beacon Press.

KALLMANN, F. J. 1938. Heredity, reproduction and eugenic procedure in the field of schizophrenia, Eugenical News, 23:105–13.

———. 1952. Genetic aspects of psychoses. In: The biology of mental health and disease, pp. 283–98. New York: Paul B. Hoeber, Inc.

KALLMANN, F. J., and SANDER, G. 1947. The genetics of epilepsy. In: HOCH, P., and KNIGHT, R. (eds.), Epilepsy, pp. 27–41. New York: Grune & Stratton.

KRZYWICKI, L. 1934. Primitive society and its vital statistics. London: Macmillan & Co., Ltd.

LENNOX, W. G. 1951. The heredity of epilepsy as told by relatives and twins, J.A.M.A., 146:529–36.

LORIMER, F. 1952. Trends in capacity for intelligence, Eugenical News, 37:17–24.

MEHROTRA, S. N., and MAXWELL, J. 1949. The intelligence of twins: a comparative study of eleven-year-old twins, Population Studies, 3:295–302.

MONTROSS, L. 1946. War through the ages. New York: Harper & Bros.

PENROSE, L. S. 1949. The biology of mental defect. New York: Grune & Stratton.

RÖLL, A., and ENTRES, J. L. 1936. Zum Problem der Erbprognosebestimmung: Die Erkrankungsansichten der Neffen und Nichten von Manisch-depressiven, Ztschr. f. d. ges. Neurol. u. Psychiat., 156:169–202.

SLATER, E. 1938. Zur Erbpathologie des manisch-depressiven Irreseins: die Eltern und Kindern von Manisch-depressiven, Ztschr. f. d. ges. Neurol. u. Psychiat., 163:1–47.

SOROKIN, P. A. 1937. Social and cultural dynamics, Vol. 3: Fluctuation of social relationships, war, and revolution. New York: American Book Co.

STENSTEDT, A. 1952. A study in manic-depressive psychosis, Acta psychiat. et neurol. Scandinav., suppl. 79, pp. v and 111.

TIETZE, C., and JOHNSON, B. S. 1950. Observations on the fertility of patients discharged from the Laconia State School, 1924 to 1934, Am. J. Ment. Deficiency, 54:551–55.

GENERAL REFERENCES

DUNN, L. C., and DOBZHANSKY, TH. 1952. Heredity, race, and society. Rev. ed. New York: Mentor Books.

FISHER, R. A. 1930. The genetical theory of natural selection. Oxford: Oxford University Press.

GRANT, M. 1921. Passing of the great race. New York: Charles Scribner's Sons.

HALDANE, J. B. S. 1938. Heredity and politics. New York: W. W. Norton & Co.

HOGBEN, L. 1946. An introduction to mathematical genetics. New York: W. W. Norton & Co.

LANDMAN, J. H. 1932. Human sterilization. New York: Macmillan Co.

LI, C. C. 1948. An introduction to population genetics. Peiping: Peking University Press.

MULLER, H. J. 1950. Our load of mutations, Am. J. Human Genetics, 2:111–76.

OSBORN, F. 1951. Preface to eugenics. Rev. ed. New York: Harper & Bros.

Report of the Departmental Committee on Sterilization. 1934. Great Britain, Parliament House of Commons, Sessional papers, Vol. XV, No. 7.

SIMPSON, G. G. 1950. The meaning of evolution. New Haven: Yale University Press.

VON HOFSTEN, N. 1951. The genetic effect of negative selection in man, Hereditas, 37:157–265.

Author Index

351

Subject Index

355

PRINTED
IN U·S·A